Effects of Adding Cement Admixtures on the Microstructure and Properties of Cement Materials

Effects of Adding Cement Admixtures on the Microstructure and Properties of Cement Materials

Guest Editors

Ying Shi
Hongwei Wang

MDPI

Basel • Beijing • Wuhan • Barcelona • Belgrade • Novi Sad • Cluj • Manchester

Guest Editors

Ying Shi
School of Resources and
Safety Engineering
Central South University
Changsha
China

Hongwei Wang
School of Resources and
Safety Engineering
Central South University
Changsha
China

Editorial Office
MDPI AG
Grosspeteranlage 5
4052 Basel, Switzerland

This is a reprint of the Special Issue, published open access by the journal *Materials* (ISSN 1996-1944), freely accessible at: https://www.mdpi.com/journal/materials/special_issues/9PXIS0WHK9.

For citation purposes, cite each article independently as indicated on the article page online and as indicated below:

Lastname, A.A.; Lastname, B.B. Article Title. *Journal Name* **Year**, *Volume Number*, Page Range.

ISBN 978-3-7258-7530-6 (Hbk)
ISBN 978-3-7258-7531-3 (PDF)
https://doi.org/10.3390/books978-3-7258-7531-3

Contents

Preface

Cementitious materials remain indispensable to modern infrastructure, yet their production entails substantial resource consumption and environmental impact. This reprint, Effect of Admixtures on the Microstructure and Properties of Cementitious Materials, was conceived to address the pressing need for sustainable advancements in this field. Its scope encompasses the development of novel admixtures, the elucidation of hydration mechanisms, and the optimization of mechanical performance and durability. The primary aim is to present a collection of studies that integrate experimental characterization with computational modeling to reveal the fundamental relationships between material composition, microstructure, and macroscopic properties. The motivation stems from the rapid progress in areas such as nanotechnology, machine learning, and the utilization of industrial byproducts, which together open new pathways for designing high-performance, low-carbon cement-based systems. This Reprint is intended for researchers, engineers, and graduate students working in materials science, civil engineering, and sustainable construction, offering both theoretical insights and practical references for advancing green innovation in the field.

Ying Shi and Hongwei Wang
Guest Editors

materials

Editorial

Current Status and Trends of the Cement Admixtures

Hongwei Wang and Ying Shi *

School of Resources and Safety Engineering, Central South University, Changsha 410083, China; hongwei.wang@csu.edu.cn
* Correspondence: shiyingfriend@csu.edu.cn

Cement-based materials are central to modern infrastructure construction. However, their production and application are accompanied by significant resource consumption and environmental emissions. To achieve sustainable development in the construction sector, optimizing cement performance, enhancing durability, and reducing environmental impact through admixture technology have become important research directions [1].

Current research on cement-based materials primarily follows two directions: investigating microscopic mechanisms and developing novel admixtures. At the micro-mechanism level, progress in experimental techniques, interdisciplinary approaches, and computational science has provided strong support for studying the hydration processes and gel properties of cement-based materials. For instance, Wang et al. [2] applied the Krstulović–Dabić kinetic model and GEMS thermodynamic analysis to show that a low-heat-release polymer retards cement hydration by "inhibiting nucleation and growth processes and delaying gypsum consumption." Golewski [3] used scanning electron microscopy to demonstrate that quaternary blended cements containing silica fume, fly ash, and nanosilica can reduce the width of microcracks in the interfacial transition zone by up to 48%, mainly due to the "homogenization of the composite structure and limitation of initial internal damage in concrete." Through X-ray spectral microanalysis, diffraction granulometric analysis, and rheological tests, Klyuev et al. [4] showed that low-cement mortars with fly ash, crushed quartzite sandstone, and PFM-NLK superplasticizer achieve a 70% reduction in cement content while reaching 28-day compressive and flexural strengths of 62.9 MPa and 7.9 MPa, respectively, exceeding conventional cement by 23% and 52%. These mortars also exhibit "suitability for 3D printing applications" owing to their optimized particle size distribution and flowability. The continuous output of research findings is further deepening this field. For example, based on a comprehensive literature review, Dahlan [5] concluded that nanotechnology, particularly the incorporation of nanoparticles such as SiO_2, TiO_2, and carbon nanotubes, "significantly enhances the performance of cement and concrete" by improving mechanical strength and durability, as well as introducing multifunctional properties like self-cleaning and piezoresistivity. Among these, the fracture mode of materials directly determines their engineering and safety performance, constituting a key area of research at the microscopic level: Golewski [6] demonstrated through SEM analysis that incorporating 20% fly ash minimizes interfacial transition zone (ITZ) crack width to 300 nm in mature concrete, thereby enhancing durability and potentially reducing hazardous fly ash disposal by 160 million tons annually.

The in-depth exploration of microstructural characteristics has directly spurred numerous interdisciplinary collaborations, leading to the emergence of novel external additives that enhance the performance of cement-based materials: Dvorkin et al. [7] demonstrate, through experimental design, regression analysis, and high-temperature testing, that the mechano-chemical activation of cement–ash binders by increasing fly ash specific surface

https://doi.org/10.3390/ma19010187

area and introducing a 1% Na_2SiF_6 additive significantly could enhances the properties of heat-resistant mortars. Yang et al. [8] studied the swelling behavior of a superabsorbent polymer in cementitious environments, finding that its absorptivity is critically influenced by factors including mix composition, superplasticizer dosage, stirring rate, and curing conditions. Whilst enhancing the performance of cementitious materials remains the core objective, their integration with solid waste and environmentally friendly materials can significantly amplify their environmental benefits. This approach has garnered widespread attention in the global drive towards industrial sustainability: Klyuev et al. [9] investigated the utilization of molten spherical basalt particles (beads), a waste product from mineral wool production, as an active mineral additive in cement paste, and found that finely ground beads, particularly at 5–10% replacement, exhibit pozzolanic activity by absorbing portlandite and contribute to a densified microstructure, thereby enhancing the composite's compressive strength. Aiken et al. [10] found that partially replacing MgO in magnesium oxychloride cement with supplementary materials, especially metakaolin, significantly enhances its water resistance, primarily through physical particle packing effects rather than new chemical reactions. Zhao et al. [11] developed a low-carbon binder for cemented paste backfill using low-reactivity blast furnace slag, superfine fly ash, and chemical additives (aluminum sulfate and carbide slag), which significantly enhanced compressive strength through both physical filling and chemical activation. Chernysheva et al. [12] confirmed through XRD, SEM, and DTA analyses that incorporating mechanically activated thermal power plant waste (fly ash/slag) and microfibers (polyamide/basalt) into gypsum–cement binders results in a densified microstructure and enhanced mechanical properties. Bio-based additives, exemplified by biochar, confer novel green benefits to cement-based materials: Wei et al. [13] demonstrate that pyroligneous acid (PA), a bio-additive from applewood pyrolysis, effectively enhances strength and retards setting in both sulfoaluminate and Portland cements at low doses ($\leq$2%), as determined through performance evaluation and microstructural characterization, demonstrating its potential as a sustainable alternative to conventional chemical admixtures. Chen et al. [14] demonstrated that incorporating carbon-negative biochars (rice husk and yard waste) as green additives in cementitious binders enhanced both hydration via internal curing/pozzolanic reactions and the immobilization efficiency of toxic elements in municipal solid waste incineration fly ash, as evidenced by multi-technique characterization (TGA, XRD, NMR, etc.), thereby advancing low-carbon stabilization/solidification technology for hazardous waste treatment. Concurrently, the rapid advancement of machine learning has opened new avenues for verifying the performance of cementitious materials: Han et al. [15] demonstrated through experimental tests and an Adaptive Neuro-Fuzzy Inference System (ANFIS) that volcanic powder effectively maintains slump flow while the combination of micro-silica and volcanic powder in ternary mixes optimally enhances compressive strength, electrical resistivity, and durability of self-consolidating green concrete with partial cement replacements.

Cement-based materials have reached a relatively advanced stage of development. Nevertheless, driven by interdisciplinary integration and emerging technological innovations, this field is once again demonstrating remarkable research vitality and innovation potential. This Special Issue, centered on the theme "Effects of Adding Cement Admixtures on the Microstructure and Properties of Cement Materials," compiles several cutting-edge studies. It covers aspects such as mechanical property enhancement, micro-mechanism analysis, development of novel additives, and environmental co-benefits, aiming to provide the latest scientific and technological trends and theoretical references for scholars and engineers in the field.

Mechanical strength is a fundamental indicator for evaluating the performance of cement-based materials, directly impacting engineering safety and service life. Multiple

https://doi.org/10.3390/ma19010187

articles in this issue focus on strength enhancement and performance prediction. For instance, Ma et al. [16] investigated the performance changes of steel slag aggregate after carbonation treatment with carbon dioxide, finding that carbonation significantly improved the compressive strength and volume stability of the aggregate, enabling it to meet Class II aggregate standards while simultaneously achieving CO_2 sequestration. This provides a new pathway to synergistically enhance material performance and environmental benefits.

Regarding performance prediction, Wang et al. [17] employed an explainable machine learning approach to establish a predictive model for the compressive strength of sustainable cement–fly ash mortar. They found that the Gradient Boosting Regressor (GBR) offered the best prediction accuracy, with curing time being the most critical factor influencing strength. The Al_2O_3 content in fly ash was more influential than its dosage or the water-to-binder ratio. Zhang et al. [18] conducted research on recycled concrete incorporating superabsorbent polymer (SAP) and manufactured sand powder (MSP) under freeze–thaw conditions. They identified shifts in the dominant influencing factors before and after freeze–thaw cycles. They established a linear prediction model for strength loss, providing a basis for concrete design in harsh environments.

The regulatory mechanisms of admixtures on the cement hydration process and microstructure are key to the targeted optimization of material properties. Konan et al. [19] revealed the complex relationship between bubble dynamics and rheological behavior in cement paste through simulation, providing theoretical support for optimizing process parameters and reducing defects. Tao and Massoudi [20] studied the influence of nano-silica and fly ash on the pulsating flow behavior of cement suspensions, finding that they had opposing effects on flow velocity and particle distribution, with nano-silica dominating in composite systems.

Regarding microstructure enhancement, Shi et al. [21] investigated the stress–strain behavior and strength development of high-volume phosphogypsum-based cement materials. They indicated that increasing the slag content could significantly improve mechanical properties, and that sulfates in phosphogypsum promote ettringite formation, thereby enhancing system strength. Jiang et al. [22] studied the synergistic effect of MgO expansive agent and steel fibers, finding their combined use produced a "superposition effect," markedly improving the crack resistance and mechanical properties of concrete, with uniform fiber distribution being a prerequisite for effective confinement and bridging.

Developing novel functional admixtures is a crucial pathway for advancing the high-performance and green development of cement-based materials. Dong et al. [23] synthesized hydrotalcite materials using steel slag as a raw material and verified their high-efficiency adsorption capacity for chloride and sulfate ions in salt-washing wastewater, demonstrating significant environmental benefits.

Furthermore, Ordonez et al. [24] investigated the regulatory effects of rice husk ash with varying reactivity on the performance of composite cements. The results showed that highly reactive rice husk ash, at appropriate dosages, could significantly improve cement performance and be combined with materials such as fly ash to prepare ternary systems with high replacement levels and high performance, providing technical support for the high-value utilization of agricultural waste.

The research compiled in this Special Issue showcases the latest progress in the field of cement admixtures from multiple dimensions: macroscopic performance enhancement, micro-mechanism interpretation, and development of novel additives. These works not only advance cement-based materials toward higher strength, durability, and functionality but also significantly enhance their environmental friendliness and resource-use efficiency. In the future, further revealing the intrinsic relationships among "composition–structure–performance–environment" through interdisciplinary approaches and intelligent methods

https://doi.org/10.3390/ma19010187

will become an important development direction in this field. It is hoped that the articles in this issue will provide valuable references for researchers and engineering practice, jointly promoting the green innovation and sustainable development of cement technology.

Funding: This research received the financial support of the Natural Science Foundation of Hunan Province (Grant No. 2024JJ6510).

Acknowledgments: The guest editors would like to acknowledge all contributors and reviewers for their excellent work.

Conflicts of Interest: The author declares no conflicts of interest.

References

1. Saedi, A.; Jamshidi-Zanjani, J.; Darban, A.K. A review of additives used in the cemented paste tailings: Environmental aspects and application. *J. Environ. Manag.* **2021**, *289*, 112501. [CrossRef] [PubMed]
2. Wang, L.G.; Wu, M.; Zhang, Y.; Ju, S.Y.; Wang, F.J.; Liu, Z.Y.; Jiang, J.Y. The influence of low temperature rise polymer on early cement hydration from the point of view of hydration kinetics and thermodynamics. *CASE Stud. Constr. Mater.* **2024**, *21*, e03466. [CrossRef]
3. Golewski, G.L. Concrete Composites Based on Quaternary Blended Cements with a Reduced Width of Initial Microcracks. *Appl. Sci.* **2023**, *13*, 7338. [CrossRef]
4. Klyuev, S.; Klyuev, A.; Fediuk, R.; Ageeva, M.; Fomina, E.; Amran, M.; Murali, G. Fresh and mechanical properties of low-cement mortars for 3D printing. *Constr. Build. Mater.* **2022**, *338*, 127644. [CrossRef]
5. Dahlan, A.S. Impact of nanotechnology on high performance cement and concrete. *J. Mol. Struct.* **2021**, *1223*, 128896. [CrossRef]
6. Golewski, G.L. Evaluation of morphology and size of cracks of the Interfacial Transition Zone (ITZ) in concrete containing fly ash (FA). *J. Hazard. Mater.* **2018**, *357*, 298–304. [CrossRef]
7. Dvorkin, L.; Zhitkovsky, V.; Tracz, T.; Sitarz, M.; Mróz, K. Mechanical-Chemical Activation of Cement-Ash Binders to Improve the Properties of Heat-Resistant Mortars. *Materials* **2024**, *17*, 5760. [CrossRef]
8. Yang, J.; Wang, J.F.; Su, Y.; He, X.Y.; Wang, F.L.; Liang, W. On the factors affecting the swelling behavior of superabsorbent polymers in cement-related environment. *Constr. Build. Mater.* **2023**, *409*, 12. [CrossRef]
9. Klyuev, A.; Klyuev, S.; Fomina, E.; Shorstova, E.; Ageeva, M.; Nedoseko, I.; Sabitov, L.; Shamanov, V.; Shayakhmetov, R.; Liseitsev, Y. Wastes from the production of heat-insulating basalt wool as an additive in cement-based materials. *Case Stud. Constr. Mater.* **2023**, *19*, 9. [CrossRef]
10. Aiken, T.A.; Kwasny, J.; Russell, M.; McPolin, D.; Bagnall, L. Effect of partial MgO replacement on the properties of magnesium oxychloride cement. *Cem. Concr. Compos.* **2022**, *134*, 16. [CrossRef]
11. Zhao, Z.Y.; Guo, K.H.; Wang, X.L. A binder prepared by low-reactivity blast furnace slags for cemented paste backfill: Influence of super-fine fly ash and chemical additives. *Constr. Build. Mater.* **2022**, *327*, 11. [CrossRef]
12. Chernysheva, N.; Lesovik, V.; Fediuk, R.; Vatin, N. Improvement of Performances of the Gypsum-Cement Fiber Reinforced Composite (GCFRC). *Materials* **2020**, *13*, 3847. [CrossRef] [PubMed]
13. Wei, G.S.; Zhang, J.Z.; Mao, Y.P.; Wang, X.J.; Li, J.W.; Pang, D.J.; Jin, Y.; Wang, W.L. Effect of pyroligneous acid as a novel bio-additive on the hydration mechanism of calcium sulfoaluminate cement and ordinary Portland cement. *Constr. Build. Mater.* **2024**, *439*, 18. [CrossRef]
14. Chen, L.; Wang, L.; Zhang, Y.Y.; Ruan, S.Q.; Mechtcherine, V.; Tsang, D.C.W. Roles of biochar in cement-based stabilization/solidification of municipal solid waste incineration fly ash. *Chem. Eng. J.* **2022**, *430*, 10. [CrossRef]
15. Han, S.Y.; Zheng, D.S.; Mehdizadeh, B.; Nasr, E.A.; Khandaker, M.U.; Salman, M.; Mehrabi, P. Sustainable Design of Self-Consolidating Green Concrete with Partial Replacements for Cement through Neural-Network and Fuzzy Technique. *Sustainability* **2023**, *15*, 4752. [CrossRef]
16. Ma, J.; Dai, G.; Jiang, F.; Wang, N.; Zhao, Y.; Wang, X. Effect of Carbonation Treatment on the Properties of Steel Slag Aggregate. *Materials* **2023**, *16*, 5768. [CrossRef]
17. Wang, H.; Ding, Y.; Kong, Y.; Sun, D.; Shi, Y.; Cai, X. Predicting the Compressive Strength of Sustainable Portland Cement-Fly Ash Mortar Using Explainable Boosting Machine Learning Techniques. *Materials* **2024**, *17*, 4744. [CrossRef]
18. Zhang, L.; Liu, R.; Jiang, F. Mechanical Properties of Recycled Concrete Incorporated with Super-Absorbent Polymer and Machine-Made Stone Powder under the Freeze-Thaw Cycle Environment. *Materials* **2024**, *17*, 5006. [CrossRef]
19. Konan, N.A.; Rosenbaum, E.; Massoudi, M. Motion of Air Bubbles in a Cement Slurry. *Materials* **2023**, *16*, 6433. [CrossRef]
20. Tao, C.; Massoudi, M. On the Flow of a Cement Suspension: The Effects of Nano-Silica and Fly Ash Particles. *Materials* **2024**, *17*, 1504. [CrossRef]

https://doi.org/10.3390/ma19010187

21. Shi, Y.; Li, Y.; Wang, H.; Ma, Y.; Lu, X. Stress-Strain Behavior and Strength Development of High-Amount Phosphogypsum-Based Sustainable Cementitious Materials. *Materials* **2024**, *17*, 4927. [CrossRef]
22. Jiang, F.; Deng, W.; Wang, Q.; Gu, Z.; Wang, J. The Influence of Fiber Dispersion on the Properties of MgO Concrete and Engineering Applications. *Materials* **2025**, *18*, 261. [CrossRef]
23. Dong, Z.; Huang, B.; Zhang, T.; Liu, N.; Mao, Z. Preparation of Steel-Slag-Based Hydrotalcite and Its Adsorption Properties on Cl^- and SO_4^{2-}. *Materials* **2023**, *16*, 7402. [CrossRef]
24. Ordonez, L.M.; Borrachero, M.V.; Monzo, J.; Soriano, L.; Paya, J. Binary and Ternary Blended Portland Cements Containing Different Types of Rice Husk Ash. *Materials* **2024**, *17*, 2923. [CrossRef]

https://doi.org/10.3390/ma19010187

materials

MDPI

Article

The Influence of Anhydrite on the Mechanical Performance of Calcium Sulfoaluminate Cement-Based Grouting Materials

Lin Liao *, Nathan Saye Teah, Zhiling Liao, Ruiqing Hao, Yubin Zhao and Wanwen Xue

Department of Underground Engineering, College of Mining Engineering, Taiyuan University of Technology, 79 Yingze West Street, Taiyuan 030024, China; nathanteah93@gmail.com (N.S.T.); lzlfight66@163.com (Z.L.); rqhao_tyut@163.com (R.H.); zhaoyubin1309@link.tyut.edu.cn (Y.Z.); xueww1112@163.com (W.X.)

* Correspondence: liaolin@tyut.edu.cn; Tel.: +86-15035181288

Abstract: Cement-based grouting reinforcement technology is an essential method to enhance the mechanical performance of fractured rock depending on the type of grouting material used. To further understand the influence of anhydrite on the calcium sulfoaluminate cement-based grouting materials, this study investigates the sample of the grouting material with (calcium sulfoaluminate cement, anhydrite, and quicklime) under different ratios with a series of experiments including compressive strength, setting time, and slurry pH test. The macro and micro mechanical characteristics of different calcium sulfoaluminate cement grouting samples were revealed using X-ray diffraction (XRD) and scanning electron microscopy (SEM). The results indicate that anhydrite enriched the system with needle-like ettringite (AFt), and plume-shaped aluminum hydroxide (AH_3), which contributed to the strength improvement. The optimal dosage of anhydrite-quicklime was 80:20 with a compressive strength of 9.5 MPa, 14.7 MPa, and 18.4 MPa at (1 d, 14 d, and 28 d), making up 38.5% of the total strength, and 80% independent of the quicklime dosage (14.2 MPa, 23.2 MPa, and 24.5 MPa), making up 52.5% of the total strength. Furthermore, X-ray diffraction and scanning electron microscopy results proved that anhydrite ratios are the main factor influencing grouting material reinforcement effectiveness and are more beneficial for improving the mechanical performance of calcium sulfoaluminate cement-based grouting material.

Keywords: calcium sulfoaluminate cement; grouting material; anhydrite; quicklime; hydration mechanism

1. Introduction

The structural anchoring technique is an essential component of structural engineering that prevents and mitigates disasters. Anchoring techniques, as an important area of structural engineering, are used extensively in many construction projects, such as tunnel engineering, slope engineering, mining engineering, and foundation supporting due to their efficient and cost-effective benefits [1–5]. Meanwhile, as an essential element of anchorage structure, grouting technology plays a vital role in improving the mechanical properties of rock masses, self-stabilizing the surrounding rock's capacity, reducing the likelihood of the surrounding rock mass instability and failure, and successfully preventing stress concentration in fracture zones [6].

In terms of cementitious material selection for anchoring grouting, PC and CSA cementitious are now used as the most common cementitious materials in anchoring grouting [7–9]. For over a century, PC hydraulic binders have been utilized as a grouting material in construction [10–12]. However, as a grouting material, it is prone to extended

 https://doi.org/10.3390/ma18071547

setting times, low early strength, and a smaller grouting range [13–15]. Additionally, it is important to recognize that PC manufacturing is not environmentally sustainable [16,17]. In the context of global warming and greenhouse gas emission reduction, it may result in a rise in the overall emissions of carbon dioxide (CO_2) during the production process [18,19]. Therefore, to guarantee the best grouting performance, it is essential to enhance the extraordinary grouting material's fluidity, setting time, expansion rate, and mechanical strength.

Calcium sulfoaluminate cement (CSA) is a hydraulic binder and an inorganic material that was developed in the 1970s and has been utilized and standardized in the construction and underground industry in China for more than a century [20,21]. Furthermore, CSA has positive effects on the environment. For instance, research shows that CSA cement can reduce carbon dioxide (CO_2) emissions in China by roughly 35–48% [22,23]. It can also be marketed as an environmentally friendly alternative to PC because of its quick setting time, early high strength, and low CO_2 emissions. However, because CSA cement has properties like low clinker sintering temperature, low limestone consumption, and strong chemical corrosion resistance, it can be considered a good grouting cementitious material and an alternative method of reducing greenhouse gas emissions in the context of global warming [24–26], and its addition to anhydrite can result in a larger concentration of ye'elimite phase minerals [27,28].

In this study, the CSA cement is mainly composed of ye'elimite ($C_4A_3\overline{S}$), and belite (C_2S). In addition, it also contains some trace minerals, such as Iron (III) oxide (Fe_2O_3), (graphite fluoride) C_2F, as shown in Table 1. In the CSA cementitious material system, ettringite ($C_6A\overline{S}_3H_{32}$) production is mostly dependent on ye'elimite ($C_4A_3\overline{S}$) . There are two ways that the hydration process of ye'elimite ($C_4A_3\overline{S}$) plays a crucial role in the ettringite formation. First, anhydrite ($C\overline{S}$) typically occurs in the hydration of ye'elimite ($C_4A_3\overline{S}$) to produce ettringite ($C_6A\overline{S}_3H_{32}$) and hydroxide aluminum (AH_3) via reaction (1). This results in high early strength and quick hardening of the CSA cement paste [29]. Second, if there is not enough anhydrite ($C\overline{S}$), ye'elimite ($C_4A_3\overline{S}$) may hydrate to produce mono-sulfoaluminate ($C_4A\overline{S}H_{12}$) via reaction (2). Because of the quick reactivity of ye'elimite ($C_4A_3\overline{S}$) and the expansion of the cement structure as ettringite ($C_6A\overline{S}_3H_{32}$) forms (Equation (2)), CSA cement causes rapid coagulation and expansion characteristics [30]. This characteristic makes it useful for precast products, self-repairing concrete, and quick repairs.

$$C_4A_3\overline{S} + 2C\overline{S}H_2 + 38H \longrightarrow C_3A\cdot 3C\overline{S}\cdot 32H + 2AH_3 \quad (1)$$

$$C_4A_3\overline{S} + 18H \longrightarrow C_3A\cdot C\overline{S}\cdot 12H + 2AH_3 \quad (2)$$

Table 1. Mineralogical composition of the CSA clinker/wt.%.

Mineral	$C_4A_3\overline{S}$	β–C_2S	C_2F	Fe_2O_3
wt.%	58.64	24.43	6.41	1.92

Although CSA cement has better material performance, its limited raw ingredients, high cost, and poor alkalinity limit its engineering use [31,32]. To improve the engineering performance of CSA cement, researchers have explored possibilities for using mixed materials and supplemental cementitious materials (SCMs). Research has indicated that incorporating fly ash, limestone powder, calcite, and anhydrite can enhance the engineering performance of CSA cement [33,34]. Currently, the usage of limestone in cement is receiving a lot of attention. In the case of CSA cement paste mixed with limestone powder, Pelletiter et al. [35] utilized limestone powder in CSA cement. It speeds up the early hydration of CSA cement because of its filler effect, thereby promoting the early age of heat evolution in CSA cement. Furthermore, the limestone powder can fill in the pores of cementitious materials, thus reducing the porosity of cement-based materials [36], and improving the strength, permeability, carbonation resistance, and corrosion resistance of reinforced concrete [37,38]. Xie et al. [39] utilized calcium carbonate, and limestone powder in PC-CSA clinker. The mixture reacted well, thereby improving the ettringite. This is because the generation of calcium carboaluminate can stabilize ettringite causing low porosity at early ages. The calcite in CSA cement can react with mono-sulphoaluminate (AFm) to form ettringite and mono-carbo aluminate, as shown in Equations (3) and (4).

$$3C4A3\overline{S} + 2C\overline{C} + 72H \longrightarrow 2C4A\overline{C}H11 + C3A \cdots 3C4\overline{S} \cdots 32H + 6AH3 \quad (3)$$

$$6C4A3\overline{S} + C\overline{C} + 135H \longrightarrow 2C4A\overline{C}\ 0.5H11 + 2C3A \cdots 3C4\overline{S} \cdots 32H + 6AH3 + 14AH3 \quad (4)$$

Additionally, in the case of anhydrite, Pelletier et al. [40] utilized anhydrite in a ternary system with PC–CSA clinker; the anhydrite promotes early hydration and allows proper hydration mechanism and reaction formation (Equation (5)). Anhydrite is a natural product, and its environmental impact improves binder hydration. Moreover, following the stoichiometry reaction of Equation (5), the hydration characteristic of the quantity of anhydrite represents around half of the ye'elimite weight (anhydrite weight/ye'elimite weight = 0.45), which had significant rapid hardening, early strength, and micro expansion characteristics as a CSA-based grouting material. In ternary blends of CSA, anhydrite, and limestone, anhydrite is more reactive with CSA clinker than limestone [41]. The production of ettringite and mono-carbonate rises with increasing anhydrite content, whereas mono-sulfoaluminate content falls.

Moreover, early hydration is accelerated as the molar ratio (M) of anhydrite to ye'elimite rises. This happens because of the reactive bonding anhydrite has with CSA cement. This bonding creates a relationship between ye'elimite and calcium sulfate, which makes them contribute immensely to the production of ettringite secondary. According to Zhou et al. [42], when excess anhydrite ($C\overline{S}$) is incorporated into a CSA–OP cement grouting material mix, rapid hydration of ye'elimite (C_4A_3) will occur, leading to the formation of ettringite (AFt) crystal and aluminum hydroxide (AH_3). Adding CSA and Portland cement will boost strength and durability. However, incorporating the proper dosage of anhydrite ($C\overline{S}$) will make the cement mortar more durable and stable enough to sustain pressure and stress during the construction phase.

$$C_4A_3\overline{S} + 2C\overline{S} + 38H \longrightarrow C_6A\overline{S}_3H_{32} + 2AH_3 \quad (5)$$

The above research findings have significantly contributed to the extensive and comprehensive application of calcium sulphoaluminate (CSA) cement as a cement-based grouting material. However, there is limited research on how the mechanical qualities and performance of CSA cement-based grouting materials are influenced by anhydrite. To meet the practical application requirements, this research explores methods to enhance the early and later age compressive strength of the grouting materials by incorporating anhydrite into the cement-based CSA, thereby creating a novel cement-based grouting material. Additionally, quicklime and anhydrite are added to influence the hydration reactions of the grouting materials, and their impact on the compressive strength, expansion, pH, and setting time is studied. The hydration process and the micro-level effects of anhydrite on the novel CSA cement-based grouting materials are investigated using X-ray diffraction (XRD) and scanning electron microscopy (SEM). The results of this research can offer valuable insights into the application of CSA-anhydrite-based grouting material.

2. Materials and Methods

2.1. Materials

The cement used in this study is a commercial CSA 42.5 cement produced by Shanxi Zhongbao Construction and Building Materials Co. of China, Ltd. (Taiyuan, China). It has a specific surface area of 450 m^2/kg and a measured density of 2.89 g/cm^3, according to the factory performance data. The CSA cement mineral composition was characterized through a Rietveld refinement X-ray diffraction analyzer (XRD) version 5.3, as shown in Table 1, and chemical components through an XRD analyzer (Rietveld refinement), as shown in Table 2. The anhydrite used in this study is off-white. It was purchased from Changzhi Zhongbao Building Material Co. of China, Ltd., Taiyuan, China, and has a density of 2.85 g/cm^3. It was formed from calcium sulfate (CS) in an aqueous solution that was deposited as gypsum crystals. Its chemical composition was determined through the Malvern Panalytical X-ray fluorescence (XRF), as shown in Table 3. In the case of CSA cement, when incorporated with anhydrite or reacted with sulfuric anhydrite (SO_3), the anhydrite serves as a regulator to control tricalcium aluminate (C_3A) hydration rate and produce more ettringite. The quicklime in this study appears white, with a density of 3.1 g/cm^3, has an effective calcium oxide content of more than 90%, and is supplied by Xilong Scientific Co., Ltd., Guangzhou, China. The quicklime easily reacts with water to form calcium hydroxide ($Ca(OH)_2$).

Table 2. Chemical composition of the CSA clinker/wt.%.

Oxide	Fe_2O_3	TiO_2	Al_2O_3	CaO	MgO	SiO_2	SO_3	LOI
wt.%	2.21	1.45	29.54	45.36	1.94	10.59	8.45	0.46

LOI = loss on ignition.

Table 3. Chemical composition of the anhydrite/wt.%.

Oxide	SiO_2	CaO	Al_2O_3	MgO	Fe_2O_3	SO_3	TiO_2	Loss
wt.%	2.83	39.5	0.23	1.36	0.03	54.16	1.17	0.75

2.2. Mix Proportions

This research intends to study the influence of anhydrite on the mechanical performance of CSA cement-based grouting materials. To achieve that, this research experiment compares the influence of anhydrite in two distant forms. First, based on the quality of the CSA clinker, the sample of CSA does not change and the anhydrite content is incorporated alone with a water-cement ratio of 0.5 to see its influence on the CSA cement grouting material. At this stage of the experiment, the content of anhydrite began from low to high (0–100%) and did not necessarily depend on the content of CSA cement to summarize the composition, as shown in Table 4.

Table 4. The influence of anhydrite content on the CSAGM.

Sample Number	1	2	3	4	5	6	7	8	9	10	11
CSA	1	1	1	1	1	1	1	1	1	1	1
anhydrite	0	10%	20%	30%	40%	50%	60%	70%	80%	90%	100%

Note: The sample content of CSA cement and anhydrite in Table 4 is changed from 10:1 to 1:1, and their ratio does not necessarily depend on each other's content to summarize the composition.

Second, anhydrite is incorporated with quicklime under the same water–cement ratio to determine how it will react to the CSA cement as a grouting material. The ratio of both materials is incorporated at equilibrium with the mass ratio of CSA cement:

(anhydrite + quicklime) is 1:1, in which the ratio of anhydrite and quicklime is changed from 1:1 to 8.5:1.5, as shown in Table 5. During these two forms of experimental procedures on anhydrite influence, the calcium sulfoaluminate cement-based grouting materials mixed mixture does not secrete water and sets the grouting material rapidly. As the test ingredients are displayed in Tables 4 and 5, the ideal ratio of the calcium sulfoaluminate cement-based grouting materials was not simply selected; rather, it was attained through extensive experimentation and natural maintenance conditions to the designated age. The quality of the anhydrite and the water-to-cement ratio were employed in the slurries to satisfy the flow of the grouting material and study the basis of the experiment.

Table 5. The influence of anhydrite + quicklime ratio on the CSAGM.

Sample Number	Slurry A	Slurry B	
	CSA	Anhydrite	Quicklime
1-A/B(50:50)	100 wt.%	50 wt.%	50 wt.%
2-A/B(75:25)	100 wt.%	75 wt.%	25 wt.%
3-A/B(80:20)	100 wt.%	80 wt.%	20 wt.%
4-A/B(85:15)	100 wt.%	85 wt.%	15 wt.%

2.3. Test Methods

2.3.1. Compressive Strength Test

According to the shape and tolerances of test specimens standard for testing the method of mechanical properties of ordinary concrete GB/T 50081-2002 [43], the specimen compressive strength was measured. As shown in Figure 1, the blended paste slurry was poured into a 50 mm × 50 mm × 50 mm mold and was prepared according to the mix proportion, as shown in Section 2.2. The paste test block compressive strength was measured for 1 d, 14 d, and 28 d using the ETM205D-TS compression testing equipment, at a loading rate of 0.3 MPa/s. According to the experiment proportion, the sample per series for each ratio is 1:4 (one proportion ratio–four test series). The average value of four replicates was taken as the experiment value, as shown in Figure 2.

Figure 1. Forming mold.

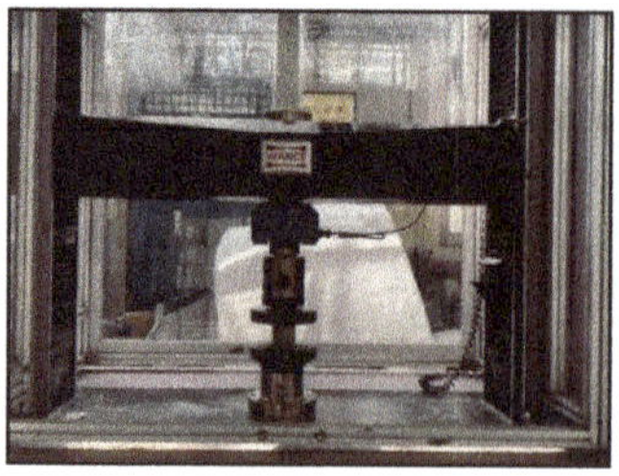

Figure 2. Compressive strength test.

2.3.2. pH Test

The pH of the slurry was tested using a PHS-3C precision pH tester produced by the Nanjing T-Bota Scietech Instruments & Equipment Co., Ltd., Nanjing, China. The slurry was stirred for 3 min to make it homogeneous. The calibrated pH meter test pen was inserted into the slurry, and readings were taken when the pH had stabilized.

2.3.3. Setting Time Test

After the blending of the slurry, it was poured into a Vicat apparatus mold. The vicat apparatus was used to measure the initial and final setting times of the blended cement paste by complying with ASTM C191 [44]. The penetration of the Vicat needle determined the initial and final setting times of the blending cement paste at 26 ± 4 mm and 0.5 ± 0.5 mm, respectively.

2.3.4. Expansion Test

The paste specimen expansion stability was studied using thermomechanical methods on solid material [45]. The slurry paste was blended for 3-mins and poured into a mold of 50 mm × 50 mm × 50 mm. After demolding, the initial lengths (L_0) of the specimens were measured as shown in Figure 3, then the specimens were cured in water at 80 °C for 24 h. Starting from the day on which the initial lengths were measured, the changes in the specimens were observed at 1 d 14 d, and 28 d, and the corresponding lengths (L_1) were measured using the triplex steel film of 25 mm × 25 mm × 280 mm as shown in Figure 4. The formula for the calculation of the expansion rates (E_X) of the specimens is given in Equation (6), and the average value of the expansion rates of the four specimens was regarded as the result for each group of specimens.

$$E_x = \frac{L_1 - L_0}{L_0 - (\Delta_1 - \Delta_2)} \times 100 \quad (6)$$

where E_X is the expansion rate of a specimen at the observed day (%); L_1 is the length of a specimen at the observed day (mm); L_0 is the initial length of a specimen (mm); and Δ_1 and Δ_2 are the lengths of the measuring heads at the left and right ends of the specimen (mm).

Figure 3. Expansion-shrinkage test device.

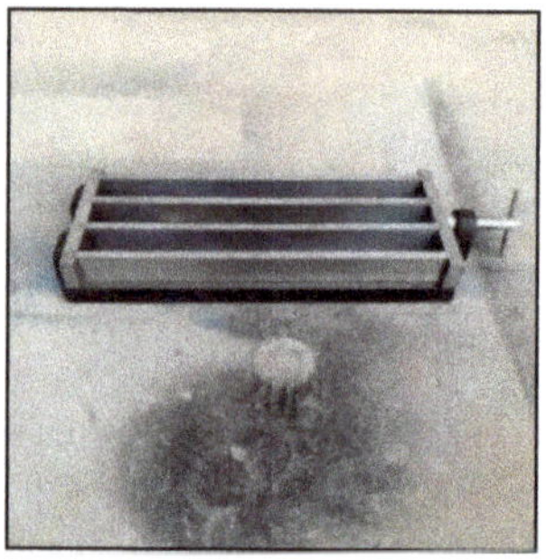

Figure 4. Triplex steel mold.

2.3.5. X-Ray Diffraction (XRD)

The hydrated samples were soaked in anhydrous ethanol for 3 d to stop the hydration. The hydrated specimen was dried in a vacuum drying oven. The temperature and the vacuum degree were 40 °C and 0.08 Mpa, respectively. After heating, the hydrated specimens were ground by hand into a powder below 0.025 mm and passed through a 200-mesh sieve. An X-ray diffraction analyzer (Rietveld refinement) was used to determine the mineralogical composition of the hydrated specimen. The test range was 5–80° at a scanning rate of 10°/min.

2.3.6. Scanning Electron Microscopy (SEM)

The samples were taken from the same hydrated sample that was immersed in anhydrous ethanol for 3 d. The hydrated samples were ground flat to a size of less than 1 cm. Then, the grounded specimen was examined by scanning electron microscopy (ZEISS Gemini SEM 300 produced in Oberkochen, Germany) using a secondary electron image and energy-dispersive X-ray (Rietveld refinement) analysis.

3. Results and Discussion

3.1. Compressive Strength

Figure 5 shows that when only CSA cement was used, the overall trend of the strength change curve fluctuated. The 1 d, 14 d, and 28 d strengths were 2.52 MPa, 122 MPa., and 10.4 MPa, respectively. As ≤30% of anhydrite dosage was incorporated, there was little increase in the early strength. The 1 d strength was 2.2 MPa at 10%, 2.72 MPa at 20%, and 3.2 MPa at 30%, respectively. However, as the curing age increases, the strength increases with a little fluctuation. The 14 d strengths are 15.2 MPa, 15.8 MPa, and 14.8 MPa, while the 28 d strengths are 19.4 MPa, 19 MPa, and 15 MPa, respectively. The cause of the decrease in the early strength was the fact that anhydrite could not participate in the early hydration because of its low dosage incorporation. However, as the curing age increases, the calcium silicate (CS) and sulfuric anhydrite (SO_3) react in the hydration reaction and produce more ettringite (AFt) to improve the strength. Following the increase in the anhydrite ratio to ≤60%, the strengths of the 1 d, 14 d, and 28 d showed a rapid increase at an early age. The 1 d strengths are 11.4 MPa at 40%, 12.05 MPa at 50%, and 15.01 MPa at 60%. The 14 d strengths are 14.7 MPa at 40%, 15.3 MPa at 50%, and 18.2 MPa at 60%, while the 28 d strengths are 15.2 MPa at 40%, 15.7 MPa at 50%, and 23.5 MPa at 60%, respectively. The cause of the increase in the early strength was the fact that anhydrite participated in the early hydration and improved the strength. As anhydrite dosage increases to ≤100%, 80% dosage is the most appropriate. The trend curve shows a more significant rise at all ages. The 1 d, 14 d, and 28 d strengths are 14.22 MPa, 23.2 MPa, and 24.52 MPa. They contribute to 10.2%, 32.3%, and 42.5% of the overall strength of anhydrite alone, respectively. At 90%

dosage, the strength curve increases continuously. The 1 d, 14 d, and 28 d strengths are 14.85 MPa, 21.5 MPa, and 23.7 MPa, respectively. At 100% dosage, the strengths of the 1 d, 14 d, and 28 d, are 14.3 MPa, 18.22 MPa, and 20.52 MPa. This shows that the impact of anhydrite on the strength of the CSA grouting material depends on the dosage.

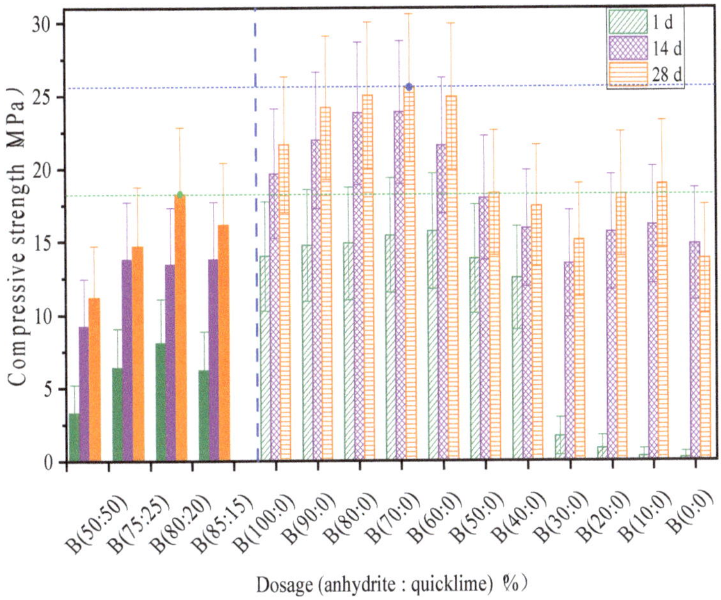

Figure 5. Effect of anhydrite–quicklime dosage on compressive strength of slurry B at different ages.

Following the incorporation of an anhydrite–quicklime ratio, the mixture with only CSA cement was used, and the overall trend of the strength change curve was the same as earlier. As anhydrite–quicklime (50:50) dosage was incorporated, the compressive strength change curve exhibited different trends. The 1 d, 14 d, and 28 d strengths are 1.7 MPa, 9.7 MPa, and 12.5 MPa which are 5% and 11.5% more than the compressive strength in the slurry with CSA alone at the early and final age. The incorporation of anhydrite-quicklime regulates the early hydration of the CSA cement that is facilitated by tricalcium aluminate (C_3A).

As the mixture ratio of anhydrite–quicklime increased to a 75:25 ratio, the strengths of the 1 d, 14 d, and 28 d were 7.5 MPa, 13.5 MPa, and 14.8 MPa. The strength was almost two times better than the 50:50 ratio at all ages. This indicates that when the anhydrite–quicklime ratio is at equilibrium, it delays the hydration which could lead to a decrease in the early strength. Following the change in the ratio of anhydrite–quicklime, the 80:20 ratio was the best. It has the highest compressive strength at all ages, reaching 9.5 MPa at 1 d, 14.7 MPa at 14 d, and 18.4 MPa at 28 d, which increases the strength to 10.2%, 20.3%, and 38.5%, respectively. At an 85:15 ratio, there was a little decrease in the early strength. The 1 d and 14 d strength curves exhibited the same trend as the 75:25 ratio. This decrease in strength was caused by the pores in the specimen and the slow participation of anhydrite in the hydration reaction. As the 85:25 ratio age increases to 28 d, the strength increases to 15.2 MPa. This shows that anhydrite has a great impact on CSAGM strength at both early and later ages when it reacts with CSA cement, and the decrease in quicklime dosage does not affect the impact on the specimen strength.

3.2. pH and Setting Time

The length of the setting time is the basis of ensuring excellent flow. In addition to the use of admixture adjustment, the influence of raw materials is also very important. The slurry pH ensures the important factor of setting time, so it is necessary to study the pH and setting time for different ratios of the raw materials. Figures 6 and 7 show the scatter diagrams of the slurry pH and the change curve of the setting time, respectively, as the CSAGM slurry increases with different dosages of anhydrite admixture.

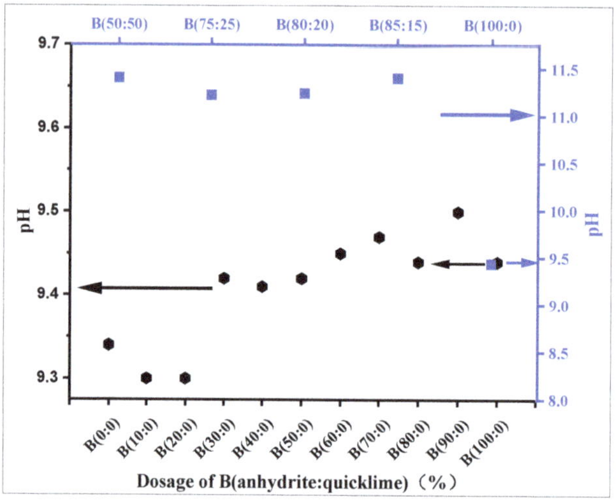

Figure 6. The pH of different anhydrite–quicklime admixtures on slurry B.

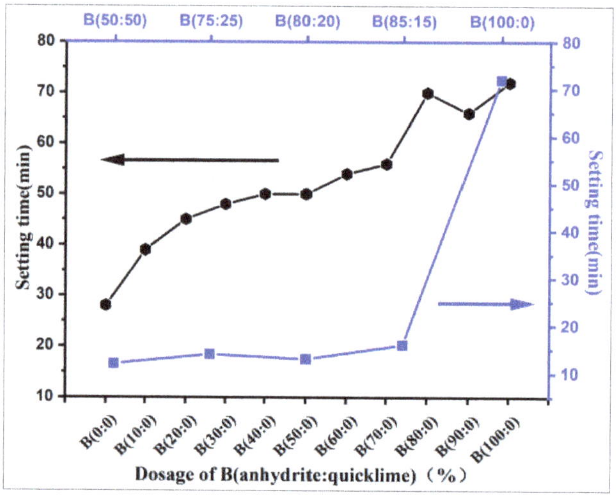

Figure 7. Slurry setting time at different anhydrite–quicklime dosages.

Figure 6 shows that in the system of only CSA clinker and anhydrite, the different amounts of anhydrite admixture have little effect on the pH value of the CSAGM slurry, and the pH value varies between 9.30 and 9.50, with no large fluctuation or change pattern. These results, combined with the strength testing results, show that the solidification strength of the slurry with a low pH can also be high. When quicklime is added, the slurry's pH value increases to approximately 11.3. The reason for this is that the quicklime dissolves in water and releases a large amount of OH-, which increases the alkalinity of the system, indirectly proving that the role of quicklime in the slurry material is mainly to increase the alkalinity of the system, thus speeding up the coagulation and promoting the hydration process [46]. This indicates that, as the relative proportion of quicklime increases, the setting time decreases rapidly, and as it becomes relatively large, it absorbs a large proportion of water because it raises the pH value of the slurry.

Following the increase in the proportion of anhydrite–quicklime admixture in slurries A and B, the setting time gradually increases and adheres to the surface of unreacted CSA clinker, thus extending the setting time and changing the overall linear pattern [47], as shown in Figure 7. In the case of CSA clinker (Slurry A) only, anhydrite admixture significantly prolongs the setting time of the clinker. This is because cement has high solubility and reactivity compared with admixture and the setting time of cement depends partly on the solubility of the cement material. In addition, this indicates that, as the amount of anhydrite increases, the hydration reaction is hindered.

3.3. Expansion Properties

Appropriate expansion can improve the practical performance of the grouting material. However, if the expansion rate is too high, it will lead to the destruction of the material, so it is necessary to consider how to achieve an appropriate expansion rate as well as a good compressive strength. Figure 8a,b shows the effect of the anhydrite–quicklime admixture on the expansion rate of the specimen.

As shown in Figure 8a, when the slurry material is mixed with CSA cement (slurry A) and anhydrite alone, the shrinkage of the test block is obvious. Following the increase in the anhydrite–quicklime ratio of slurry B material to 50:50, the expansion and cracking damage could not be measured for the specimens at the age of 1 d (Figure 8b). The reason for this is that the increase in the anhydrite–quicklime dose is at an equilibrium ratio. Incorporating too much quicklime can affect the early and later age expansion performance of the specimen, which can cause the specimen hydration and strength growth to be slow, with cracking occurring on the specimen as a result of the system not effectively controlling the expansion performance, as shown in Figure 5 The change in expansion properties after the addition of quicklime is caused by the calcium alumina product on the one hand and the expansion of quicklime on the other (Figure 8b). This indicates that, as the quicklime content increases, the expansion rate decreases [48].

When the incorporation of anhydrite is less than 40%, the slurry material exhibits a certain degree of shrinkage at all ages, as shown in Figure 8b. The volumetric unsetting mechanism of traditional cement grouting material is observed in the fluid phase, due to the long setting time. This causes cement slurry to delaminate the water separation under the action of gravity of its particles, thus decreasing the amount of water absorbed. During the delamination process of the early hardening phase, the large rapid hydration reaction dissolves a large amount of mixing water, resulting in insufficient hydration and material shrinkage deformation. In addition, the low-anhydrite admixture appears to form alumina, ferric oxide (AFm) from ettringite (AFt) conversion, which causes the volume of the cured specimen to shrink.

When the anhydrite admixture exceeds 40%, the expansion rate gradually increases at all ages because of the increase in calcium alumina generated with cumulative expansion. The addition of anhydrite significantly increased the AFt content, but the expansion effect is related not only to the ettringite (AFt) content but also to the setting time of the cement. The analysis of the setting time of the specimens (Figure 7) showed that the final setting time was prolonged after the addition of anhydrite, and the formation of ettringite (AFt) probably occurred before the hardening of the paste, resulting in the consumption of expansion energy in plastic deformation. When the anhydrite admixture continued to increase, in addition to the expansion of AFt, the remaining anhydrite also expanded, but the amount of expansion was smaller compared to that when quicklime was added [49].

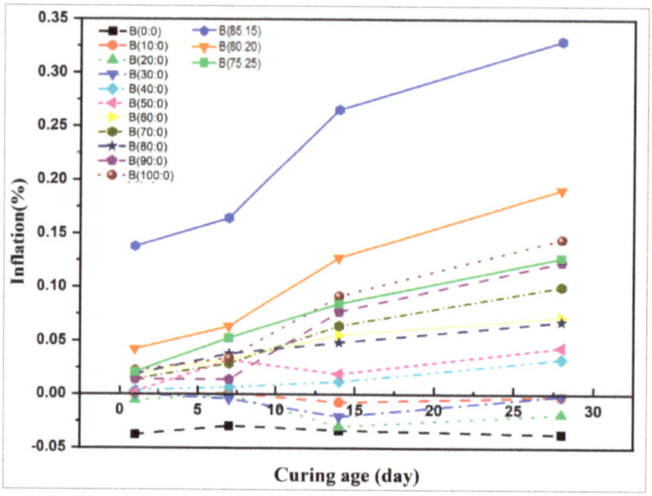

(**a**)

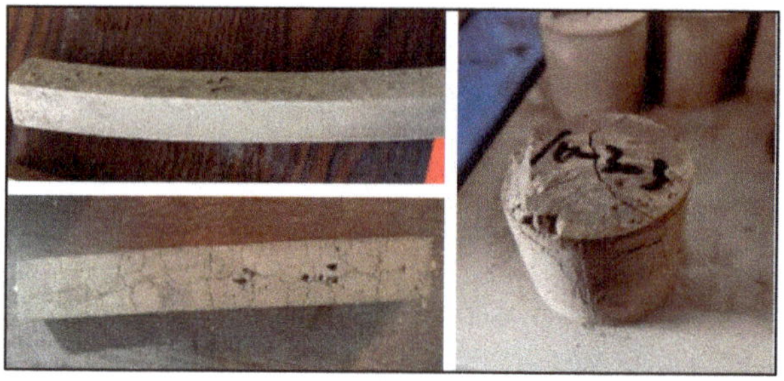

(**b**)

Figure 8. (**a**) Variation curve of the expansion rate at different anhydrite–quicklime dosages. (**b**) Sample damaged by expansion and cracking.

This indicates that different morphologies of ettringite (AFt) can produce different expansion effects. This result is in agreement with Bizzozero et al.'s research [47], which proposed that when the anhydrite admixture ratio is larger, it follows the threshold effect of expansion. This indicates that ettringite's (AFt) capacity to expand and enter the system of interconnected pores is linked to its supersaturation. It can only apply pressure in a small number of isolated pores below the critical sulfate threshold. Once the critical sulfate threshold is reached, the pores will enlarge, increasing the total volume pressure produced by calcium alumina and causing expansion.

3.4. XRD Material Phase Hydration of the CSAGM

Figure 9 shows the XRD of the CSAGM with different anhydrite–quicklime ratios at the age of 28 d. The decrease in the specimen strength of the grouting material with the increase in anhydrite admixture is due to the excess anhydrite generating a low-crystalline-phase gel similar to the coating stabilized on the surface of hydrated particles, which promoted calcium hydroxide (CH) and Sulfate (S) in the early phase, and inhibits the formation of ettringite (AFt) in the later phase. As the second product of anhydrous calcium sulfoaluminate hydration, aluminum hydroxide (AH_3) is visible in form and co-exists as $Al(OH)_3$ (bayerite and gibbsite). XRD hydration peaks of dicalcium silicate (C_2S) were detected in the XRD pattern 32.026°, but the peak variation is not obvious, indicating that the anhydrite–quicklime admixture has no significant effect on the hydration of dicalcium silicate (C_2S) in sulfoaluminate cement-based grouting materials.

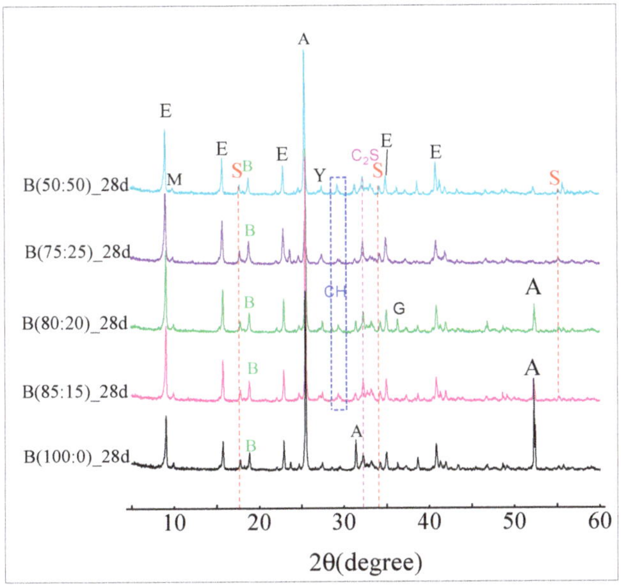

Figure 9. XRD at 28 d of age at different anhydrite-quicklime dosages (E: ettringite, Y: ye'elimite, A: anhydrite, M: alumina ferric-oxide, B: bayerite, G: gypsum, S: strätlingite).

In addition, the trace amounts of the hydration products katoite (C_2ASH_8) and calcium hydroxide ($Ca(OH)_2$) (Equation (1)) were detected when quicklime was involved in the reaction due to the dissolution of quicklime into calcium hydroxide($Ca(OH)_2$), which inhibited the hydration of dicalcium silicate (C_2S) [Equation (2)] from forming calcium silicate hydrate (C-S-H), while dicalcium silicate (C_2S) reacted with aluminum hydroxide (AH_3) to form strätlingite (C_2ASH_8). This explains why there are not any hydration products visible in the XRD diagram behind the weak calcium silicate hydrate (C-S-H) filling in the calcium alumina structure, as seen in Figure 6 and Equations (7) and (8).

$$C_2S + 2H_2O \rightarrow \text{C-S-H} + Ca(OH)_2 \tag{7}$$

$$C_2S + AH_3{\cdot}5H_3O\text{-}H + C_2\ ASH_8 \tag{8}$$

Figures 10 and 11 show the XRD spectra of the hydration products of CSA clinker with different dosages of anhydrite admixtures at 1 d and 28 d of curing age. When anhydrite was not added, the formation of ettringite (AFt) was detected in the XRD of 1 d hydration of CSA clinker accompanied by the formation of AFm (Figure 11), but the reaction was not complete, and there was remaining CSA clinker in the system. At 28 d, almost no CSA clinker was monitored (Figure 12), the main hydration product was, ettringite (AFt) and the hydration reaction was complete.

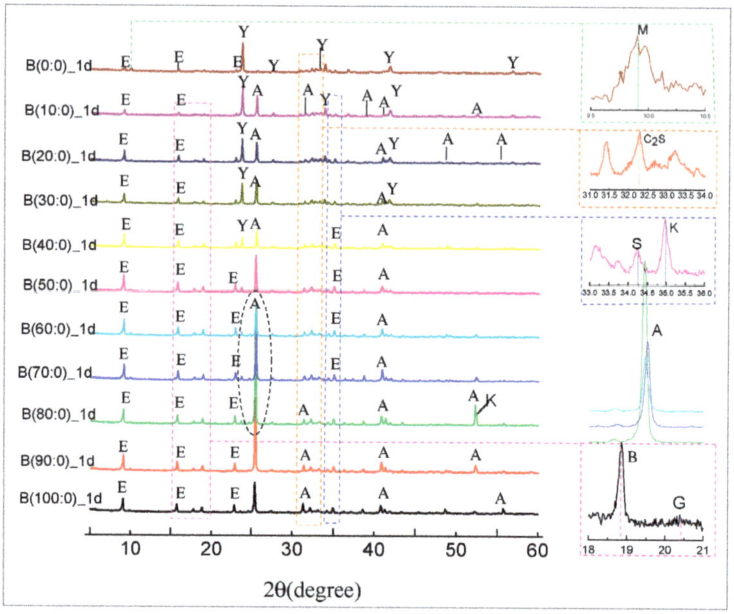

Figure 10. XRD at 1 d of age at different anhydrite dosages (E: ettringite, Y: ye'elimite, A: anhydrite, K: katoite, M: alumina ferric-oxide, B: bayerite, C: tricalcium aluminate, S: strätlingite, G: gypsum).

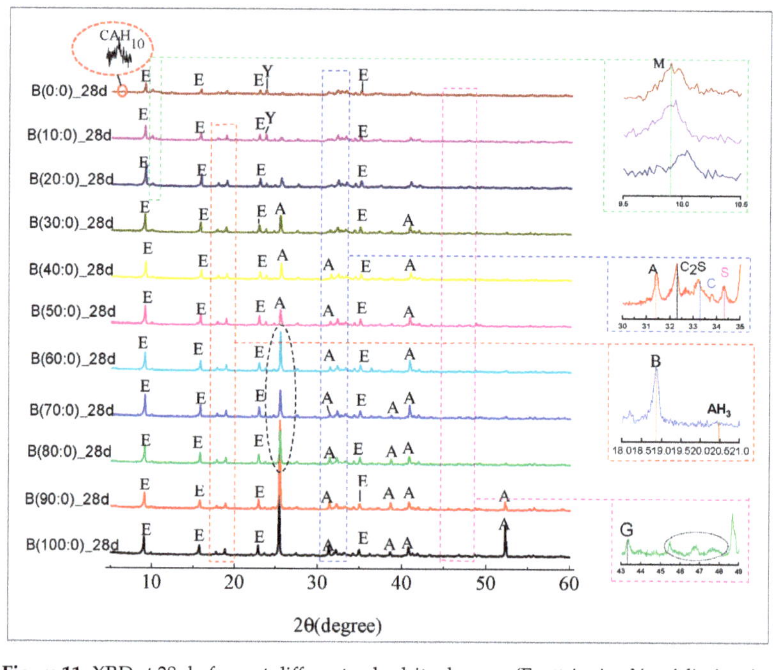

Figure 11. XRD at 28 d of age at different anhydrite dosages (E: ettringite, Y: ye'elimite, A: anhydrite, M: alumina ferric-oxide, B: bayerite, C: tricalcium aluminate G: gypsum, S: stratlingite).

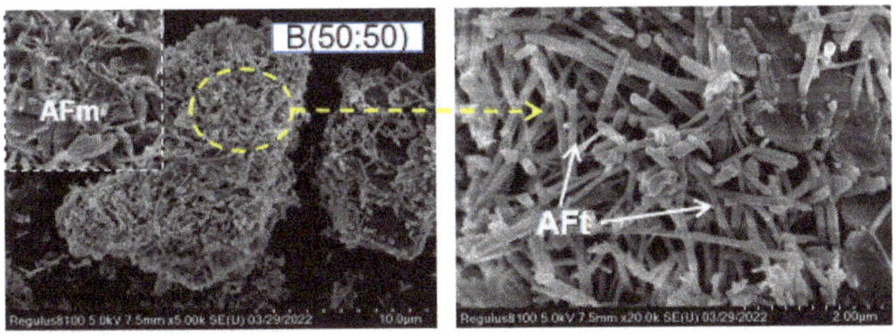

(**a**)

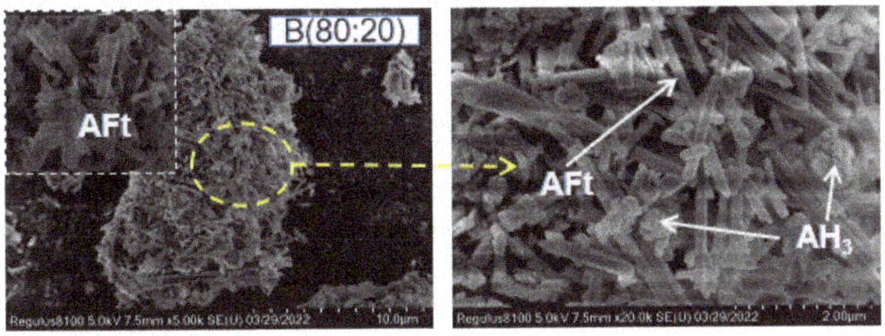

(**b**)

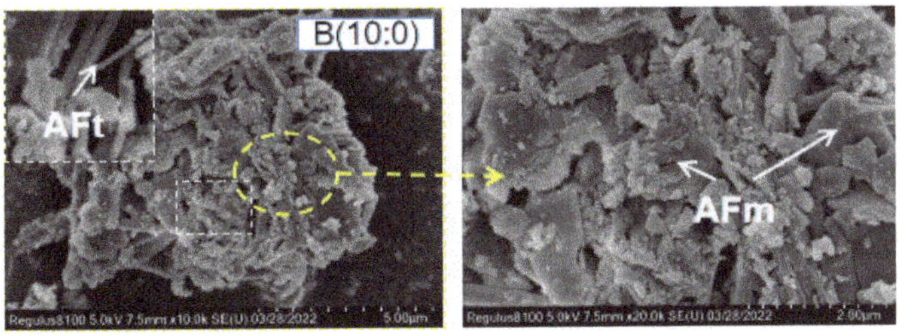

(**c**)

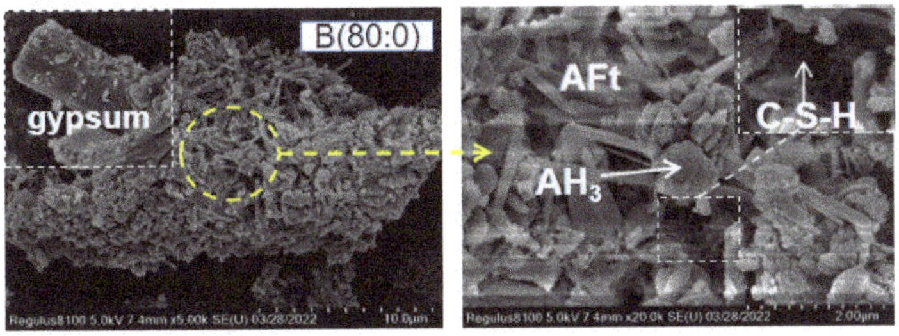

(**d**)

(**e**)

Figure 12. SEM of 28 d with different anhydrite dosage ((**a**): B(50:50); (**b**): B(80:20); (**c**): B(10:0); (**d**): B(80:20); (**e**): B(100:0)).

As shown in Figure 10, AFm was detected at 28 d of reaction when 10 to 30 wt.% anhydrite was incorporated, indicating that the hydration continued with an increase in age. AFt was decomposed and transformed to alumina ferric-oxide (AFm) due to the lack of anhydrite incorporation or insufficient anhydrite incorporation. Following the increase in anhydrite ratio to 80% or more, the early phase was slow; the main hydration product was gypsum (G) and no longer CSA clinker at 1 d, indicating that anhydrite promoted the hydration reaction, and after the CSA clinker was completely reacted, there was also a slight increase in the peak intensity of ettringite (AFt) at the later stage, as seen in Figure 10.

Figure 10 shows that no ettringite (AFt) was found in the early phase of hydration when sulfoaluminate cement was used alone. However, the sulfoaluminate adsorption promoted the formation of calcium aluminate decahydrate (CAH_{10}). The generation of strätlingite ($Ca_3A_{12}(OH)_{12}$) in the hydration product was detected at 35.053° after mixing with anhydrite, which has a significant effect on the strength improvement. This is because the generation of strätlingite ($Ca_3A_{12}(OH)_{12}$) is related to the amount of anhydrite. When the hydration of CSA clinker has been mostly completed, it is accompanied by the generation of strätlingite and aluminum hydroxide (AH_3) because the remaining anhydrite and the straitlingite ($Ca_3A_{12}(OH)_{12}$) do not continue to hydrate into ettringite (AFt) [50].

The calcium silicate hydrate (C-S-H) gel was not detected (Figures 10 and 11) in the XRD pattern during the whole reaction process because it is an amorphous substance and could not be detected by XRD. In addition, because of the high concentration of anhydrite in the hydrated environment, which induces the transition from alumina ferric-oxide (AFm), dicalcium silicate (C_2S) to ettringite (AFt), the excess anhydrite will consume calcium hydroxide ($Ca(OH)_2$) by reacting with calcium hydroxide ($Ca(OH)_2$) and aluminum hydroxide (AH_3). No diffraction peak of calcium hydroxide ($Ca(OH)_2$) was detected in the XRD when anhydrite was mixed alone. Calcium hydroxide ($Ca(OH)_2$) was detected (Figure 9) when quicklime was mixed in because the hydrated calcium hydroxide of quicklime was hardly involved in the hydration reaction. Tricalcium aluminate ($3CaO{\cdot}A_{12}O_3$) and the sulfoaluminate type of hydrates were detected by XRD. Following the increase in anhydrite ratio and hydration process, aluminum hydroxide (AH_3) gel is another hydration product that contributes to the strength after AFt to a certain extent, which is also verified by the highest production of aluminous gum at an 80% anhydrite dosage. Combined with their respective swelling rates at 28 d (Figure 8), excessive anhydrite dosage leads to expansion and thus strength reduction (Figure 5). The presence of gypsum can be significantly detected when the anhydrite dosage is 80% (Figure 11). The amount of dicalcium silicate (C_2S) (Figures 10 and 11) does not vary much, indicating that the anhydrite admixture has less effect on the degree of hydration of dicalcium silicate (C_2S) in CSAGM.

3.5. SEM Material Phase Microstructure of the CSAGM

To study the effects of different anhydrite–quicklime admixture proportions on the microstructure of CSAGM, the morphology of the specimens at a specific age (28 days) was examined using SEM, and the representative micrographs were selected (Figure 12). The selected SEM images of different dosages of CSAGM slurry B are represented with B(50:50), B(80:20), B(10: 0), B(80: 0), and B(100:0). These were tested, and the main conclusions after comparative analysis are as follows:

Figure 12a shows that the anhydrite and quicklime ratio of 50:50 has a large number of flaky single sulfosalicate stacks and long needle-like AFt staggered crosses. Because of the large amount of quicklime, the hydration reaction of the slurry is rapid and consists of a large number of ettringite (AFt) crystals winding through the anhydrite and CSA clinker that have not had time to participate in the reaction, although the amount of anhydrite is not sufficient, and the ettringite (AFt) is not sufficiently stable to be converted into a large number of stacked alumina ferric-oxide (AFm) and other monosulfate crystals.

Figure 12b shows that the AFt skeletal structure contains a significant quantity of aluminum hydroxide (AH3) gel for an anhydrite and quicklime mixture ratio at 80:20 and needle-like ettringite (AFt) crystals appear in the form of nucleation points emitting multiple clusters ettringite (AFt) which contribute to the strength improvement to a certain extent and thus better mechanical properties (Figure 5).

As shown in Figure 12c, when the anhydrite dosage was 10%, the formation of pin-rods of ettringite (AFt) was observed, as well as a large number of flakes of alumina ferric-oxide (AFm), which was consistent with the physical phase detected in the XRD peaks. However, the hydration products were poorly developed and not tightly connected, as shown in the figure, so macroscopically, the mechanical properties were poorer at this low anhydrite dosage.

When the anhydrite is 80%, the XRD pattern of Figure 12d shows that the system is rich in ettringite (AFt) crystals and plume-shaped alumina hydroxide clusters, which exhibit higher strength, and a large number of radiating ettringite (AFt) crystal clusters that mechanically compact with each other to form the structure. The overall structure is denser and is filled inside the skeleton, thus forming a dense structure and increasing its hardness. In addition, the appearance of rod-shaped gypsum inside the calcium alumina skeleton structure, which is also detected in XRD (Figure 11), is due to the partial hydration of the remaining anhydrite to form gypsum, which is explained by the fact that when the concentrations of calcium ion (Ca^{2+}) and sulfate ion (SO_4^{2-}) in the solution exceeds the supersaturation of gypsum, the calcium ion (Ca^{2+}) and sulfate ion (SO_4^{2-}) in the liquid phase becomes more concentrated. They and water molecules will spontaneously combine to form gypsum and precipitate crystals, which enhances the strength.

When anhydrite dosage is 100%, the XRD pattern of Figure 12e shows that there is a large amount of well-developed columnar ettringite (AFt) generated that is not involved in the reaction of anhydrite, as in the local enlargement. The large number of cracks is one of the reasons for the strength being slightly reduced with increasing anhydrite content. Anhydrite at this dosage has the greatest effect on the expansion rate (Figure 8) of CSAGM and causes rapid generation of a large amount of calcium ettringite (AFt) in the slurry after solidification expansion and no new material generation, so it is easy to reduce its strength due to expansion.

4. Conclusions

An early-stren grouting material was developed using calcium sulfoaluminate cement clinker (CSA), anhydrite, and quicklime. To determine the influence of anhydrite on the mechanical performance of CSA cement-based grouting material, laboratory experiments were conducted (within 1 d, 14 d, and 28 d) to evaluate the change in the setting time, expansion rate, slurry pH, and compressive strength. The hydration mechanism and microstructure properties of the grouting material were analyzed using XRD and SEM. Based on these experimental procedures, the following conclusions were drawn:

(1) Due to the relatively fast dissolution rate of anhydrite, the high strong and ettringite formation can be completed before 3 h of hydration when anhydrite is selected at a ratio ($\geq$40%), causing non-expansion characteristic on the specimen hardened body in the later stage of hydration. The optimal dose of anhydrite in the CSA grouting material paste is 80%. (80% when it is incorporated alone, and 80:20 when it is incorporated with quicklime). This dose increases the ultimate strength at all ages and makes up 42.5% and 38.5% of the compressive strength independent or dependent on quicklime.

(2) The incorporation of anhydrite alone does not affect the pH of the grouting material paste compared to anhydrite–quicklime dosage. It increases the pH value of the grouting material paste and releases a large amount of hydroxide (OH^-), which increases the alkalinity of the system This significantly prolongs the setting time of the CSA grouting material paste but does not have a direct effect on the ettringite production, nor the early or final strength.

(3) The equilibrium increase in the incorporation of the anhydrite–quicklime ratio slows the dissolution rate of the paste, leads to the expansion rate, and constrains the strength, thus leading to unified ettringite production. Increasing the anhydrite dosage inhibited the expansion performance and increased the strength. Besides, a uniformly distributed ettringite (AFt), calcium silicate hydrate, (C-S-H), and aluminum hydroxide (AH_3) were produced.

(4) The incorporation of anhydrite has a positive significant effect on the dicalcium silicate (β-C_2S) hydration products of CSAGM. The XRD and SEM characterizations confirmed that the hydration products of the CSAGM system are rich in needle-like AFt crystals, and plume-shaped aluminum hydroxide (AH3), which contribute to strength improvement. The hydration of dicalcium silicate (β-C_2S) is inhibited by forming calcium silicate hydrate (C-S-H), due to its dissolution into calcium hydroxide, which produces trace amounts of katoite (C_2ASH_8) as the hydration product.

Author Contributions: Conceptualization, Z.L.; Software, Z.L. and Y.Z.; Formal analysis, W.X.; Investigation, N.S.T.; Resources, Y.Z.; Data curation, R.H.; Writing—original draft, N.S.T.; Writing—review & editing, L.L.; Visualization, Y.Z.; Supervision, L.L. and R.H.; Project administration, W.X. All authors have read and agreed to the published version of the manuscript.

Funding: The authors greatly appreciate the financial support from the Key Laboratory of Highway Construction & Maintenance Technology in the Loess Region, Ministry of Transport, PRC. (Grant number KLTLR-Y23-4) and the Joint Funding Project of the Basic Research Program of Shanxi Provincial Department of Science and Technology (Shan Xi Transportation Holdings Group Co., Ltd.), (Grant number 202303011221002).

Institutional Review Board Statement: Not applicable.

Informed Consent Statement: Not applicable.

Data Availability Statement: The data presented in this study are available on request from the corresponding author due to privacy.

Acknowledgments: First and foremost, the authors thank God Almighty for providing them with the wisdom and expertise necessary to finish this research. The authors gratefully acknowledge the economic support for this research from Shanxi Transportation Holdings Group Co., Ltd. Likewise, the authors want to acknowledge Shanxi Zhongbao Construction and Building Materials Co., Ltd. for their materials supporting carrying out the experimental campaigns.

Conflicts of Interest: The authors declare that this study received funding from Shan Xi Transportation Holdings Group Co., Ltd. The funder was not involved in the study design, collection, analysis, interpretation of data, the writing of this article, or the decision to submit it for publication.

Abbreviations

The following abbreviations are used in this manuscript:

CSA	Calcium sulfoaluminate
CSAGM	Calcium sulfoaluminate grouting mater
wt.%	Weight percentage
pH	Potential of hydrogen
XRD	X-ray diffraction
SEM	Scanning electron microscopy
E_X	Expansion rate
L_X	Length (test block)
L_1	Initial length (test block)

References

1. Wang, Q.; Jiang, Z.; Jiang, B.; Gao, H.; Huang, Y.; Zhang, P. Research on an automatic roadway formation method in deep mining areas by roof cutting with high-strength bolt-grouting. *Int. J. Rock Mech. Min. Sci.* **2020**, *128*, 104264. [CrossRef]
2. Ding, W.; Wang, Z.; Huang, X.; Chen, L.; Zheng, Y. Influence of corrosion on anchoring bond behavior of jointed rock mass. *KSCE J. Civ. Eng.* **2022**, *26*, 1914–1928. [CrossRef]
3. Ma, S.; Zhao, Z.; Nie, W.; Gui, Y. A numerical model of fully grouted bolts considering the tri-linear shear bond-slip model. *Tunn. Undergr. Space Technol.* **2016**, *54*, 73–80. [CrossRef]
4. Dai, H.L.; Wang, X.; Xie, G.X.; Wang, X.Y. Theoretical model, and solution for the rheological problem of anchor-grouting a soft rock tunnel. *Int. J. Press. Vessel. Pip.* **2004**, *81*, 739–748. [CrossRef]
5. Bobet, A.; Einstein, H.H. Tunnel reinforcement with rockbolts. *Tunn. Undergr. Space Technol.* **2011**, *26*, 100–123. [CrossRef]
6. He, S.; Lai, J.; Wang, L.; Wang, K. A literature review on properties and applications of grouts for shield tunnel. *Constr. Build. Mater.* **2020**, *239*, 117782. [CrossRef]
7. Zhang, J.; Li, G.; Yang, X.; Ren, S.; Song, Z. Study on a high-strength ternary blend containing calcium sulfoaluminate cement/calcium aluminate cement/ordinary Portland cement. *Constr. Build. Mater.* **2018**, *191*, 544–553. [CrossRef]
8. Gu, P.; Beaudoin, J.J.; Quinn, E.G.; Myers, R.E. Early strength development and hydration of ordinary Portland cement/calcium aluminate cement pastes. *Adv. Cem.-Based Mater.* **1997**, *6*, 53–58. [CrossRef]
9. Trauchessec, R.; Mechling, J.-M.; Lecomte, A.; Roux, A.; Le Rolland, B. Impact of anhydrite proportion in a calcium sulfoaluminate cement and Portland cement blend. *Adv. Cem. Res.* **2014**, *26*, 325–333. [CrossRef]
10. Lu, J.; Ling, S.; Wang, X.; Li, X.; Liu, D. Research on Portland Cement Grouting Material in Sandy Pebble Soil. *Sci. Adv. Mater.* **2019**, *11*, 1027–1036. [CrossRef]
11. Mollamahmutoğlu, M.; Avci, E. Ultrafine Portland cement grouting performance with or without additives. *KSCE J. Civ. Eng.* **2015**, *19*, 2041–2050. [CrossRef]
12. Moosavi, M.; Bawden, W.F. Shear strength of Portland cement grout. *Cem. Concr. Compos.* **2003**, *25*, 729–735. [CrossRef]
13. Shi, C.; Qu, B.; Provis, J.L. Recent progress in low-carbon binders. *Cem. Concr. Res.* **2019**, *122*, 227–250. [CrossRef]
14. Sabbah, A.; Zhutovsky, S. Effect of sulfate content and synthesis conditions on the phase composition of belite-ye'elimite-ferrite (BYF) clinker. *Cem. Concr. Res.* **2022**, *155*, 106745. [CrossRef]
15. Fu, J.; Hu, B.; Guo, W.; Hu, Y.; Jiang, C. Effect of additive compounding on early properties of belite sulfoaluminate cement. *Adv. Cem. Res.* **2024**, *36*, 164–178. [CrossRef]
16. Habert, G. 1—Environmental impact of Portland cement production. In *Eco-Efficient Concrete*; Pacheco-Torgal, F., Jalali, S., Labrincha, J., John, V.M., Eds.; Woodhead Publishing: Cambridge, UK, 2013; pp. 3–25. [CrossRef]
17. Ige, O.E.; Olanrewaju, O.A.; Duffy, K.J.; Collins, O.C. Environmental Impact Analysis of Portland Cement (CEM1) Using the Midpoint Method. *Energies* **2022**, *15*, 2708. [CrossRef]
18. Gartner, E. Industrially interesting approaches to "low-CO_2" cements. *Cem. Concr. Res.* **2004**, *34*, 1489–1498. [CrossRef]
19. Giergiczny, Z.; Król, A.; Tałaj, M.; Wandoch, K. Performance of concrete with low CO_2 emission. *Energies* **2020**, *13*, 4328. [CrossRef]
20. Popescu, C.D.; Muntean, M.; Sharp, J.H. Industrial trial production of low energy belite cement. *Cem. Concr. Compos.* **2003**, *25*, 689–693. [CrossRef]
21. Sherman, N.; Beretka, J.; Santoro, L.; Valenti, G.L. Long-term behaviour of hydraulic binders based on calcium sulfoaluminate and calcium sulfosilicate. *Cem. Concr. Res.* **1995**, *25*, 113–126. [CrossRef]
22. Mohamad, N.; Muthusamy, K.; Embong, R.; Kusbiantoro, A.; Hashim, M.H. Environmental impact of cement production and Solutions: A review. *Mater. Today Proc.* **2022**, *48*, 741–746. [CrossRef]

23. Soomro, M.; Tam, V.W.; Evangelista, A.C.J. Production of cement and its environmental impact. In *Recycled Concrete*; Elsevier: Amsterdam, The Netherlands, 2023; pp. 11–46. [CrossRef]
24. Juenger, M.C.G.; Winnefeld, F.; Provis, J.L.; Ideker, J.H. Advances in alternative cementitious binders. *Cem. Concr. Res.* **2011**, *41*, 1232–1243. [CrossRef]
25. Park, S.; Jeong, Y.; Moon, J.; Lee, N. Hydration characteristics of calcium sulfoaluminate (CSA) cement/portland cement blended pastes. *J. Build. Eng.* **2021**, *34*, 8650. [CrossRef]
26. Liu, Y.J.; Xu, Y.M.; Geng, C.L. Sulfoaluminate cement: An alternative to portland cement. *Adv. Mater. Res.* **2012**, *368–373*, 478–484. [CrossRef]
27. Bullerjahn, F.; Zajac, M.; Ben Haha, M. CSA raw mix design: Effect on clinker formation and reactivity. *Mater. Struct.* **2015**, *48*, 3895–3911. [CrossRef]
28. Gu, X.; Tan, H.; He, X.; Smirnova, O.; Zhang, J.; Luo, Z. Utilization of carbide slag by wet grinding as an accelerator in calcium sulfoaluminate cement. *Materials* **2020**, *13*, 4526. [CrossRef]
29. Winnefeld, F.; Lothenbach, B. Hydration of calcium sulfoaluminate cements—Experimental findings and thermodynamic modelling. *Cem. Concr. Res.* **2010**, *40*, 1239–1247. [CrossRef]
30. Ma, J.; Yu, Z.; Ni, C.; Shi, H.; Shen, X. Effects of limestone powder on the hydration and microstructure development of calcium sulphoaluminate cement under long-term curing. *Constr. Build. Mater.* **2019**, *199*, 688–695. [CrossRef]
31. Shen, B.; Poulsen, B. Investigation of overburden behaviour for grout injection to control mine subsidence. *Int. J. Min. Sci. Technol.* **2014**, *24*, 317–323. [CrossRef]
32. Wu, T.; Gao, Y.; Zhou, Y. Application of a novel grouting material for prereinforcement of shield tunnelling adjacent to existing piles in a soft soil area. *Tunn. Undergr. Space Technol.* **2022**, *128*, 104646. [CrossRef]
33. Guo, X.; Shi, H.; Hu, W.; Wu, K. Durability and microstructure of CSA cement-based materials from MSWI fly ash. *Cem. Concr. Compos.* **2014**, *46*, 26–31. [CrossRef]
34. Martin, L.H.; Winnefeld, F.; Müller, C.J.; Lothenbach, B. Contribution of limestone to the hydration of calcium sulfoaluminate cement. *Cem. Concr. Compos.* **2015**, *62*, 204–211. [CrossRef]
35. Pelletier-Chaignat, L.; Winnefeld, F.; Lothenbach, B.; Müller, C.J. Beneficial use of limestone filler with calcium sulphoaluminate cement. *Constr. Build. Mater.* **2012**, *26*, 619–627. [CrossRef]
36. Wang, D.; Shi, C.; Farzadnia, N.; Shi, Z.; Jia, H. A review on effects of limestone powder on the properties of concrete. *Constr. Build. Mater.* **2018**, *192*, 153–166. [CrossRef]
37. Elgalhud, A.A.; Dhir, R.K.; Ghataora, G. Limestone addition effects on concrete porosity. *Cem. Concr. Compos.* **2016**, *72*, 222–234. [CrossRef]
38. Tsivilis, S.; Tsantilas, J.; Kakali, G.; Chaniotakis, E.; Sakellariou, A. The permeability of Portland limestone cement concrete. *Cem. Concr. Res.* **2003**, *33*, 1465–1471. [CrossRef]
39. Xie, Y.; Qian, C. Improved ettringite stabilization by calcium carbonate and calcium nitrate additions in ternary PC-CSA-C $ systems. *Cem. Concr. Res.* **2024**, *175*, 107383. [CrossRef]
40. Pelletier, L.; Winnefeld, F.; Lothenbach, B. The ternary system Portland cement–calcium sulphoaluminate clinker–anhydrite: Hydration mechanism and mortar properties. *Cem. Concr. Compos.* **2010**, *32*, 497–507. [CrossRef]
41. Long, Z.; Long, G.; Tang, Z.; Shangguan, M.; Zhang, Y.; Wang, L.; Peng, L.; Yi, M. Hydration, strength, and microstructure evolution of Portland cement-calcium sulphoaluminate cement-CSH seeds ultra-early strength cementitious system. *Constr. Build. Mater.* **2024**, *430*, 136492. [CrossRef]
42. Zhou, Z.; Li, Z.; Gao, X.; Liu, C.; Yang, L. Study on properties of Portland cement-sulfoaluminate cement-based grouting materials modified by in-situ polymerization of calcium acrylate. *Constr. Build. Mater.* **2023**, *409*, 134087. [CrossRef]
43. *GB 2002;50081:3–5*; Rong, J.M.; Lu, J.W. Shape and Tolerances of Test Specimens. The Standard for Test Methods of Mechanical Properties of Ordinary Concrete. China Building Industry Press: Beijing, China, 2003.
44. *ASTM C191-08 2008*; Standard Test Methods for the Time of Setting of Hydraulic Cement by Vicat Needle. ASTM International: West Conshohocken, PA, USA, 2008.
45. Wetton, R.E. Thermomechanical methods. In *Handbook of Thermal Analysis and Calorimetry*; Elsevier B.V.: Amsterdam, The Netherlands, 1998; pp. 363–399. [CrossRef]
46. Gabrisova, A.; Havlica, J.; Sahu, S. Stability of calcium sulphoaluminate hydrates in water solutions with various pH values. *Cem. Concr. Res.* **1991**, *21*, 1023–1027. [CrossRef]
47. Bizzozero, J.; Gosselin, C.; Scrivener, K.L. Expansion mechanisms in calcium aluminate and sulfoaluminate systems with calcium sulfate. *Cem. Concr. Res.* **2014**, *56*, 190–202. [CrossRef]
48. Mehta, P.K. Mechanism of expansion associated with ettringite formation. *Cem. Concr. Res.* **1973**, *3*, 1–6. [CrossRef]

49. Odler, I.; Colán-Subauste, J. Investigations on cement expansion associated with ettringite formation. *Cem. Concr. Res.* **1999**, *29*, 731–735. [CrossRef]
50. Jeong, Y.; Hargis, C.W.; Chun, S.-C.; Moon, J. The effect of water and gypsum content on strätlingite formation in calcium sulfoaluminate-belite cement pastes. *Constr. Build. Mater.* **2018**, *166*, 712–722. [CrossRef]

Disclaimer/Publisher's Note: The statements, opinions and data contained in all publications are solely those of the individual author(s) and contributor(s) and not of MDPI and/or the editor(s). MDPI and/or the editor(s) disclaim responsibility for any injury to people or property resulting from any ideas, methods, instructions or products referred to in the content.

materials

MDPI

Article

The Influence of Fiber Dispersion on the Properties of MgO Concrete and Engineering Applications

Feifei Jiang [1], Wencong Deng [2], Qi Wang [2], Zihan Gu [1,*] and Jialei Wang [1,*]

[1] School of Civil Engineering, Nantong Institute of Technology, Nantong 226000, China; 999620140019@just.edu.cn
[2] China State Construction Engineering (Macau) Co., Ltd., Macau 999078, China; dengwencong@hotmail.com (W.D.); wangqi_2019@cohl.com (Q.W.)
* Correspondence: 18752791818@139.com (Z.G.); wangjl@nit.zju.edu.cn (J.W.)

Abstract: Adding expansion agents to compensate for concrete shrinkage is a common crack resistance technique, but excessive expansion can also increase the porosity of concrete and reduce its strength. The addition of fibers can reduce expansion and improve the compactness of concrete. However, too little fiber will not be effective in inhibition, while too much fiber will cause aggregation. In this study, steel fiber and MgO expansive agent were used at the same time, and the effect of fiber on the mechanical properties of MgO concrete was studied. The results showed that the appropriate amount of MgO (8%) could compensate for the shrinkage of concrete and slightly improve the strength of concrete. When the content reached 10%, MgO produced excessive expansion under free conditions, which reduced the strength of the concrete. After using MgO and steel fiber at the same time, steel fiber could restrain the expansion of MgO, improve the compactness of concrete, produce a "super superposition" benefit, and increase the strength of concrete by 20%. In addition, the reinforcing effect of steel fiber on MgO was closely related to its distribution. In the composite system, steel fiber not only played a "bridge role" but also needed steel fiber to effectively restrain the expansion of MgO and produce self-stress. Only when the steel fibers were evenly distributed could reliable bonding be ensured between the fibers and the matrix, and at this time, the fibers could restrain the expansion of MgO. Considering the uniformity of steel fiber distribution and construction cost, adding 8% MgO and 1% steel fiber has the maximum benefit.

Keywords: MgO expansion agent; fiber; fiber distribution; crack resistance; compressive strength

1. Introduction

In modern engineering construction, concrete is the most commonly used building material. But at the same time, the problem of cracking in the hardening and service process of concrete is becoming increasingly prominent. The emergence of cracks not only affects the aesthetics of the structure but also may lead to a decrease in the durability of the concrete, which in turn affects the overall safety and service life of the engineering structure [1–3]. The main causes of concrete cracking include factors such as drying shrinkage, autogenous shrinkage, temperature stress, and external environmental changes [4–6]. Especially in modern high-strength concrete, due to the low water–cement ratio, high strength, and concentrated hydration heat, drying shrinkage and temperature shrinkage are more pronounced [7,8]. These shrinkages generate tensile stress inside the concrete, ultimately leading to cracks [9,10].

Adding an expansion agent is a commonly used method for crack resistance. The hydration of the expansion agent produces expansion, which can counteract the drying shrinkage and self-shrinkage during the early hardening process of concrete and reduce the risk of cracks caused by shrinkage stress. Traditional expansion agents, such as sulphoaluminate and CaO expansion agents, can provide some expansion compensation. However, due to their fast reaction speed, large early expansion, and small expansion in the middle and late stages, they cannot match the long-term shrinkage process of concrete [11–13]. On the other hand, a MgO expansion agent has become a more ideal compensating shrinkage material due to designable expansion [14], low water demand [15], and stable hydration products [16], especially for structures requiring durability and crack resistance, such as underground engineering, bridges, and water conservancy projects [17].

However, some people doubt that the MgO expansive agent will reduce the compressive strength of concrete. They believe that MgO hydration will not only compensate for the shrinkage and produce macro expansion but also increase the porosity of concrete, which will have an adverse impact on the strength. Wang [18] found that after adding 8% MgO, the expansion of MgO increased the porosity of concrete and reduced the compressive strength of 3d, 28d, and 180d by 17.2%, 16.4%, and 14.6%, respectively. What is more serious is that the excessive expansion will produce harmful expansion inside the concrete, damaging the microstructure and integrity of the concrete. Ye [19] found that adding 10% MgO resulted in an excessive expansion of MgO under high-temperature conditions, leading to expansion cracks and a 37% decrease in the compressive strength of concrete.

In order to avoid excessive expansion of MgO causing a decrease in concrete strength, researchers restricted the expansion of MgO concrete by applying constraints. Zhou [20] studied the application of MgO in concrete-filled steel tubes and found that the compensation of MgO compensated for the shrinkage of concrete avoided the void between the inner wall and concrete. In addition, the steel tube restrained the expansion and improved the density of the concrete, thereby increasing the bearing capacity by 19%. However, the steel tube–concrete composite structure is not common in engineering. Therefore, researchers studied common reinforced concrete structures, where steel bars also restrain the expansion of concrete. Yu [21] found that in the reinforced concrete structure, the longitudinal reinforcement restrained the expansion of MgO, improved the compactness of concrete, reduced the 28-day porosity by 6.8%, and improved the strength of concrete. However, as shown in Figure 1, because the longitudinal stressed reinforcement is distributed in the direction of the principal stress generated by the external load, it can effectively restrain the expansion in the vertical direction, but there is no constraint in the parallel direction, so the constraint effect is limited.

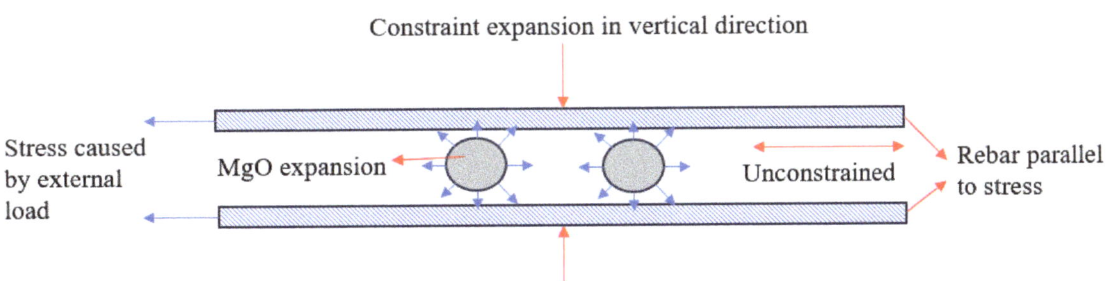

Figure 1. Expansion of MgO in reinforced concrete structures.

Based on the above analysis, the performance of MgO concrete can be improved by the constraint, but both the steel tube and the stressed reinforcement have certain limitations. In the previous study, our group attempted to use steel fibers to constrain the expansion of MgO. The randomly distributed steel fibers can constrain the expansion in any direction, thereby enhancing the strength of the concrete [22]. However, it has also been found that the distribution of fiber has a great influence on the performance of concrete. When the fibers are unevenly distributed and agglomerated, the expansion cannot be limited, and the "strengthening and toughening effect" cannot be played. Previous studies have shown that the distribution of steel fibers has a significant impact on the mechanical and durability properties of concrete. When the fibers are unevenly distributed, or oriented perpendicular to the principal tensile stress, the reinforcement and toughening effects of the fibers and the effect of restraining expansion are greatly reduced [23,24]. In the composite system, the distribution of steel fibers will not only affect its "bridging effect" but also affect its constraint effect on MgO expansion. Therefore, the study of fiber dispersion is more valuable for fiber—MgO composite systems than ordinary fiber concrete.

This paper takes the super high-rise buildings in Macao as the research object and analyzes in detail the influence of the coupling effect of MgO and fiber on the performance of mass platform concrete. The basement platform of high-rise buildings usually has a large area and thickness. After concrete is poured, there will be significant temperature drop shrinkage and drying shrinkage, which can easily lead to shrinkage cracks. Therefore, the MgO expansion agent is used to compensate for shrinkage, and steel fibers are added to further improve the strength and crack resistance of concrete. The internal mechanism of the influence of fiber dispersion on the performance of MgO concrete is studied in detail through the layered cutting experiment. Finally, the optimal dosage of steel fibers is confirmed, and fiber-reinforced MgO concrete is successfully applied in engineering construction. The research findings are of significant importance in addressing the issue of MgO reducing the strength of concrete.

2. Materials and Methods

2.1. Materials

The cement is PO 52.5 grade ordinary Portland cement, with a specific surface area of 348 m^2/kg and a specific density of 3142 kg/m^3. The fly ash is Class I fly ash with a specific surface area of 409 m^2/kg and a sulfate content of 1.2%. The detailed chemical composition of cement and fly ash is shown in Table 1. The fine aggregate is river sand, with an apparent density of 2623 kg/m^3 and a fineness modulus of 2.93. Coarse aggregate is 5–20 mm limestone, with an apparent density of 2728 kg/m^3 and a bulk density of 1478 kg/m^3. The steel fiber is a wavy steel fiber with a tensile strength of 520 MPa, a density of 7.8 kg/m^3, a diameter of 0.58 mm, and a length of 38 mm. The activity of the MgO expansion agent is 105 s. The water reducer is a polycarboxylic acid high-performance water reducer with a water reduction rate of 28.1%. To meet the construction and workability of pumping during later on-site construction, the amount of water-reducing agent is adjusted to ensure that the concrete slump of each mixing ratio is always within a reasonable range (100–140 mm) in the test. MgO is added in three different concentrations: 6%, 8%, and 10% by mass. The steel fiber is set at three different concentrations: 0.5%, 1%, and 1.5% by volume. Detailed mix proportions of concrete are shown in Table 2. In order to make the fiber distribution as uniform as possible, the mixing process of secondary feeding is adopted. First, put the coarse and fine aggregates, cement and fly ash into the mixer and mix for 3 min; then, add steel fiber and mix again for 3 min. Finally, add water and water-reducing agent, and stir for another 5 min.

Table 1. Chemical composition of cement and fly ash/%.

No.	CaO	MgO	Al_2O_3	SiO_2	Fe_2O_3	SO_3	K_2O	Na_2O	Loss	Total
Cement	60.9	2.3	6.4	22.4	3.0	1.6	0.58	0.16	2.1	99.4
Fly ash	5.9	1.4	33.9	47.9	6.1	1.4	0.42	0.17	1.9	99.1

Table 2. Detailed mix proportions of concrete/kg.

No.	Cement	Fly Ash	Water	Water Reducer	Crushed Stone	Sand	MgO	Steel Fiber
MF-0-0	450	50	160	7	1025	713	0	0
MF-6-0	450	50	160	7	1025	713	30 (6%)	0
MF-8-0	450	50	160	7.5	1025	713	40 (8%)	0
MF 10-0	450	50	160	8.5	1025	713	50 (10%)	0
MF-0-0.5	450	50	160	6	1025	713	0	0.5
MF-0-1	450	50	160	6	1025	713	0	1
MF-0-1.5	450	50	160	6.5	1025	713	0	1.5
MF-6-0.5	450	50	160	7	1025	713	30 (6%)	0.5
MF-6-1	450	50	160	7.5	1025	713	30 (6%)	1
MF-6-1.5	450	50	160	8.5	1025	713	30 (6%)	1.5
MF-8-0.5	450	50	160	7	1025	713	40 (8%)	0.5
MF-8-1	450	50	160	7.5	1025	713	40 (8%)	1
MF-8-1.5	450	50	160	8.5	1025	713	40 (8%)	1.5
MF-10-0.5	450	50	160	7	1025	713	50 (10%)	0.5
MF-10-1	450	50	160	7.5	1025	713	50 (10%)	1
MF-10-1.5	450	50	160	8.5	1025	713	50 (10%)	1.5

2.2. Methods

2.2.1. Compressive Strength Test and Splitting Tensile Strength Test

The compressive strength test and splitting tensile strength test are conducted in accordance with the test method described in the "Test Method for Mechanical Properties of Ordinary Concrete" (GB/T 50081-2016) [25], using a specimen size of 150 mm × 150 mm × 150 mm. After demolding, each fresh concrete test block is placed in a curing room and tested for compressive strength when it reaches the corresponding testing age. The curing temperature is 20 °C ± 2 °C, and the curing humidity is above 95% RH. The loading speed is 0.5 MPa/s.

2.2.2. Fiber Distribution Test Process

(1) Cutting of test piece. First, as shown in Figure 2, drill and core the steel fiber concrete to obtain a cylinder with a diameter of 70 mm and a height of 150 mm. Next, divide the cylinder into five equal parts along the height direction, numbered 1–5 (1 for the top surface and 5 for the bottom surface). The height of each small test block is 30 mm, and the cutting diagram is shown in Figure 3;

Figure 2. Drilling of concrete.

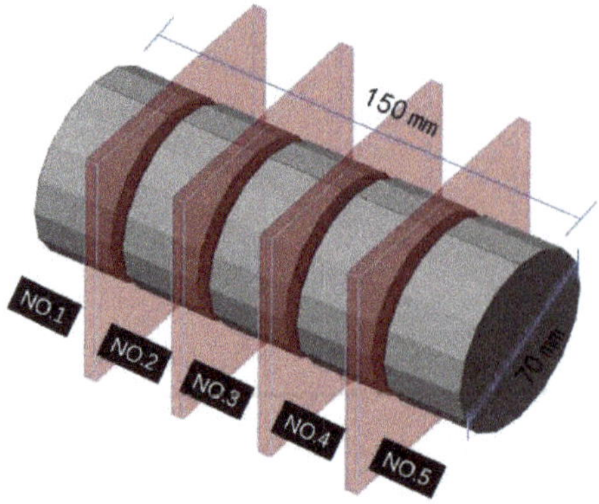

Figure 3. Schematic diagram of concrete cutting.

(2) Crushing of test pieces. Place the cut small test block flat on the workbench and slowly tap it with a hammer to break it;
(3) Weighing of steel fibers. Use a strong magnet to suck out all steel fibers from the crushed small test block and weigh them, recording the weight as mi (where i is the test piece number).

2.2.3. Calculation Method

After weighing the steel fibers in each section of the test piece, the steel fiber content of the five parts from the top to the bottom of the test piece is calculated. The fiber content of each type of concrete height is taken as the average value of eight test pieces. Two methods are used to analyze the uniformity of steel fiber distribution at different locations.

(1) Direct calculation method.

Directly compare the fiber volume fraction (x_i) of specimens with different heights to evaluate the distribution. The specific calculation method is shown in Formula (1).

$$x_i = \frac{m_i/\rho_F}{V_i} \quad (1)$$

where x_i is the volume fraction of steel fiber in the ith small test block, m_{Fi} is the mass of steel fiber in the ith small test block, ρ_F is the density of steel fiber, and V_i is the measured volume of the ith test block

(2) The fiber variation coefficient ($\psi_{(x)}$) and fiber distribution coefficient γ are used to evaluate the distribution of fibers. The calculation method is shown in Formulas (2) and (3).

$$\psi(x) = \sqrt{\frac{\Sigma(x_i/\overline{x} - 1)^2}{n}} \quad (2)$$

$$\gamma = exp[-\psi(x)] \quad (3)$$

where n is the number of test pieces divided, and n = 5 in this experiment; x_i is the measured volume fraction of steel fibers in the ith small test block; $\overline{x}$ is the average value of the measured volume fraction of steel fibers in 5 small test blocks; and γ is the distribution coefficient of steel fibers in the test piece. When γ is closer to 1, it indicates that the steel fibers are more uniform. When γ is closer to 0, it indicates that the steel fiber is less uniform.

2.2.4. Microstructure Testing

A scanning electron microscope (SU3500, Hitachi, Chiyoda, Japan) is used to scan the fiber–matrix interface transition zone to study the effects of steel fiber distribution and MgO expansion on the microstructure of concrete. In order to reduce the influence of surface laitance on the test, the sample 5 cm below the surface is selected for microstructure test.

3. Results and Discussion

3.1. Mechanical Properties of Concrete

The compressive strength of 3 days and 28 days concrete is shown in Figures 4 and 5. From Figure 4, it can be found that in the early stage, the effect of steel fiber on the strength of concrete is not obvious, and even some concrete strength has decreased, such as the strength of MF-0-5 is reduced by 5.7%. There are two main reasons for this. The first is that in the early stage of concrete (within 3 days), the hydration of cement is not yet complete, and the matrix has not yet formed sufficient strength. Steel fibers come into contact with an incompletely hydrated matrix, resulting in low interfacial bonding strength. The second reason is due to the shrinkage of the cement matrix. After the cement is hardened, drying shrinkage and chemical shrinkage occur, and tiny gaps and internal tensile stresses are generated at the interface between the steel fibers and the matrix, weakening the fiber–matrix bond and forming microstructural defects, resulting in a decrease in compressive strength. This results in M-10-0.5 having a strength of 4.8 MPa higher than MF-0-0.5. The addition of MgO compensates for shrinkage, resulting in higher strength at equal fiber content.

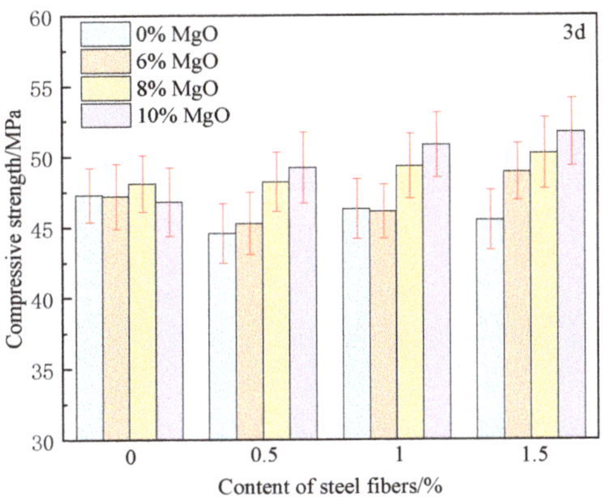

Figure 4. Compressive strength of concrete at 3 days.

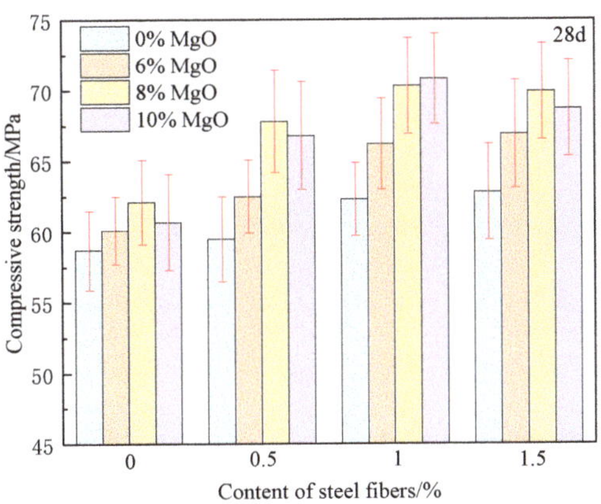

Figure 5. Compressive strength of concrete at 28 days.

On the other hand, when there is no constraint (no steel fiber added), the harmful expansion occurs when the MgO content reaches 10%. The strength of M-10-0 is 1.4 MPa lower than that of M-8-0, indicating that in engineering, the amount of MgO cannot be increased blindly. Under free conditions, excessive MgO can cause harmful expansion, increase the porosity of concrete, reduce its strength, and even lead to expansion cracks. This is also the main reason why structural designers currently oppose the use of MgO in engineering construction.

In order to further analyze the influence of the composite effect of MgO and steel fibers on compressive strength, the compressive strength enhancement coefficient μ_c of concrete under various mix proportions was calculated according to Formula (4), and the results are shown in Tables 3 and 4.

$$\mu_c = \frac{f_{ci} - f_{co}}{f_{co}} \times 100\% \quad (4)$$

where f_{ci} is the compressive strength of fiber-reinforced expansive concrete with different dosages, and f_{co} is the compressive strength of ordinary concrete.

Table 3. Reinforcement factor of compressive strength at 3 days.

ID	0% Fiber	0.5% Fiber	1% Fiber	1.5% Fiber
0% MgO	0.0%	−5.7%	−2.1%	−3.8%
6% MgO	−0.2%	−4.2%	−2.5%	3.4%
8% MgO	1.7%	1.9%	4.2%	6.1%
10% MgO	−1.1%	4.0%	7.4%	9.3%

Table 4. Reinforcement factor of compressive strength at 28 days.

ID	0% Fiber	0.5% Fiber	1% Fiber	1.5% Fiber
0% MgO	0.0%	1.4%	6.1%	7.0%
6% MgO	2.4%	6.5%	12.8%	14.0%
8% MgO	5.8%	15.5%	19.8%	19.1%
10% MgO	3.4%	13.8%	20.6%	17.0%

From Table 4, it can be observed that after 28 days, the hydration reaction of cement is basically completed, and the strength of the concrete matrix significantly increases. After hydration, the matrix becomes denser, and the strength of the interface transition zone between steel fibers and the matrix is enhanced, which helps to fully utilize the reinforcing effect of steel fibers. Therefore, after adding steel fibers, the 28 days reinforcement coefficient is greater than that of 3 days. When the steel fiber content is 0.5%, the strength is only 1.4% higher than that of M-0-0. This is mainly due to insufficient fiber content and uneven distribution, which can not give full play to the "bridge" role of fibers. When the dosage increases from 0.5% to 1.0%, the reinforcement coefficient rapidly increases from 1.4% to 6.1%, mainly due to the uniform distribution. When the dosage continues to increase from 1.0% to 1.5%, the reinforcement coefficient only increases from 6.1% to 7.0%. This is mainly because although the fiber dosage increases, the fiber uniformity deteriorates, and even agglomeration occurs. Fiber aggregation leads to a significant decrease in the bonding strength between the matrix and fibers, and the residual strength of the concrete decreases after the matrix cracks, making it impossible to fully utilize the reinforcing effect of fibers. Considering the high price of steel fiber, it is recommended that the dosage of steel fiber should not exceed 1% in engineering applications.

On the other hand, after using both MgO and steel fibers simultaneously, the strength of concrete increased by 20.6%, which is greater than the sum of the increments when using

MgO and steel fibers alone (7.0% + 5.8%), indicating that MgO and steel fibers have produced a super superposition effect. MgO expansion agent produces an expansion effect during the hydration process. When the expansion is limited by steel fibers, compressive stress (self-stress) is generated between the fibers and the matrix. This self-stress effectively counteracts the shrinkage tensile stress in the early stage of concrete hardening, avoiding the generation of microcracks caused by shrinkage. The pre-stress applied by the MgO expansion agent to the matrix can generate a high bonding force in the interface transition zone, making the interface bonding between steel fibers and cement matrix more stable, reducing the occurrence of interface microcracks, and thus better exerting the reinforcement effect of steel fibers. Of course, this strengthening effect is closely related to the distribution of fibers, and only a uniform distribution of fibers can better constrain the expansion of MgO.

From the previous analysis, it can be seen that the dosage of steel fibers and MgO expansion agent can both affect the compressive strength of concrete. By conducting regression analysis using Origin analysis software 2022 on the strength of different dosages, a formula for predicting the compressive strength of concrete under the dual factor effect is obtained to quantitatively calculate the influence of both on the strength of concrete. This provides a calculation method for predicting the strength of fiber-reinforced MgO concrete in engineering design. The prediction formula is calculated based on two raw materials—MgO with an activity of 105 s and steel fiber with a length of 38 mm—used in this experiment. If the raw materials change, the parameters of the model can be corrected through a small amount of supplementary experiments to improve the prediction accuracy. The fitting results are shown in Figure 6, and the calculation method is shown in Formula (5).

$$f_c = -4.08F^2 - 0.08M^2 + 9.19F + 0.54M + 0.26MF + 57.4 \quad (5)$$

where f_c is the compressive strength of concrete at 28 days, M is the content of MgO, and F is the content of steel fiber.

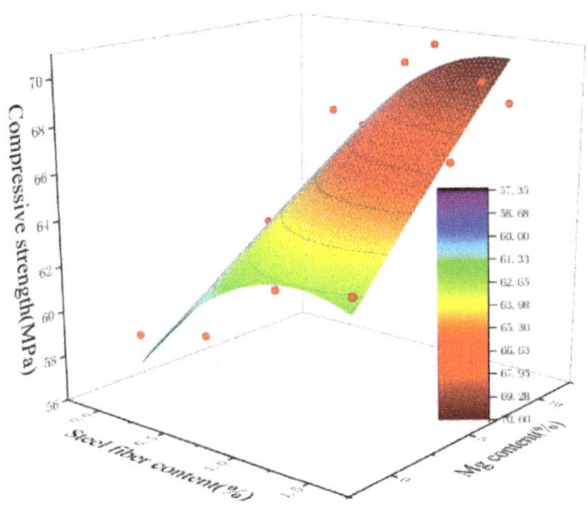

Figure 6. Regression analysis of the compressive strength of fiber-restrained expansive concrete.

Figures 7 and 8 show the splitting tensile strength of concrete at 3 days and 28 days, respectively. It can be found that when steel fibers and MgO expansion agent are used in combination, MgO further stimulates the ability of steel fibers, and the tensile strength of concrete is improved by 33.6–65.0%. When the MgO content is 8%, the tensile strength of 1% and 1.5% steel fiber-reinforced concrete reaches 6.63 MPa and 6.76 MPa, respectively. Compared to ordinary concrete, it has increased by 61.3% and 64.5%, respectively, which is much higher than the sum of the increments when used separately. Moreover, the composite effect enhances the splitting tensile strength more than the compressive strength. This is mainly due to the transfer of load to adjacent steel fibers after the matrix cracks. Due to the composite effect, the bond strength between the matrix and fibers is improved, which enhances the residual strength of the concrete after cracking and allows it to continue to withstand tensile stress.

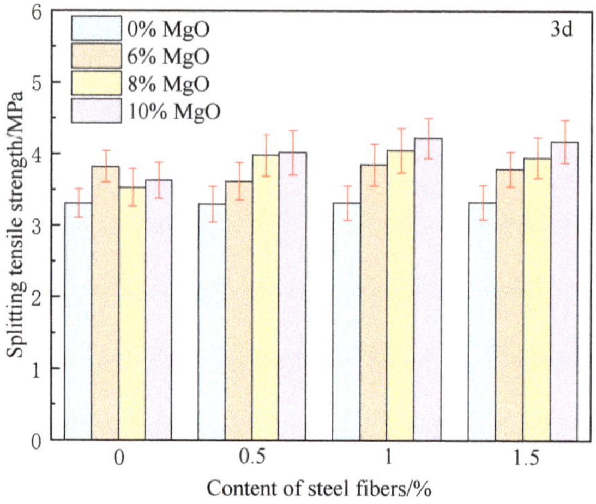

Figure 7. Splitting tensile strength of concrete at 3 days.

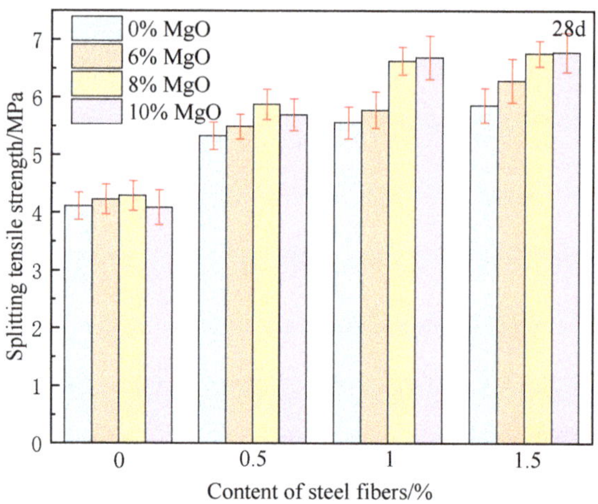

Figure 8. Splitting tensile strength of concrete at 28 days.

3.2. Analysis of Steel Fiber Distribution at Different Depths

The changes in fiber volume fraction of concrete slices with different MgO contents at different depths are shown in Tables 5–8. Based on the measured volume fractions of each slice, the variation coefficient (ψ) and distribution coefficient (γ) can be calculated.

Table 5. Fiber volume fraction of 0% MgO concrete.

Depth	0.5% Fiber	1% Fiber	1.5% Fiber
1	0.36	0.88	1.26
2	0.49	1.14	0.98
3	0.55	0.93	1.73
4	0.62	1.1	2.31
5	0.71	1.15	1.68
Average value	0.55	1.04	1.59
Variation coefficient	0.217	0.108	0.285
Distribution coefficient	0.805	0.897	0.752

Table 6. Fiber volume fraction of 6% MgO concrete.

Depth	0.5% Fiber	1% Fiber	1.5% Fiber
1	0.33	1.08	1.56
2	0.35	0.86	1.28
3	0.45	1.06	1.01
4	0.58	1.11	1.68
5	0.71	1.12	2.29
Average value	0.48	1.05	1.56
Variation coefficient	0.297	0.091	0.275
Distribution coefficient	0.743	0.913	0.759

Table 7. Fiber volume fraction of 8% MgO concrete.

Depth	0.5% Fiber	1% Fiber	1.5% Fiber
1	0.35	0.89	1.21
2	0.40	1.08	0.93
3	0.42	1.17	1.51
4	0.52	0.98	1.13
5	0.56	1.18	2.13
Average value	0.45	1.06	1.38
Variation coefficient	0.173	0.105	0.302
Distribution coefficient	0.841	0.900	0.739

Table 8. Fiber volume fraction of 10% MgO concrete.

Depth	0.5% Fiber	1% Fiber	1.5% Fiber
1	0.39	0.85	1.26
2	0.33	0.83	0.93
3	0.42	0.91	1.28
4	0.56	1.06	1.42
5	0.63	1.02	2.16
Average value	0.47	0.93	1.41
Variation coefficient	0.239	0.098	0.289
Distribution coefficient	0.787	0.907	0.749

The variation of the steel fiber volume fraction in samples of different heights is shown in Figure 9. From Figure 9, it can be observed that the volume fraction of steel fibers in each section does not remain constant but varies along the height direction. When the steel fiber content is small (0.5%), gravity plays a dominant role. As the depth increases, the volume content of steel fibers shows a significant increase trend, rapidly increasing from 0.36% to 0.71%. This drastic change in content along the depth position can easily lead to local defects in the cement matrix. When the steel fiber content increases to 1%, due to the mutual friction between steel fibers and the obstruction of coarse aggregates, the fiber settlement caused by gravity is no longer significant, and the fiber content is evenly distributed along the height direction. When the steel fiber content continues to increase to 1.5%, as shown in Figure 10, there is a significant agglomeration of steel fibers. The steel fiber content varies greatly at different heights, with a maximum of 2.31% and a minimum of only 0.98%. This also explains why the strength of concrete decreases instead of increasing when the steel fiber content increases to 1.5%. After fiber agglomeration, the following three adverse consequences will occur: (1) Fiber agglomeration reduces the uniformity of concrete, reduces the bond strength between fiber and matrix, and the weak interface between fiber and matrix will become a channel for harmful ions to penetrate, thus reducing the durability of concrete. (2) Agglomeration reduces the constraint of fiber on MgO expansion and reduces self-stress, which is unfavorable to crack resistance. (3) Due to the increase in fiber content, it also increases the construction cost, resulting in a waste of resources. On the other hand, only after the fibers are evenly distributed can a reliable connection between the fibers and the matrix be ensured. After the matrix cracks under applied load, the fibers can continue to bear tensile stress and avoid crack propagation. This can improve the residual strength of concrete after cracking. Moreover, reducing the number and width of cracks can effectively decrease the penetration rate of harmful ions, which has a high value in improving the durability of concrete. Meanwhile, uniform fiber distribution can better constrain the expansion of MgO, generate significant self-stress, and improve the volume stability of concrete.

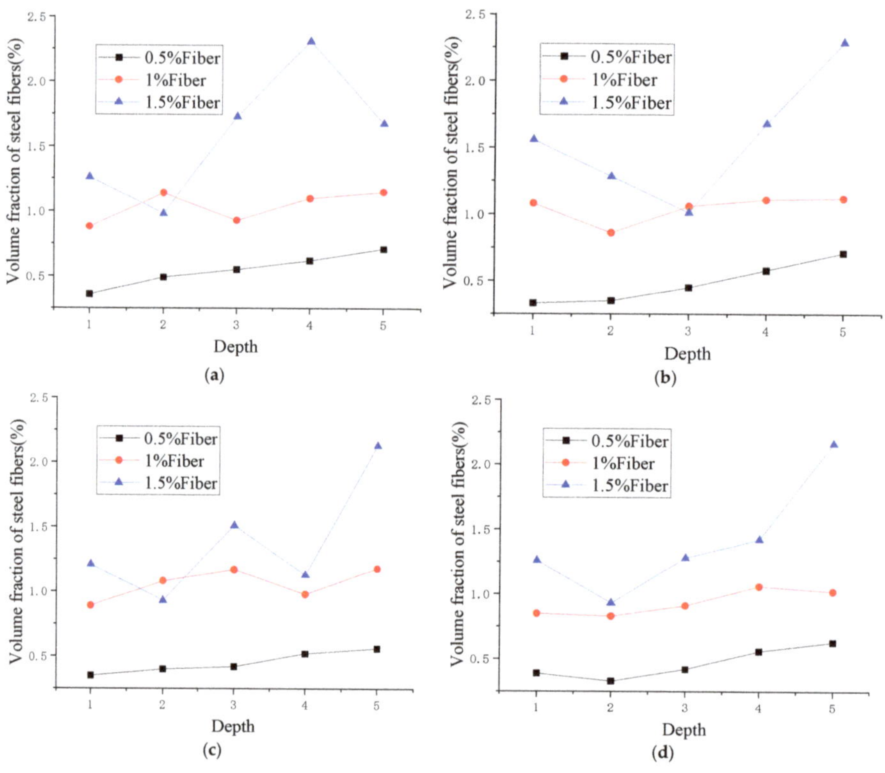

Figure 9. Changes in steel fiber content at different depths: (**a**) 0% MgO; (**b**) 6% MgO; (**c**) 8% MgO; (**d**) 10% MgO.

Figure 10. Morphology of concrete with 10% steel fiber content.

Figure 11 shows the fitting surface of the volume fraction of steel fibers with different dosages. For specimens with different steel fiber dosages, there are significant differences in the fiber volume fraction. When the steel fiber content is 1%, the volume fraction of fibers is most stable and is minimally affected by depth and MgO content. When the steel fiber content is 0.5%, the stability of the fiber volume fraction slightly decreases and is less affected by the MgO content but more affected by depth. When the content of steel fiber is 1.5%, the stability of the fiber volume fraction is poor, and the fiber agglomerates and fluctuates violently with depth.

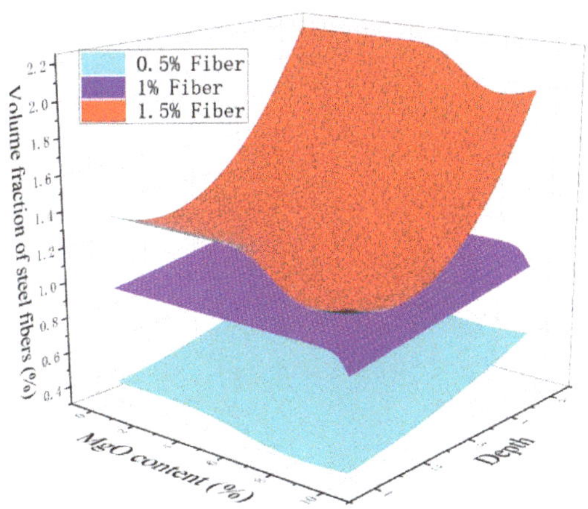

Figure 11. Fitting of volume fraction of steel fibers with different depths and fibers.

Distribution Coefficient of Steel Fibers

The fiber distribution coefficients of different steel fibers and MgO content are shown in Figure 12. From Figure 12, it can be observed that for concrete specimens without MgO, as the fiber content increases from 0.5% to 1%, the distribution coefficient increases from 0.81 to 0.90, and the fiber distribution gradually becomes more uniform. When the fiber content increases from 1% to 1.5%, the distribution coefficient rapidly decreases to 0.75, causing strong agglomeration of fibers and seriously affecting the uniformity of distribution. Due to the consumption of water during the hydration of MgO, theoretically, as the dosage of the MgO expansion agent gradually increases, the flowability of concrete greatly decreases, and the viscosity also increases, which will have a certain impact on the distribution of fibers. However, in this experiment, in order to ensure the pumpability of concrete construction, the amount of water-reducing agent was increased to ensure the construction performance of concrete. Therefore, the MgO expansion agent has a relatively small impact on fiber dispersion. Overall, when the steel fiber content is 1%, the volume fraction of fibers is the most stable and is minimally affected by depth and MgO content. When the steel fiber content is 0.5%, the stability of the fiber volume fraction slightly decreases and is less affected by the MgO content but more affected by depth. The steel fiber content is 1.5%, and the stability of the fiber volume fraction is poor. Due to random fiber aggregation, the content of steel fibers fluctuates violently. Therefore, in terms of fiber distribution, 1% of steel fibers have the most uniform spatial distribution, which can effectively enhance the reinforcing effect of steel fibers and also have a stronger restraining effect on MgO expansion.

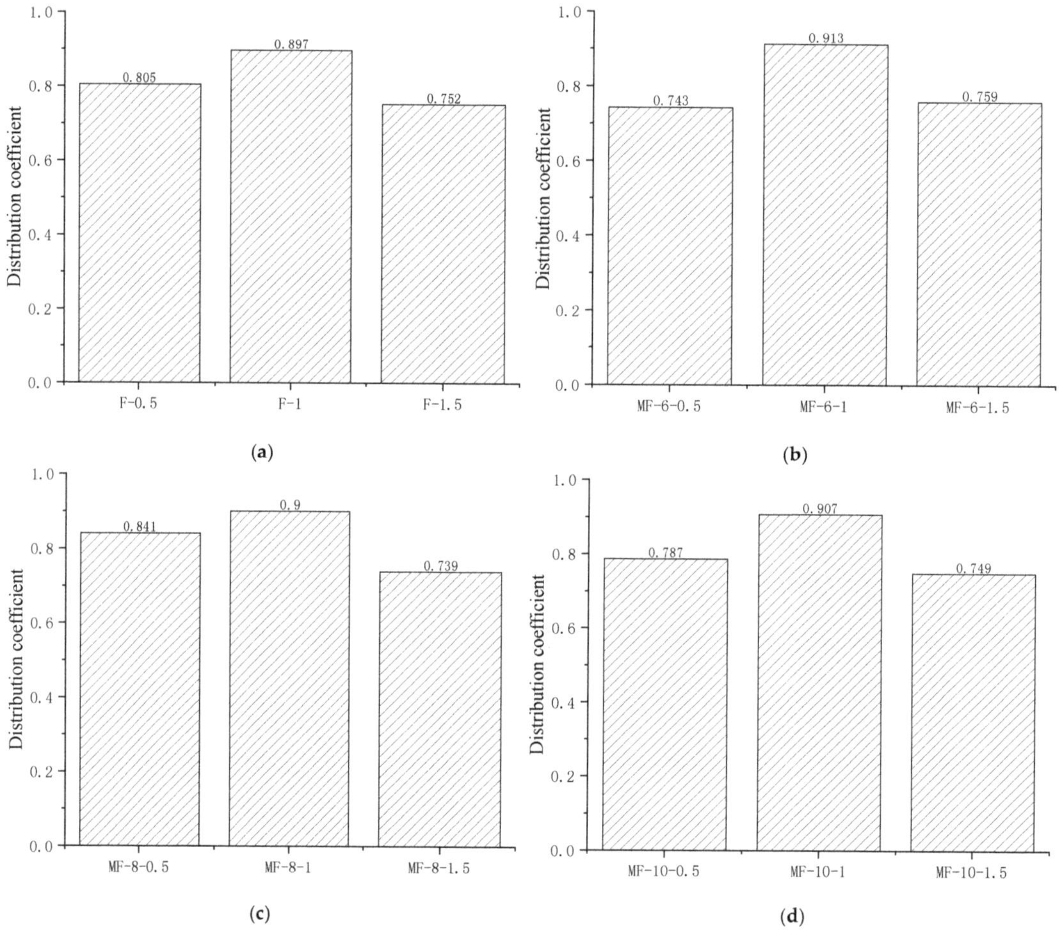

Figure 12. Distribution coefficient of steel fibers in concrete: (**a**) 0% MgO; (**b**) 6% MgO; (**c**) 8% MgO; (**d**) 10% MgO.

3.3. Microstructure of Concrete

As shown in Figure 13a, when steel fibers are used alone, due to the hydrophobicity of steel fibers and the "sidewall benefit", the water/binder ratio on the fiber surface is greater than that of the matrix [26], resulting in some holes in the transition zone. On the other hand, due to the continuous hydration and shrinkage of cement after concrete pouring, there is a gap between the fiber and the matrix. The existence of holes and gaps reduces the strength of concrete.

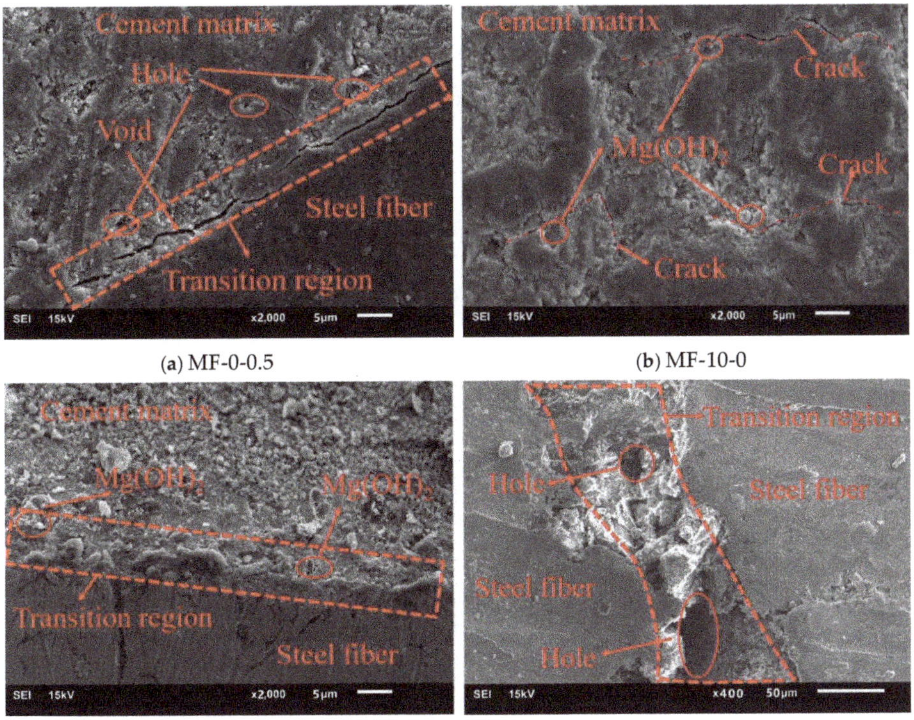

(a) MF-0-0.5 (b) MF-10-0

(c) MF-8-1 (d) MF-8-1.5

Figure 13. Microstructure of concrete.

It can be seen from Figure 13b that when MgO is used alone and the content is large, reaching 10%, a large number of expansion cracks are generated in the matrix. From the energy spectrum, it can be found that there is a large amount of Mg $(OH)_2$ in the cracks, indicating that this is the harmful expansion caused by excessive MgO. The relevant literature [18] indicates that MgO expansion increases the porosity of concrete, thereby reducing its compactness. As a result, the 28-day compressive strength of the concrete decreases by 16.4%. Therefore, when using MgO to compensate for concrete shrinkage, attention should be paid to avoid the negative effect of excessive expansion on strength and durability.

It can be found from Figure 13c that when 8% MgO and 1% steel fiber are reasonably added, the steel fiber is evenly distributed without agglomeration. And the steel fiber can restrain the expansion of MgO from producing self-stress, which improves the compactness of the fiber–matrix interface transition zone and greatly improves the "bridge effect" and "expansion limiting effect" of the fiber.

As shown in Figure 13d, when the steel fiber content reaches 1.5%, the fibers are locally agglomerated, further increasing the water–binder ratio in the fiber transition zone. After the matrix hardens, the excess water dries, resulting in a large number of holes, which seriously reduces the fiber–matrix bond strength and weakens the role of steel fibers.

3.4. Engineering Application

3.4.1. Project Overview

This construction project is a super high-rise building in Macao, with a total of six residential buildings of 149.90 m, 50 floors high, and one basement. The project covers an area of about 20,000 square meters and a construction area of about 330,000 square meters. The foundation structure mainly uses bored piles and pre-stressed concrete piles, which are connected to each pile cap by a base plate. The basement in this project is 267.9 m long and 139.7 m wide.

3.4.2. Analysis of the Performance of the Concrete in the Basement Platform

(1) Huge size: In super high-rise buildings, the concrete platform has a large surface area and substantial thickness. The large surface area leads to rapid drying shrinkage of concrete, resulting in significant drying shrinkage. And the substantial thickness results in the slow dissipation and accumulation of heat generated by the hydration reaction, leading to a significant temperature drop and shrinkage at later stages. When the shrinkage is restrained, tensile stress develops, which can cause cracking in the concrete;

(2) Complex load: The platform needs to withstand the combined effects of static loads from the upper structure and dynamic loads from vehicles. Therefore, it requires concrete to have high strength;

(3) High waterproof requirements: The basement floor concrete, as the bottom structure of the building, needs to have good waterproof performance to prevent damage to the building caused by groundwater leakage;

(4) Durability requirements: Due to the long service life of super high-rise buildings, the concrete of the platform needs to have good durability and crack resistance to cope with long-term loads and environmental factors.

Through the above analysis, it is found that the platform concrete has high requirements for crack resistance, strength, and durability, so 8% MgO and 1% steel fiber are added during construction. Detailed construction coordination is shown in Table 2. The main construction process includes cushion construction, steel bar binding, concrete pouring, surface finishing, and concrete curing, as shown in Figure 14.

3.4.3. On-Site Testing

To avoid the accumulation of hydration heat in the concrete and reduce the accumulation of shrinkage deformation, the concrete is poured into separate sections. After the curing reaches the specified age, the strength of the concrete is tested using the rebound method, and its deformation is measured using strain gauges. The detailed results are shown in Figure 15. From Figure 15, it can be found that in the early stage, due to the heat release of cement hydration, the concrete undergoes thermal expansion, reaching 143 $\mu\varepsilon$ at 1d. After that, as the temperature decreases, the volume shrinks. At this time, the hydration expansion of MgO compensates for the temperature drop shrinkage and drying shrinkage, and the concrete remains in a slightly expanded state during the testing period of 14 days without experiencing shrinkage cracking problems. And the strength also exceeds the design strength (50 MPa). Therefore, the use of MgO and steel fiber composite can meet the construction requirements.

Figure 14. Platform concrete construction: (**a**) Reinforcement binding; (**b**) concrete pouring; (**c**) surface finishing; (**d**) maintenance.

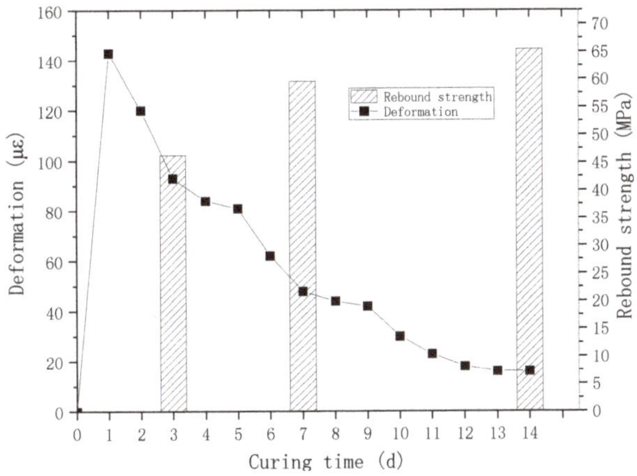

Figure 15. On-site measured deformation and strength of concrete.

3.4.4. Economic Analysis

(1) Initial material cost

For every cubic meter of concrete, 8% MgO (by mass) and 1% steel fiber (by volume) are added, equivalent to 40 kg of MgO and 78 kg of steel fibers. The unit price of MgO is 1 RMB/kg, and the unit price of steel fiber is 3 RMB/kg. Therefore, the additional composite material cost per cubic meter of concrete is 40 × 1 + 78 × 3 = 274 RMB.

(2) Post-Maintenance cost analysis

(a) Crack resistance

MgO provides a micro-expansion effect that offsets shrinkage stress, significantly reducing the risk of concrete cracking. In the long term, fewer cracks help prevent waterproof layer failures and leakage risks, saving substantial costs on crack repairs. Steel fibers enhance the toughness and tensile strength of concrete, which is especially effective in preventing the propagation of cracks, thereby reducing costs associated with repairing structural damage caused by cracks. When they are used simultaneously, steel fibers constrain the self-stress generated after MgO expansion, further improving the concrete's strength and crack resistance.

(b) Durability

The combined use of MgO and steel fibers improves the density of concrete, enhancing its impermeability. For underground structures, the improved waterproofing and corrosion resistance reduces the long-term repair costs caused by environmental degradation.

(3) Economic comparison

The maintenance cost is calculated based on the regular maintenance frequency of the super high-rise basement platform. The cost of repairing cracks in ordinary concrete is 100 RMB per square meter per year. After using MgO expansive agent and steel fiber, the cracking probability of concrete can be reduced by more than 50%, and the later repair cost can be reduced to 50 RMB per square meter or even lower. Using a platform area of 10,000 square meters as an example, the maintenance cost for 10 years can be calculated. The repair cost for cracks in ordinary concrete is RMB 1 million, while the repair cost for composite material concrete is only RMB 500,000.

(4) Overall recommendations

The simultaneous use of MgO expansion agents and steel fiber can significantly reduce crack-related maintenance costs and water seepage losses, especially for complex underground structural environments, with significant economic benefits in the later stage. From an economic perspective, for mega projects like the one in this paper, if the initial budget allows, the combination of MgO expansion agent and steel fiber is worth promoting, especially in the basement environment of super high-rise buildings, which can effectively reduce the total cost in the later stage of use. But for small projects, if the harm caused by cracking is not serious and easy to repair, then MgO and steel fiber may not be applicable.

4. Conclusions

This study addresses the cracking issues in floor concrete and the possible performance deficiencies in the application of MgO expansion agents. The research results indicate that the synergistic effect of steel fibers and MgO expansion agents can effectively improve the mechanical properties of concrete, and the uniform dispersion of fibers is the core factor in improving strength. The main conclusions are as follows:

(1) When steel fiber is used alone, the bond strength between steel fiber and incomplete hydration matrix is low at the early stage, and the fiber–matrix interface transition zone becomes the weak point of concrete, so the fiber can not improve the compressive strength. In the later stage, with the continuous hydration of cement, the matrix–fiber bond strength gradually increased, the fiber gradually played a "bridging effect", and the compressive strength increased by 7.0%. Without restraint (fiber), harmful expansion will occur when the MgO content reaches 10%, and the strength of M-10-0 is 1.4mpa lower than that of M-8-0, indicating that the MgO content cannot be blindly

increased. It is necessary to avoid harmful expansion after excessive MgO under free conditions. In engineering practice, the use of steel fiber and MgO alone has some negative effects;

(2) After using MgO and steel fiber at the same time, the strength of concrete increases by 20.6%, which is greater than the sum of increments when using MgO and steel fiber alone (7.0% + 5.8%), indicating that MgO and steel fiber have a super superposition effect. When the expansion is limited by the steel fiber, it will introduce compressive stress (self-stress) between the fiber and the matrix. This kind of self-stress effectively offsets the shrinkage tensile stress at the initial stage of concrete hardening and reduces the generation of micro-cracks caused by shrinkage. The preloading stress applied by MgO to the matrix can produce a higher bonding force in the interface transition zone, which makes the interface between steel fiber and cement matrix more stable and reduces the occurrence of interface micro-cracks so as to give better play to the reinforcement effect of steel fiber. Of course, this strengthening effect is closely related to the distribution of fibers, and only the uniform distribution of fibers can better restrain the expansion of MgO;

(3) The distribution of fiber has a great influence on the properties of MgO concrete. When the fiber content is small, and the distribution is uneven, it can not play the role of "bridge" and "limiting expansion". When the fiber content is large, it will produce agglomeration, which has a serious impact on the performance of MgO concrete. Firstly, fiber agglomeration reduces the uniformity of concrete and the bond strength between fiber and matrix. The weak interface between fiber and matrix will become a channel for harmful ions to penetrate, thus reducing the durability of concrete. Secondly, agglomeration reduces the constraint of fiber on MgO expansion and self-stress, which is unfavorable to crack resistance. Finally, due to the increase in fiber content, it also increases the construction cost, resulting in a waste of resources;

(4) When the content of steel fiber is 0.5%, the stability of fiber volume fraction is poor, which is less affected by the content of MgO but more affected by the depth (gravity). When the content of steel fiber is 1.5%, the stability of the fiber volume fraction is very poor. Due to the random fiber agglomeration, the content of steel fiber fluctuates violently. Therefore, in terms of fiber distribution, 1% steel fiber is the most evenly distributed in space, which can better play the role of steel fiber in strengthening and toughening and has a stronger restraining effect on MgO expansion. At this time, the strength of concrete is also higher. When the steel fiber continues to increase to 1.5%, the cost increases, but the strength is no longer significantly improved. It is most reasonable to add 1% steel fiber during construction;

(5) Different from the conventional fiber-reinforced concrete, the uniformity of fiber is more important for the composite system using MgO and steel fiber at the same time. In the composite system, steel fiber not only plays the role of "bridge" but also needs to effectively restrain the expansion of MgO to produce self-stress. Only under the premise of uniform distribution and reliable fiber–matrix bonding can the fiber restrain the expansion. Considering the high price of steel fiber and MgO, adding 8% MgO and 1% steel fiber has the maximum benefit. If the dosage continues to increase, the cost of purchasing raw materials will be significantly increased, but the strength cannot be significantly improved.

Author Contributions: Methodology, F.J. and W.D.; Formal analysis, F.J., Q.W. and Z.G.; Investigation, W.D.; Resources, F.J.; Data curation, W.D., Z.G. and J.W.; Writing—original draft, Z.G.; Writing—review & editing, Q.W. and J.W. All authors have read and agreed to the published version of the manuscript.

Funding: The project was funded by Basic Science (Natural Science) Research Project of Higher Education Institutions in Jiangsu Province: 24KJD560004; Jiangsu University 'Blue Project': 2024.

Institutional Review Board Statement: Not applicable.

Informed Consent Statement: Not applicable.

Data Availability Statement: The original contributions presented in the study are included in the article, further inquiries can be directed to the corresponding author.

Acknowledgments: The authors would like to thank Zhongyang Mao from Nanjing Tech University for his precious contribution to the revision of this work.

Conflicts of Interest: Authors Wencong Deng and Qi Wang were employed by the company China State Construction Engineering (Macau) Co., Ltd. The remaining authors declare that the research was conducted in the absence of any commercial or financial relationships that could be construed as a potential conflict of interest.

References

1. Bogas, J.A.; Ahmed, H.H.; Diniz, T. Influence of Cracking on the Durability of Reinforced Concrete with Carbon Nanotubes. *Appl. Sci.* **2021**, *11*, 1672. [CrossRef]
2. Srinivasan, S.S.; Ferron, R.D. Effect of Crack Width, Density, and Depth on Strength and Durability of Concrete-Equivalent Mortar. *Aci Mater. J.* **2021**, *118*, 65–77.
3. Li, K.F.; Li, L. Crack-altered durability properties and performance of structural concretes. *Cem. Concr. Res.* **2019**, *124*, 105811. [CrossRef]
4. Deysel, R.C.; Boshoff, W.P.; Smit, M.S. Implementing capillary pressure control measures to prevent plastic shrinkage cracking in concrete. *Constr. Build. Mater.* **2023**, *397*, 132407. [CrossRef]
5. Heweidak, M.; Kafle, B.; Al-Ameri, R. Influence of Discrete Basalt Fibres on Shrinkage Cracking of Self-Compacting Ambient-Cured Geopolymer Concrete. *J. Compos. Sci.* **2023**, *7*, 202. [CrossRef]
6. Miao, Y.C.; Lu, Z.Y.; Wang, F.J.; Wang, H.; Li, Y.L.; Lin, J.L.; Jiang, J. Shrinkage cracking evolvement in concrete cured under low relative humidity and its relationship with mechanical development. *J. Build. Eng.* **2023**, *72*, 106670. [CrossRef]
7. Zhang, M.M.; Gao, S.; Liu, T.; Guo, S.Y.; Zhang, S.T. High-Performance Materials Improve the Early Shrinkage, Early Cracking, Strength, Impermeability, and Microstructure of Manufactured Sand Concrete. *Materials* **2024**, *17*, 2465. [CrossRef]
8. Du, J.; Tan, X.; Wang, Y.H.; Bao, Y.; Meng, W.N. Reducing the cracking potential of ultra-high-performance concrete (UHPC) with the prewet expansive agent. *Constr. Build. Mater.* **2024**, *431*, 136597. [CrossRef]
9. Meyer, D.M.; Combrinck, R. Utilising microCT scanning technology as a method for testing and analysing plastic shrinkage cracks in concrete. *Constr. Build. Mater.* **2022**, *317*, 125895. [CrossRef]
10. Liu, J.P.; Tian, Q.; Wang, Y.J.; Li, H.; Xu, W. Evaluation Method and Mitigation Strategies for Shrinkage Cracking of Modern Concrete. *Engineering* **2021**, *7*, 348–357. [CrossRef]
11. Mo, L.W.; Deng, M.; Tang, M.S.; Al-Tabbaa, A. MgO expansive cement and concrete in China: Past, present and future. *Cem. Concr. Res.* **2014**, *57*, 1–12. [CrossRef]
12. Guo, J.N.; Zhang, S.W.; Qi, C.G.; Cheng, L.; Yang, L. Effect of calcium sulfoaluminate and MgO expansive agent on the mechanical strength and crack resistance of concrete. *Constr. Build. Mater.* **2021**, *299*, 123833. [CrossRef]
13. Zhang, J. Recent advance of MgO expansive agent in cement and concrete. *J. Build. Eng.* **2022**, *45*, 103633. [CrossRef]
14. Kabir, H.; Hooton, R.D. Evaluating soundness of concrete containing shrinkage-compensating MgO admixtures. *Constr. Build. Mater.* **2020**, *253*, 119141. [CrossRef]
15. Choi, S.W.; Jang, B.S.; Kim, J.H.; Lee, K.M. Durability characteristics of fly ash concrete containing lightly-burnt MgO. *Constr. Build. Mater.* **2014**, *58*, 77–84. [CrossRef]
16. Li, F.X.; Chen, Y.Z.; Long, S.Z.; Wang, B.; Li, G.G. Research on the preparation and properties of MgO expansive agent. *Adv. Cem. Res.* **2010**, *22*, 37–44. [CrossRef]
17. Jiang, F.F.; Deng, M.; Mo, L.W.; Wu, W.Q. Effects of MgO Expansive Agent and Steel Fiber on Crack Resistance of a Bridge Deck. *Materials* **2020**, *13*, 3074. [CrossRef]
18. Wang, L.; Song, X.; Yang, H.; Wang, L.; Tang, S.; Wu, B.; Mao, W. Pore Structural and Fractal Analysis of the Effects of MgO Reactivity and Dosage on Permeability and F–T Resistance of Concrete. *Fractal Fract.* **2022**, *6*, 113. [CrossRef]
19. Ye, Q.; Yu, K.; Zhang, Z. Expansion of ordinary Portland cement paste varied with nano-MgO. *Constr. Build. Mater.* **2015**, *78*, 189–193. [CrossRef]

20. Zhou, X.J.; Hou, D.S.; Chen, T.; Wang, X.P. The development of concrete filled steel tube with enhanced performance via the use of expansive ultra high performance concrete. *J. Build. Eng.* **2023**, *79*, 107793. [CrossRef]
21. Yu, L.; Deng, M.; Mo, L.; Liu, J.; Jiang, F. Effects of Lightly Burnt MgO Expansive Agent on the Deformation and Microstructure of Reinforced Concrete Wall. *Adv. Mater. Sci. Eng.* **2019**, *2019*, 1948123. [CrossRef]
22. Jiang, F.F.; Liu, R.G.; Mao, Z.Y.; Deng, M. The combined effect of steel fiber and MgO on the deformation and mechanical properties of high- strength concrete. *J. Mater. Res. Technol.* **2023**, *26*, 4296–4309. [CrossRef]
23. Zhang, S.L.; Liao, L.; Song, S.Z.; Zhang, C.S. Experimental and analytical study of the fibre distribution in SFRC: A comparison between image processing and the inductive test. *Compos. Struct.* **2018**, *188*, 78–88. [CrossRef]
24. Omidinasab, F.; Moghadam, A.S. Effect of Purposive Distribution of Fibers to Prevent the Penetration of Bullet in Concrete Walls. *KSCE J. Civ. Eng.* **2021**, *25*, 843–853. [CrossRef]
25. *GB/T 50081-2016*; Standard for Test Method of Mechanical Properties of Ordinary Concrete. China Standards Press: Beijing, China, 2016.
26. Min, B.Z.; Chen, G.L.; Sun, Y.Y.; Li, K.; Chen, X.; Wang, Z.Q. Enhancing the fracture properties of carbon fiber-calcium silicate hydrate interface through graphene oxide. *Mater. Des.* **2024**, *241*, 112916. [CrossRef]

materials

MDPI

Article

The Role and Mechanism of Rice Husk Ash Particle Characteristics in Cement Hydration Process

Jialei Wang [1,2], Xiaoqing Hu [3], Feifei Jiang [4,*] and Haoyu Chen [1]

1 School of Civil Engineering and Architecture, NingboTech University, Ningbo 315100, China; wangjl@nit.zju.edu.cn (J.W.); chy1416386018@163.com (H.C.)
2 School of Mechanics & Civil Engineering, China University of Mining and Technology, Xuzhou 221116, China
3 School of Civil Engineering and Architecture, Chongqing Institute of Engineering, Chongqing 400056, China; hu812255201@hotmail.com
4 School of Civil Engineering, Nantong Institute of Technology, Nantong 226000, China
* Correspondence: 999620140019@just.edu.cn

Abstract: Reactive rice husk ash (RHA) is used as a supplementary cementitious material (SCM) to prepare cement composite pastes. The impact of RHA content and the internal curing effect on the hydration process of the cementitious system was studied. The hydration heat, degree, and product content of the cement–RHA composite system at 3, 7, and 28 d were analyzed using hydration microcalorimetry, thermogravimetry, and XRD (Rietveld) analysis. The results show that with the increase in RHA, the main exothermic peaks move forward, and the values increase. The induction period is prolonged, and the acceleration period is shortened. The induction period of 15% RHA is extended to 3 h. The hydration heat of cement composite pastes is mainly divided into three stages. Namely, the first stage (0–18 h) is the superposition of the RHA nucleation effect and chemical effect, the second stage (18–51 h) is the superposition of the dilution effect and internal curing effect, and the third stage (51–72 h) is the internal curing effect with the water-release. The internal curing effect of RHA has a certain periodicity, which is related to its content. The water-release age in the early stage (24 h) advances with the increase of content, and the water-release effect in the later stage (7–28 d) is also significant with the increase of content. The higher the content, the more significant the promotion of the internal curing effect on cement hydration and the pozzolanic reaction of RHA.

Keywords: rice husk ash; hydration heat; internal curing effect; water-release

1. Introduction

Improving the comprehensive utilization rate of solid waste resources is of great significance for environmental improvement and promoting comprehensive green development in society. As a category of solid waste, crop straw is widely sourced and generated in large quantities. Rice husk is one of the main high-yield crop straws, with enormous raw material capacity at a global level. Due to its high silica content, rice husk is difficult to degrade naturally, and the traditional utilization rate is low. If it is used as garbage for landfill, it will increase additional land occupation and soil pollution, which will cause potential harm to the environment. However, energy such as heat and electricity can be generated by burning rice husks. The residual ash, namely RHA, can be used as SCM in cement-based materials, which will greatly increase their added value [1]. The RHA obtained under certain calcination conditions contains a large amount of amorphous SiO_2 porous powder [2,3]. Its particles can provide nucleation sites for cement hydration, promote the reaction of cement, and improve the hydration degree of cement [4]. The C-S-H gel formed by the pozzolanic reaction further fills the pores and improves the compactness of the pastes. The redistribution of water by the porous structure of RHA can promote the continuous hydration reaction of cementitious materials, as well as the development of mechanical and long-term properties [5–7]. Therefore, RHA has excellent physicochemical properties, and its feasibility as an SCM has been fully confirmed [8,9]. Currently, RHA has been researched and applied in many countries such as Asia, Africa, South America, and Western Europe, etc. [10–12].

However, compared with traditional pozzolanic materials, the reactivity of RHA and the adsorption of water by its porous structure make the hydration process of the cementitious material more complex. The impact of RHA on cement hydration primarily manifests as nucleation effect, dilution effect, chemical effect [13], and internal curing effect [14,15]. The chemical effect and internal curing effect are related to the amount, reactivity, and porous structure of RHA [16]. The relevant literature [17] studied the effect of RHA on the exothermic reaction of cement hydration. Compared with the control group, 25% RHA significantly accelerated the hydration, advanced the main exothermic peak, and shortened the induction period.

Feng [18] obtained the same conclusion that the adsorption of water by RHA reduced the effective water–cement ratio, and the cement hydration entered an earlier acceleration period. Nguyen [19] found that the impact of RHA on the cement hydration process was related to the water–binder ratio and the amount of RHA. With a constant water–binder ratio, increasing the amount of RHA delays the main exothermic peak, which is different from the conclusion mentioned above. This is due to the different methods of incorporating RHA, that is, the internal incorporation method where RHA replaces cement, or the external incorporation method where the amount of cement remains unchanged. Moreover, when the RHA content is 20%, extra peaks appear in the deceleration period of the exothermic curve for cement pastes with water–binder ratios of 0.4 and 0.6, though the cause of the changes is unclear. The research [20] found that the hydration degree of cement blended with RHA increased significantly after 7 d and was higher than that of pure cement. The study suggests that the internal curing effect of RHA promotes cement hydration by releasing water, thereby increasing the hydration degree. However, most of the previous literature had qualitatively analyzed the results based on relevant theories without discussing the roles and effects of RHA's reactivity and water-release in the various hydration stages of the cementitious system.

This study aims to investigate the influence of RHA's reactivity and internal curing effect on different stages of the hydration process in cement-based composite materials. The impact of RHA content on the exothermic rate, hydration degree, and product amount in the cementitious system was analyzed by calorimetric test, TG-DTG analysis, and XRD phase quantification. Additionally, the water-release effect of RHA during the hydration process was examined by testing the relative humidity (RH) within the samples. This study seeks to clarify the influence of RHA's reactivity and internal curing effect on the hydration process of cementitious materials, providing a reference for the mechanical properties and durability of the composite system.

2. Experimental

2.1. Materials and Characterization

The cement paste was prepared by P·I 42.5 Portland cement, which was produced by China United Cement Group Co., Ltd, and it was recorded as P. The physical and mechanical properties of cement are shown in Table 1. RHA was prepared according to references as follows [3,21]: rice husks were crushed and finely ground into rice husk powder. A dilute hydrochloric acid solution with a mass concentration of 2.5% was used as a pretreatment agent to soak the rice husk powder at a solid–liquid ratio of 1:10 for 1 h, and the rice husk powder was stirred once every 20 min to ensure thorough acidification. After the pretreatment, the acidified rice husk powder was filtered through an 80-mesh stainless steel sieve, washed, and filtered with clean water 3 times. The acidified rice husk powder was dried in an oven until it formed blocks, then crushed into powder using a crusher. Subsequently, a crucible containing rice husk powder was placed in a muffle furnace for heating, raised from room temperature to 600 °C for 1 h, and maintained for 0.5 h, then the white RHA would be obtained by repeated calcination 3 times, recorded as R. The combustion program and preparation process are shown in Figure 1, and the XRD pattern of RHA is shown in Figure 2. The density of RHA is measured to be 2.21 g/cm^3.

Table 1. The physical and mechanical properties of the cement.

Fineness (%)	Density (g·cm^{-3})	Specific Surface Area (m^2·kg^{-1})	Setting Time (min)		Flexural Strength (MPa)		Compressive Strength (MPa)	
			Initial	Final	3 d	28 d	3 d	28 d
0.6	3.15	349	130	195	5.8	8.8	29.2	50.2

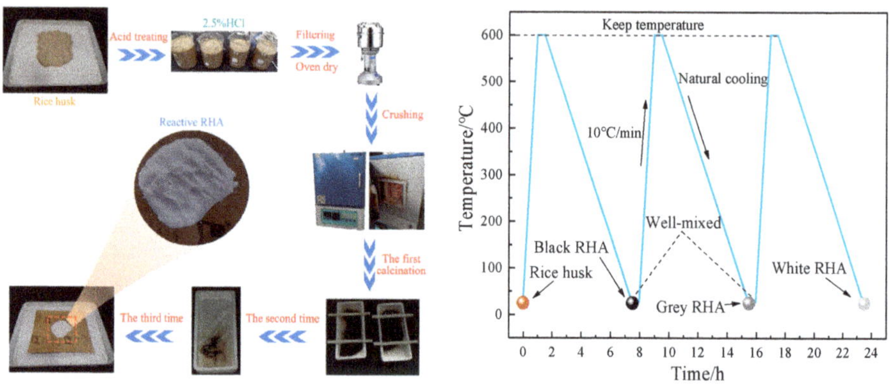

(**a**) (**b**)

Figure 1. The (**a**) preparation process and (**b**) combustion program of RHA.

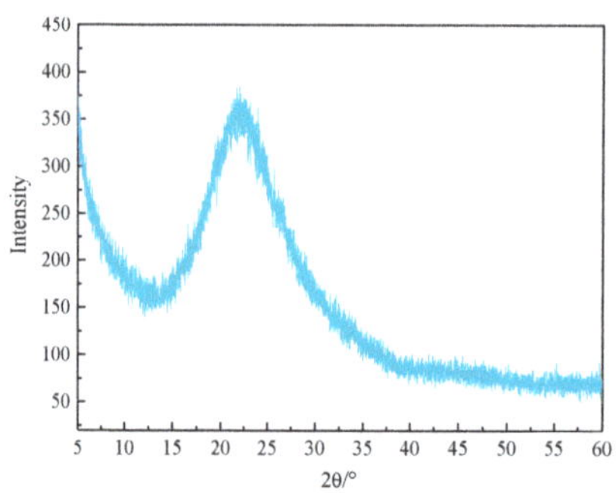

Figure 2. The XRD pattern of RHA.

2.1.1. X-Ray Fluorescence (XRF)

XRF analysis was used to determine the types and content of elements in the powder, which can obtain the chemical composition of RHA. The sample to be tested was ground using an agate mortar, and an Axios mAX type XRF spectrometer produced by PANalytical B.V. in Netherlands was employed for testing, yielding the elemental mass ratio or the mass ratio of oxides in the material. The chemical compositions of cement and RHA are shown in Table 2, and the mineral composition of cement is shown in Table 3. It can be observed that the primary component of RHA is SiO_2, and a wide peak appears at the position corresponding to the diffraction angle of SiO_2 in Figure 2, which indicates that the main phase of amorphous SiO_2 is active in the RHA.

Table 2. The chemical compositions of the cement and RHA (wt%).

	SiO_2	Al_2O_3	Fe_2O_3	CaO	MgO	SO_3	Na_2O_{eq}	f-CaO	Cl^-
Cement	20.75	4.42	3.16	62.59	2.91	2.83	0.53	0.71	0.011
RHA	99.33	0.09	0.05	0.07	0.03	0.08	0.02	0	0

Table 3. The mineral composition of cement (wt%).

Minerals	C_3S	C_2S	C_3A	C_4AF
Cement	56.58	21.76	6.45	10.70

2.1.2. Particle Size Distribution (PSD)

The PSD test was conducted using a Mastersizer 2000 type laser particle size analyzer produced by Malvern instruments Ltd in UK, which can obtain the particle size. The refractive index of solid particles was set at 0, and the refractive index of the dispersant was set at 1.33, with opacity ranging from 4% to 8%. The testing range was from 0.02 to 2000 μm. Ethanol was used as the dispersant for the testing of cement and RHA. The particle size distribution is shown in Figure 3. It can be observed that the volume average particle diameter of RHA is 6.51 μm, which is smaller than cement.

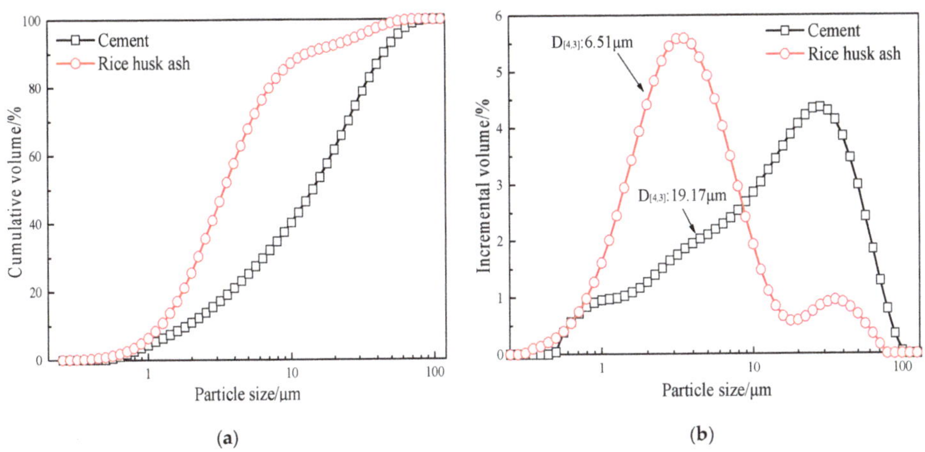

Figure 3. The (**a**) cumulative distribution and (**b**) incremental distribution of cement and RHA.

2.1.3. Scanning Electron Microscope (SEM)

A JSM-6700F type SEM produced by JEOL in Japan was used to observe the surface morphology and pore characteristics of RHA powder particles. To prevent the powder from being removed under high vacuum and to enhance the conductivity of the powder particles, a gold sputtering treatment was applied to the RHA. The vacuum condition was set to 5×10^{-5} Pa, and the operating voltage was 20 kV. The RHA obtained by calcination is a porous powder, where the combustion and separation of organic matter leads to the formation of a porous structure within the RHA particles [13]. The microscopic morphology characteristics of RHA are shown in Figure 4.

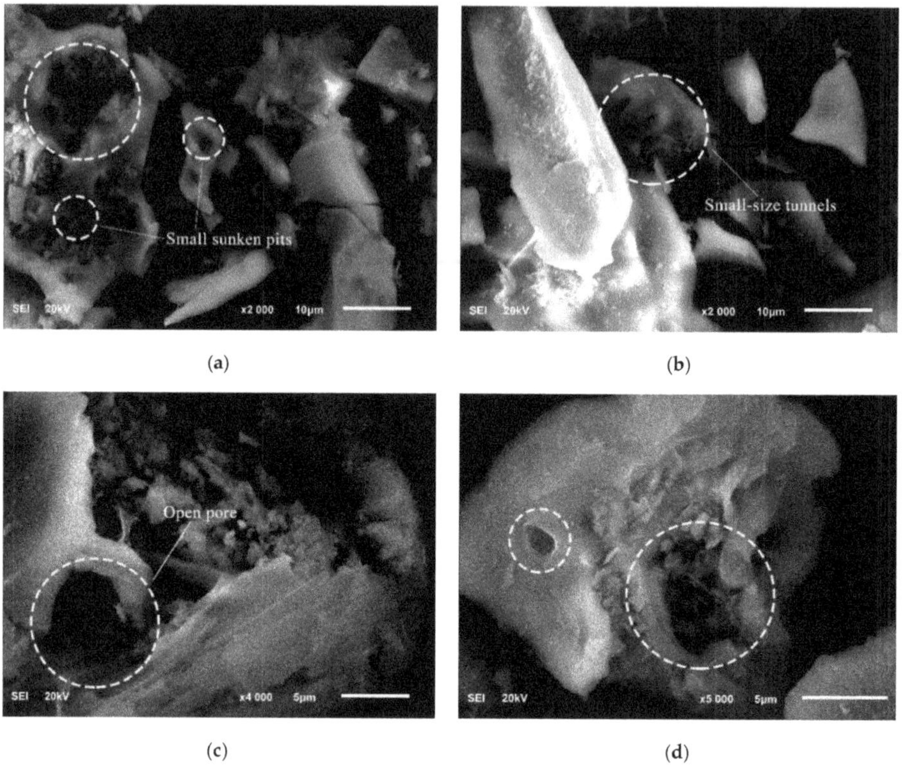

Figure 4. The SEM images (**a**), (**b**) ×2000, (**c**) ×4000, and (**d**) ×5000 of RHA particles.

It can be seen that RHA particles exhibit polygonal and angular shapes, and the porous surfaces and concave structures represent higher SSA and porosity. The pore structure of RHA particles is mainly characterized by small-sized pits, with some open pores and small-scale tunnels also present.

2.1.4. Porous Structural Characteristics

In order to verify the porous structural characteristics of RHA, the water absorption characteristics of RHA were tested according to the method in references [17,22]. Specifically, the RHA was dried at 110 ± 5 °C for 48 h until a constant weight was reached. A certain quantity of RHA was taken to lay a thin layer in a glass petri dish, which was then placed on an electronic scale and put inside a constant temperature and humidity chamber set at 20 °C and 100% RH. Changes in sample mass with time were recorded, as shown in Figure 5. The specific surface area of RHA was measured to be 207.8×10^3 m^2/kg using BET-N_2 adsorption. The pore size distribution of RHA was calculated using the BJH method, as shown in Figure 6. Based on the calculation of the cumulative pore volume per unit mass, the porosity of RHA was 67.04%.

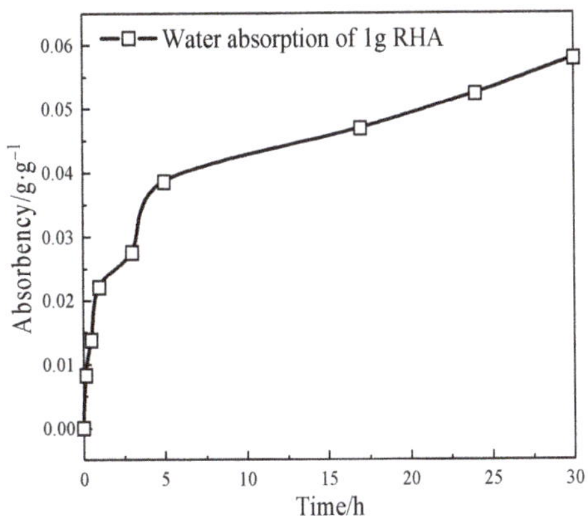

Figure 5. Moisture absorption characteristics of RHA.

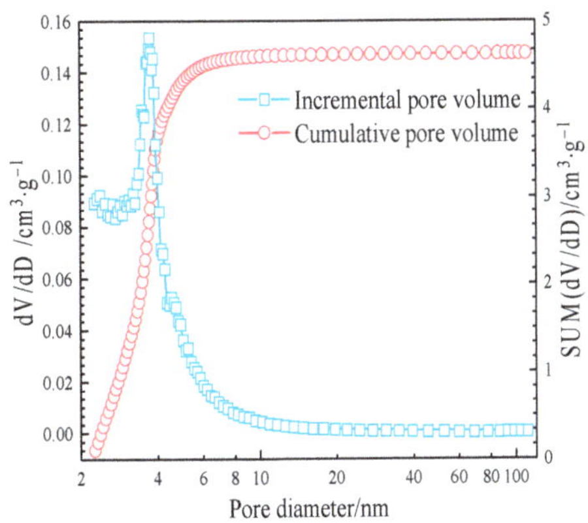

Figure 6. Pore size distribution of RHA.

The specific surface area measured by the N_2 adsorption method typically represents the internal surface area for porous cementitious materials and the external surface area for powder materials. Due to the porous structure of RHA providing more internal surfaces, the measured specific surface area is the sum of the internal and external surface areas. Therefore, the specific surface area of RHA is much greater than that of cement. It can be observed that the micropores of RHA are mainly concentrated in the range of 2–20 nm, as shown in Figure 6.

2.2. Experimental Methods

The cement–RHA composite pastes were prepared by using RHA as a substitution for cement in the experimental mix proportions. The water–binder ratio is 0.4, and the RHA replacement rates are 0%, 5%, 10%, 15%, and 20%, respectively, recorded as PC, PR5, PR10, PR15, and PR20. Due to the porous characteristics and high-water storage function of RHA particles, a large dosage of superplasticizer needs to be added when preparing cement composite pastes, as shown in Figure 7. The dosage of superplasticizer with 180–200 mm slump flow of pastes increases with the content of RHA. The more RHA is added, the greater the increase in the superplasticizer dosage. In order to avoid the impact of superplasticizer on the early hydration of cement and ensure the stir-ability and flowability of the pastes, the same dosage (0.4%) can be uniformly used for 5%, 10%, and 15% RHA. The use of 0.4% superplasticizer for 20% RHA pastes cannot be stirred and formed, and excessive superplasticizer may mislead the analysis of hydration heat data. Therefore, the RHA contents for 0%, 5%, 10%, and 15% are selected for the calorimetric test.

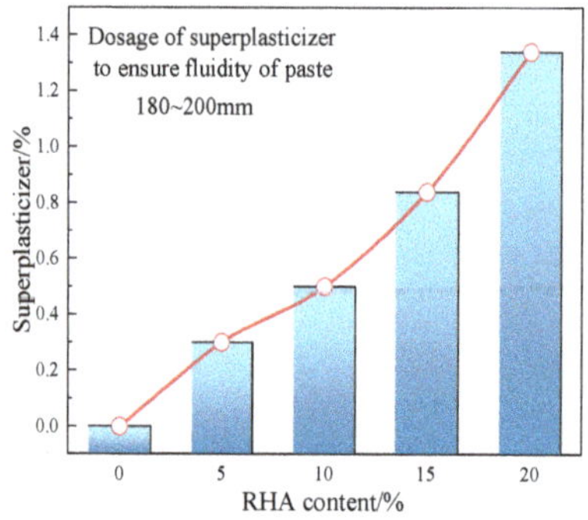

Figure 7. The dosage of superplasticizers required by RHA.

2.2.1. Calorimetric Test

TAM Air 8-channel isothermal calorimeter produced by TA instruments Ltd was used to obtain the heat flow curve of cement composite pastes. Samples were prepared outside the calorimeter in an air-conditioned room with a temperature of 20 ± 2 °C. The temperature of the chamber was set at 20 °C. The paste was mixed with an electric stirrer with a rotation speed of 1500 rpm for 3 min and 11.67 g of the sample was weighed for the test. The testing duration is 72 h, with a 10 min interval between the data collection. The calorimetric test of the RHA pozzolanic reaction uses a solid–liquid mass ratio of 1:2, a molar ratio of RHA (SiO_2) to $Ca(OH)_2$ of 1:1, and a testing duration of 50 h.

2.2.2. TG-DTG Analysis

TGA3 thermogravimetric analyzer produced by Mettler Toledo instruments Ltd in Switzerland was used to obtain the TG-DTG curve. Before testing, a portion of the samples was taken after terminated hydration and vacuum drying, then crushed and ground into powder (<100 μm). About 30–40 mg of the powder was placed into a 70 μL corundum crucible, which was inserted into the instrument. The temperature was raised from room temperature to 1000 °C at a rate of 10 °C/min with an N_2 flow of 20 mL/min. The decomposition temperature range of the phases can be determined from the DTG curves, and the mass loss rate within this range can be obtained from the TG curves. The content of $Ca(OH)_2$ and the chemically bound water can be quantified and the degree of cement hydration can be calculated by the TG-DTG analysis.

2.2.3. XRD-Rietveld Analysis

The Rietveld method can achieve quantitative analysis of various phases in the XRD pattern by fitting the whole spectrum of multiple known phases [23]. The sample from the middle position of a crushed specimen was taken and immersed in isopropanol to terminate hydration for 7 d (changing the isopropanol every 3 d). The sample was vacuum-dried in a desiccator for 3 d, ground into powder, and 2 g of the powder was mixed with 0.5 g of analytical pure ZnO (as internal standard). The mixtures were ground for an additional 15 min to ensure uniform mixing, and the particle size was controlled to less than 10 μm by sieving. The mixed powder was then pressed into tablets on a glass slide. The test was performed using a PANalytical X'Pert Pro MPD diffractometer produced by PANalytical B.V in Netherlands, with CuKα radiation, operating at 40 kV and 40 mA, with a scanning speed of 1.56 °/min and a scanning angle (2θ) of 5° to 65°. The quantitative phase analysis was conducted using the internal standard with 20% ZnO. The XRD-Rietveld analysis used X'Pert High Score Plus 3.0e software. The calculation of phase contents is given by

Equation (1), while the degree of hydration (DoH) of the cement is provided by Equation (2).

$$W_i(t) = W_q(t)/0.8 \quad (1)$$

$$(DoH)_t = 1 - \frac{w_{C_3S}(t) + w_{C_2S}(t) + w_{C_3A}(t) + w_{C_4AF}(t)}{w_{C_3S}(t_0) + w_{C_2S}(t_0) + w_{C_3A}(t_0) + w_{C_4AF}(t_0)} \quad (2)$$

where $W_q(t)$ represents the mass fraction of the phase obtained by XRD–Rietveld analysis; $W_i(t)$ denotes the mass fraction of the phase in the cementitious material; $w_i(t)$ is the mass fraction of the main phases of cement clinker at the hydration age t; $w_i(t_0)$ is the mass fraction of the main phases at the age of initial non-hydration.

2.2.4. Relative Humidity Test

In order to understand the water-release effect of RHA, the RH within the paste was measured during the hydration process [24]. A PVC pipe was used to reserve a hole with a diameter of 20 mm and a depth of 80 mm in the center of a 10 cm × 10 cm × 10 cm sample. Multiple pre-drilled holes for the PVC pipe at the central position of the sample were made at a depth of 50 mm (refilled the PVC pipe with filler during mortar forming), which aimed to ensure proper moisture exchange between the sample and the pore, and effectively avoid inaccurate humidity data caused by the sensor contacting with the pastes. To prevent the humidity inside the hole from being affected by the external environment, the reserved hole was sealed with hot-melt adhesive, which also fixed the sensor in place. The outer surface of the sample was sealed with plastic film. The ambient temperature was maintained at 20 ± 2 °C, and the internal RH was recorded every 5 min by using a multichannel data tester, as shown in Figure 8.

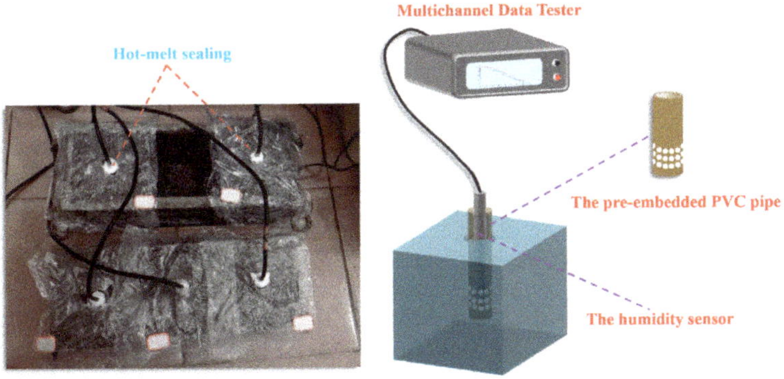

Figure 8. Test of internal RH in the cement–RHA mortar.

3. Results and Discussion

3.1. The Role of RHA in Hydration Process and Its Mechanism

Figure 9 shows the exothermic curve of the pozzolanic reaction of RHA. The exothermic reaction rate of RHA decreases rapidly with time and gradually stabilizes after approximately 36 min. The initial reaction already reaches the peak exothermic value of the pozzolanic reaction, which indicates that RHA can react rapidly with $Ca(OH)_2$ within a short period. The reactivity of RHA was determined by the conductivity method of saturated $Ca(OH)_2$ solution [25]. Figure 10 shows the effect of different RHA masses on the conductivity variation of the saturated $Ca(OH)_2$ solution within 0 to 20 min.

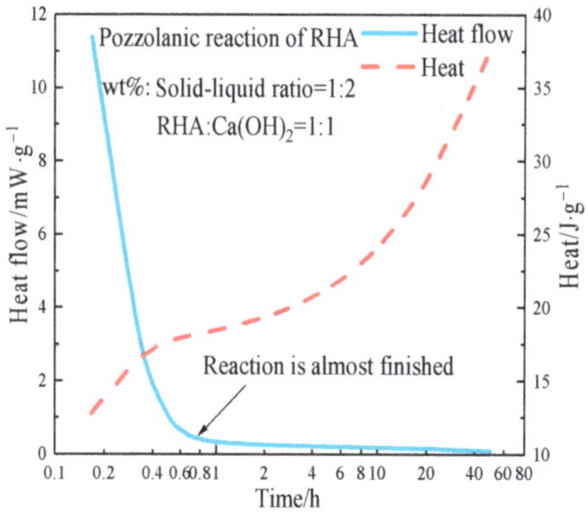

Figure 9. Exothermic curve of pozzolanic reaction of RHA.

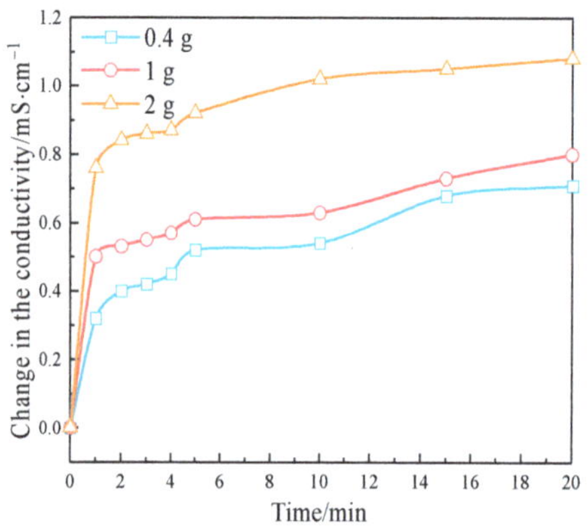

Figure 10. The conductivity variation of saturated $Ca(OH)_2$ solution.

After adding RHA to saturated $Ca(OH)_2$ solution, the amorphous SiO_2 reacts with $Ca(OH)_2$, which reduces the Ca^{2+} concentration in the solution and thereby decreases the conductivity. The greater the variation in conductivity, the higher the reaction activity of RHA. As shown in Figure 9, the RHA quickly produces a pozzolanic reaction in the saturated $Ca(OH)_2$ solution, with the reaction being more intense in 0–5 min. After 5 min, the change in conductivity tends to be gentle, indicating a slowdown of the pozzolanic reaction. The increase in the amount of RHA intensifies the reaction degree. It can be seen that RHA has a high reactivity and will react with the $Ca(OH)_2$ released by cement hydration in the early stage. However, cement hydration is complicated, and Figures 9 and 10 just represent the reaction degree between RHA and $Ca(OH)_2$ in solution, and cannot characterize the start and end times of the pozzolanic reaction in cement paste. Figure 11 shows the effect of RHA on the exothermic reaction rate of the cement paste.

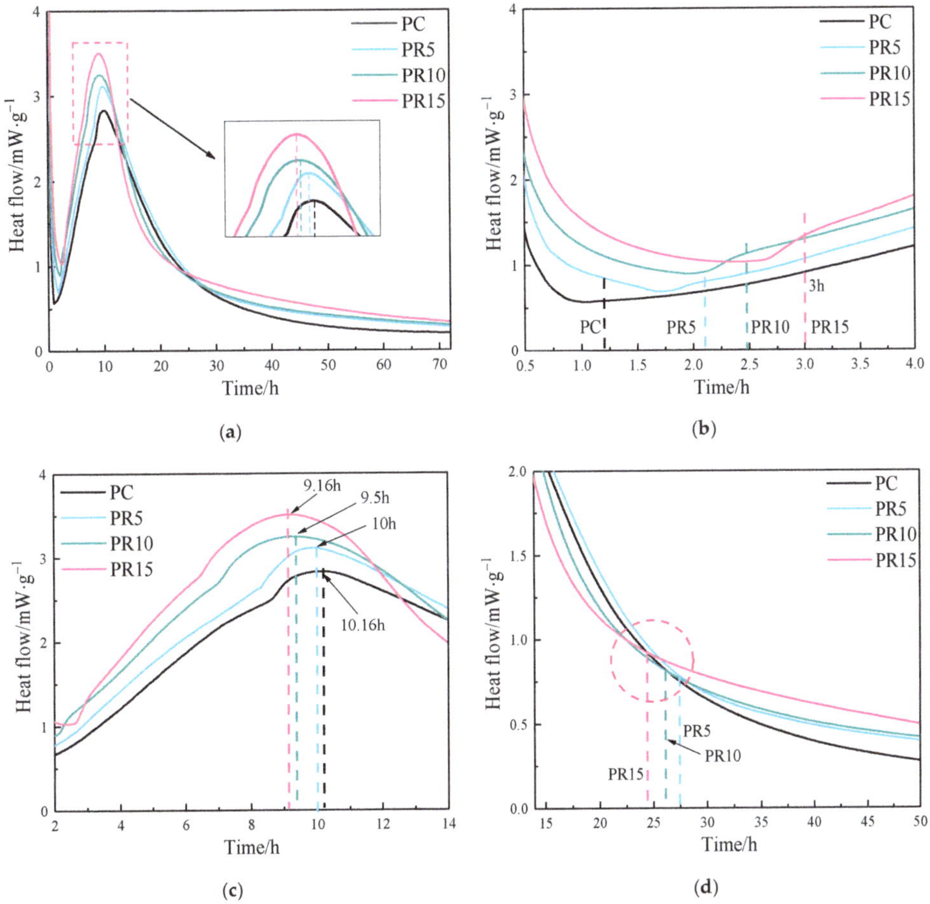

Figure 11. Hydration heat flow (**a**) 0–72 h, (**b**) 0.5–4 h, (**c**) 2–14 h, and (**d**) 14–50 h of cement–RHA pastes.

It can be observed that the addition of RHA shifts the main exothermic peak to an earlier time and increases its intensity. The induction period extends with the increase of RHA content, and 15% RHA extends the induction period to 3 h before entering the acceleration period. Figure 11b,c show that the exothermic rate increases with RHA, and the acceleration period delays the start and ends earlier. It indicates that the higher the RHA content, the shorter the acceleration period and the more intense the hydration reaction. During the deceleration period, 10% and 15% RHA reduce the exothermic rate, which is lower than that of the reference group. After reaching the stabilization period (around 24 h), the exothermic rate of the cement paste blended with RHA suddenly changes compared to the reference group and is significantly higher.

Based on the physicochemical properties of RHA particles, it can be clarified that there are two reaction mechanisms of RHA in the cement system [26], as shown in Figure 12, which can fully analyze the influence of RHA on the hydration heat of the composite pastes. Reaction mechanism I: Si-O bonds in RHA are disconnected by OH^- ions [27], and hydrolyzed to form the $H_3SiO_4^-$ ions [28,29], as shown in Equation (3) [30]. The Ca^{2+} ions from the cement hydration react with the $H_3SiO_4^-$ ions to produce the C-S-H gel. Reaction mechanism II: the surface of partially slow-dissolving RHA particles will be enriched with a large amount of Ca^{2+} ions and OH^- ions, which will react with SiO_2 molecules to form a C-S-H gel layer surrounding the particle surface as the nucleation site [31]. With the processing of hydration, the ions will gradually penetrate through the surface shell and mass transfer to the interior of the particles and continue to the diffusion reaction.

$$SiO_2(s) + H_2O + OH^- = H_3SiO_4^- \quad (3)$$

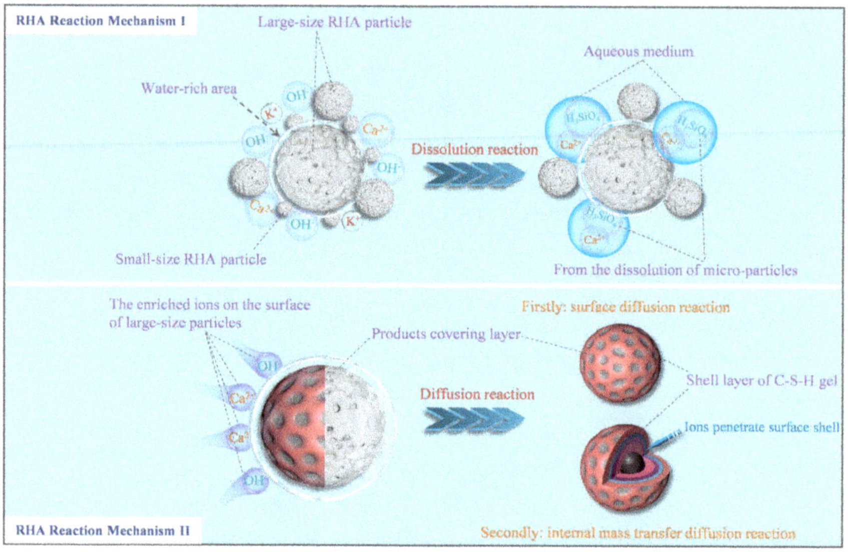

Figure 12. The mechanism of dissolution and diffusion reactions of RHA particles.

According to the RHA reaction mechanism, it is known that the hydrolysis of the micro-particles produces $H_3SiO_4^-$. $H_3SiO_4^-$ reacts with Ca^{2+}, which inhibits the super-saturated crystallization of $Ca(OH)_2$, delays the nucleation of $Ca(OH)_2$ [32], and prolongs the induction period. Ca^{2+} and OH^- released by slowly dissolved C_3S will gradually accumulate and reach super-saturation, thus leading to the accelerated period, as shown in Figure 11b. Due to the water absorption properties of RHA, the increase of RHA reduces the amount of water available for cement hydration. In the first few minutes of hydration, the amount of dissolved SiO_2 will decrease. As a result, the induction period is prolonged with the increasing RHA content. Another important reason is that the cementitious system incorporating RHA used an equal dosage of superplasticizer (0.4%). The addition of the superplasticizer can also delay the hydration of the cement and lead to an extended induction period [33].

Due to the filler effect, the addition of any fine particles will increase the reaction rate of C_3S [34]. The average particle size of RHA is much smaller than that of cement, and the porous structure of the particles provides a large internal and external surface area. This induces the nucleation sites for hydration products to increase with the RHA content. Both the internal and external surfaces of RHA can produce the nucleation effect, facilitating nucleation of the hydration products, and accelerating the hydration reaction of cement and RHA. Thereby, the exothermic rate increases, and the main exothermic peak shifts forward. The earlier end of the acceleration period also indicates that the accelerated hydration reaction causes the rapid nucleation and the growth of hydration products on the surfaces of cement and RHA particles, which encapsulates the unhydrated particles [35], and leads to the beginning of the deceleration period, as shown in Figure 11c.

Around 24 h of hydration, the exothermic rates of the cement paste blended with RHA suddenly changed, as shown in Figure 11d. According to the hydration mechanism of cement [36], the hydration reaction of cement after entering the stabilization period is mainly the diffusion reaction mechanism under the effect of liquid-phase mass transfer. The relevant literature found that the RH inside the samples blended with RHA was higher than that in samples without RHA after 1 d [37]. The water-release of the internal curing effect is the main reason for the higher RH. The release of water promotes cement hydration, and the Ca^{2+} and OH^- released by hydrolysis further induce the pozzolanic reaction of RHA, thereby increasing the instantaneous exothermic rate of the composite cementitious system [24].

The water-release time of internal curing advances with the increase of RHA content. The higher the RHA content, the greater the water storage capacity, and the less water is in the cement paste. The consumption of water by the hydration reaction causes the RH in the system to become unbalanced more quickly, which promotes the RHA to release water earlier. Internal curing of RHA with high content releases relatively more water, which promotes cement hydration and the pozzolanic reaction of RHA more significantly and increases the instantaneous hydration exothermic rate. It can be seen that the water-release time of RHA and the impact on the exothermic rate during 72 hare closely related to the amount of RHA. Figure 13 shows the cumulative heat curves for cement–RHA paste.

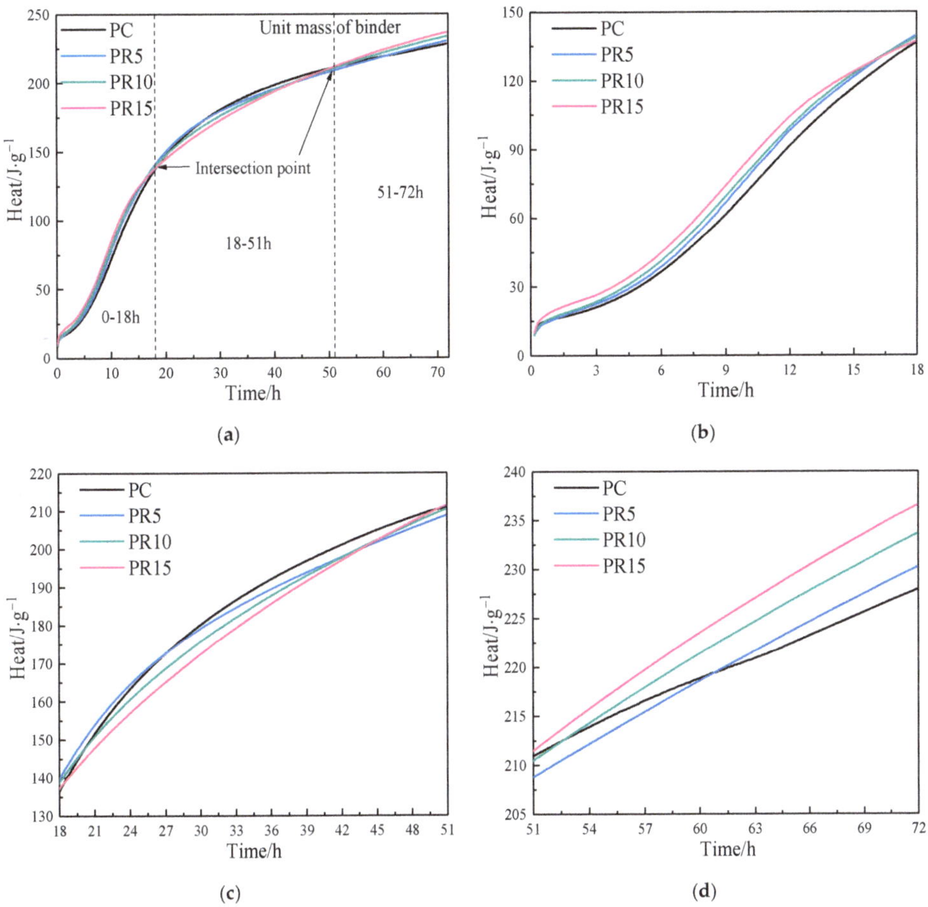

Figure 13. The cumulative heat (**a**) unit mass of binder, (**b**) 0–18 h, (**c**) 18–51 h, and (**d**) 51–72 h of cement–RHA pastes.

Figure 13a shows the cumulative heat of per unit mass cementitious materials. The effect of RHA on the cumulative heat changes twice during the hydration process of the composite pastes, as shown in Figure 13b–d. In the range of 0–18 h, the higher the RHA content, the greater the cumulative heat, which indicates a higher cumulative hydration reaction degree of cement and RHA. The stage includes the initial period, induction period, acceleration period, and deceleration period. The variation in heat release is caused by the superposition of nucleation and chemical effects of RHA. The dissolution of SiO_2 reduces the OH^-concentration, and more Ca^{2+} ions enter the solution. With the increase of RHA, the dissolution rate of cement particles accelerates after the induction period, resulting in higher heat release. In the range of 18–51 h, there is not only a decrease in cumulative heat caused by the reduced cement content but also an increase in cumulative heat caused by the furtherance of hydration reaction from the water released by RHA around 24 h. This stage is the superposition of dilution and internal curing effects of RHA. The stage of 51–72 h shows an increase in cumulative heat due to the internal curing effect of RHA.

3.2. Hydration Degree

The TG-DTG curves can be divided into two stages [38]: the first stage is the decomposition of hydration products such as C-S-H gels, hydrated calcium aluminate, and AFm phase from room temperature to 300 °C. The second stage is the decomposition of $Ca(OH)_2$ and carbonates [39]. The TG-DTG curves of cement–RHA pastes at different ages are shown in Figure 14. The decomposition temperature of $Ca(OH)_2$ typically ranges from 400 to 550 °C. However, due to the variations in testing parameters and conditions, the DTG curve should be based on actual testing. From the DTG curve, it can be determined that the decomposition of $Ca(OH)_2$ starts at 400 °C and ends at 484 °C. Therefore, 400–484 °C is selected as the temperature range for calculating the $Ca(OH)_2$ content. The main mass variation in the DTG curve is the loss of H_2O below 550 °C.

According to the stoichiometric ratio, 1 mass fraction of $Ca(OH)_2$ releases 0.243 mass fraction of H_2O upon decomposition, which can be used to obtain the $Ca(OH)_2$ content. In Figure 14f, the sample of PC hydrated for 28 d shows a mass loss at 550–720 °C, indicating the carbonation of the sample. Based on the molecular weights of $CaCO_3$ and $Ca(OH)_2$, the sample was corrected to calculate the actual $Ca(OH)_2$ content, with a correction factor of 0.595.

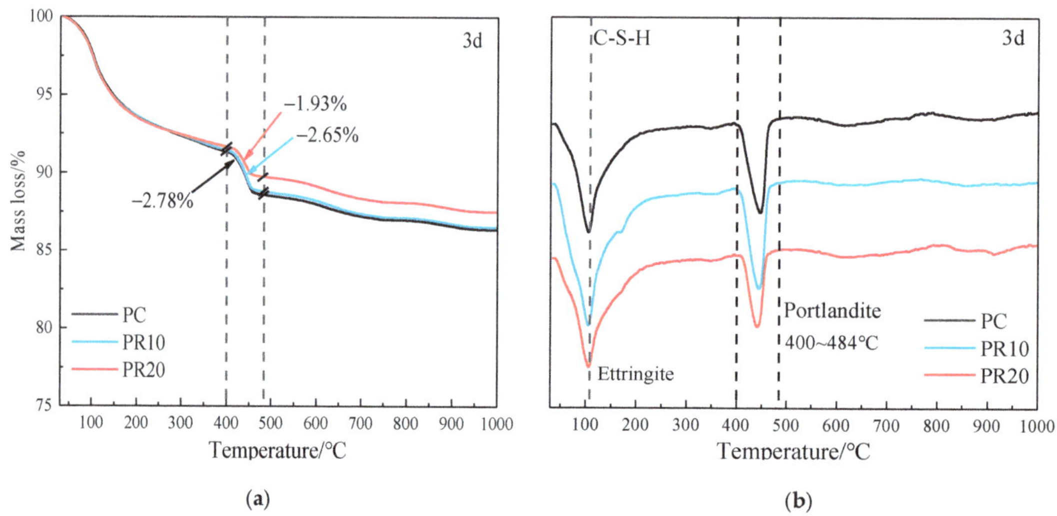

Figure 14. *Cont.*

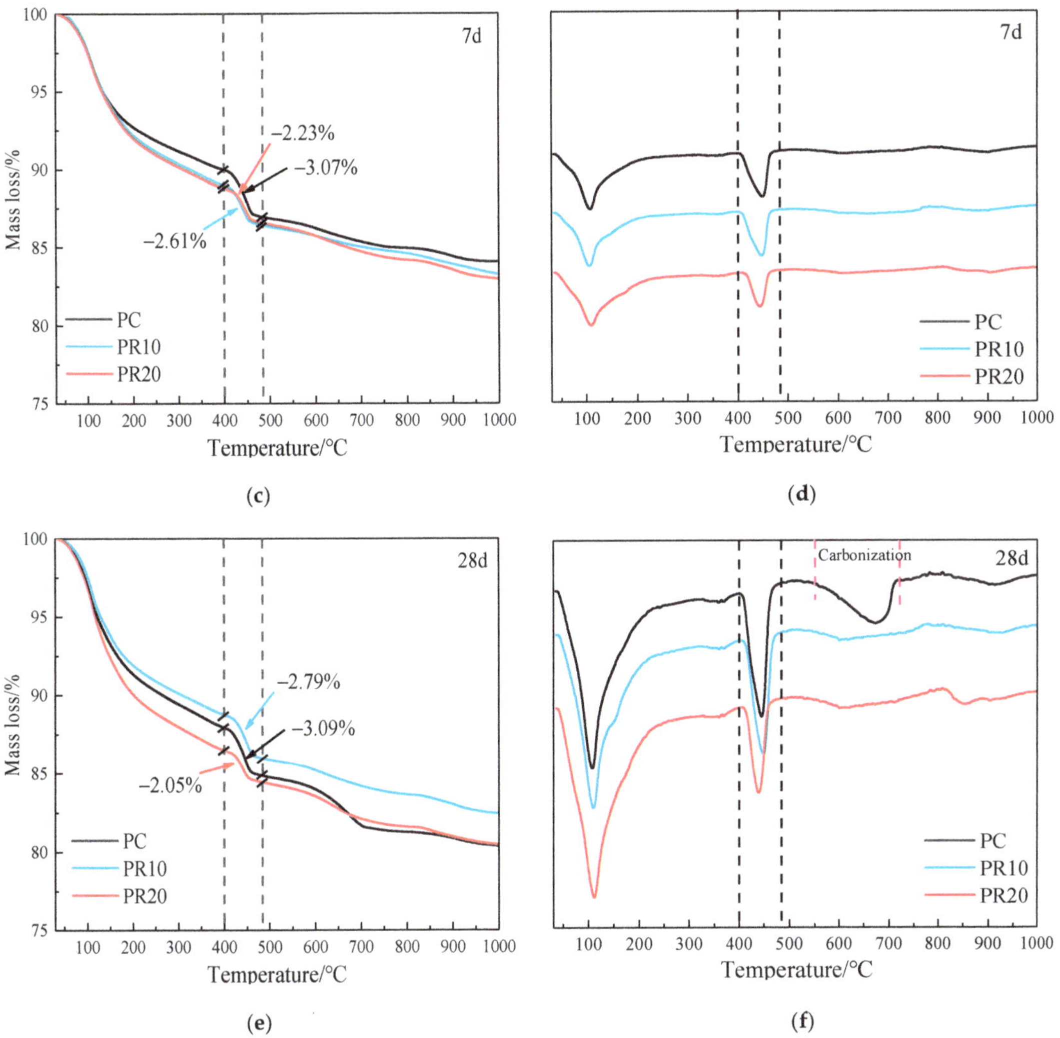

Figure 14. TG-DTG curves of composite pastes hydrated for (**a**,**b**) 3 d, (**c**,**d**) 7 d, and (**e**,**f**) 28 d.

The chemical-bound water mainly includes non-evaporable water, the water in hydration products such as AFt, AFm, $Ca(OH)_2$, and the water among interlayer pores of C-S-H gels. The amount of chemical-bound water can effectively reflect the content of hydration products to a certain extent. Although RHA will participate in the pozzolanic reaction and consume $Ca(OH)_2$, the bound water in the consumed $Ca(OH)_2$ will still be present in the C-S-H gel structure [20]. Therefore, the amount of chemical-bound water can accurately reflect the hydration degree of the composite pastes. The phases in the hydration products, except C-S-H, have specific decomposition temperature ranges. The decomposition of C-S-H almost covers the entire temperature range, especially in the range of 100–400 °C [23]. The tested samples were treated with isopropanol to terminate hydration and then vacuum dried, which eliminated the effect of free water loss from capillary pores. The mass loss at 550 °C is selected as the amount of chemical-bound water. The content of $Ca(OH)_2$ and chemical-bound water are shown in Figure 15.

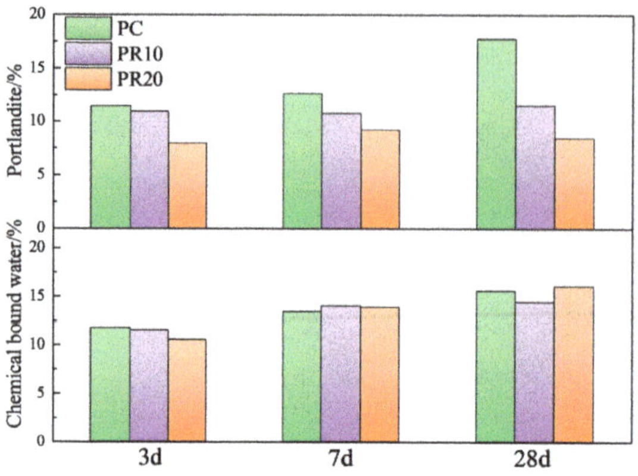

Figure 15. Content of $Ca(OH)_2$ and chemical-bound water in cement–RHA pastes at different ages.

From Figure 15, it can be seen that the $Ca(OH)_2$ content of PR10 and PR20 is lower than that of PC at all ages, and the reduction in $Ca(OH)_2$ content increases with age. The chemically bound water shows different trends. At the age of 3 d, the increase of RHA significantly reduces the $Ca(OH)_2$ content, and the chemical-bound water decreases with RHA. At the age of 7 d, the $Ca(OH)_2$ content in PR20 shows a slight increase compared to 3 d. With a higher RHA content, the effective water–cement ratio decreases due to the water absorption characteristic, which leads to a decrease in the hydration degree at 3 d. With the progress of the age (3–7 d), a higher RHA content shows a more significant internal curing effect and releases more water, which promotes cement hydration and produces more $Ca(OH)_2$. The $Ca(OH)_2$ content in PR10 at 7 d is slightly lower compared to 3 d, and there is little change in $Ca(OH)_2$ content at 7–28 d. The chemically bound water increases and exceeds that of PC at 3–7 d, while the change is minor at 7–28 d. This indicates that the 10% RHA consumes $Ca(OH)_2$ through the pozzolanic reaction in the early stage and improves the hydration degree. However, PR20 shows more significant $Ca(OH)_2$ consumption and hydration promotion from 7 to 28 d. Comparative analysis shows that an appropriate amount of RHA mainly consumes $Ca(OH)_2$ and improves hydration in the early stage by the pozzolanic reactions. An excessive amount of RHA mainly shows the internal curing effect and pozzolanic reaction, then promotes hydration of the cementitious system in the later stage.

The chemically bound water comes from the hydration products of cement and RHA. Both the hydration reaction of cement and the pozzolanic reaction of RHA jointly affect the hydration degree of the composite system. Additionally, the water absorption of RHA reduces the amount of water available for cement hydration, impacting the hydration of cement. Due to the multiple factors, the chemically bound water cannot accurately analyze the hydration degree of cement, thus requiring further study combined with XRD quantitative analysis.

3.3. *Quantitative Analysis of Hydration Products*

In the cement–RHA composite cementitious system, the hydration degree of cement at different ages cannot be determined by TG-DTG analysis. The Rietveld method can be used to quantitatively analyze the content of a single mine in the samples, and then the hydration degree of cement at each age can be calculated. The XRD patterns of the cementitious materials are shown in Figure 16.

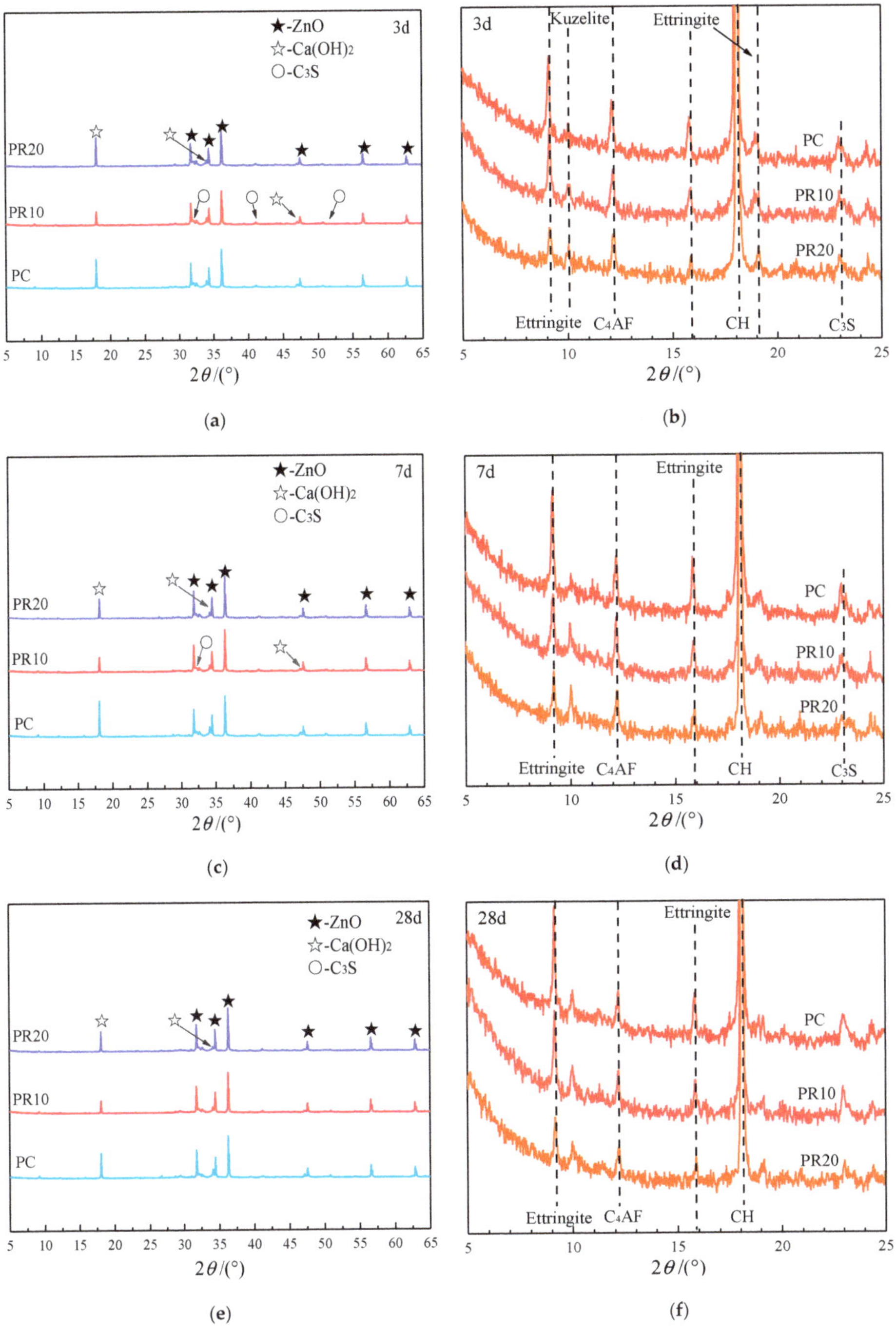

Figure 16. The XRD patterns (**a**,**b**) 3 d, (**c**,**d**) 7 d, and (**e**,**f**) 28 d of cementitious materials at different ages.

From Figure 16, it can be seen that the main phases in the samples are the internal standard ZnO, hydration product $Ca(OH)_2$, and the single mine C_3S. There are no significant differences in phase types with age. The results of the Rietveld quantitative analysis are converted using Equation (1), and the mass fractions of each phase in the cementitious material can be obtained, as shown in Table 4. The early hydration of Portland cement primarily involves the dissolution and precipitation reactions of C_3S and C_3A. The C_2S shows very low reactivity in the early stages, and a relatively obvious hydration reaction appears after 10 d [40]. Thus, the phase content analysis for hydration at 3 and 7 d does not consider C_2S.

Table 4. The quantitative analysis of main phases in samples (wt%).

	3 d			7 d			28 d			
	C_3S	$Ca(OH)_2$	Goodness of Fit	C_3S	$Ca(OH)_2$	Goodness of Fit	C_3S	C_2S	$Ca(OH)_2$	Goodness of Fit
PC	11.38	12.63	2.08	8.5	13.38	2.07	5.13	16.25	15.50	2.73
PR10	10.69	8.11	1.69	9.67	8.38	1.8	6.62	16.13	13.75	2.04
PR20	12.22	7.38	2.55	9.5	8.13	1.98	3.62	14.32	9.75	2.21

Figure 17 compares the $Ca(OH)_2$ content obtained from Rietveld quantitative analysis and TG-DTG analysis. It can be seen that a good consistency between the two methods in the $Ca(OH)_2$ content changes with the RHA content. According to the single mine content in cement from Table 3 and the calculations in Table 4, the consumption (as a percentage of the cementitious material mass) of single mine at hydration ages of 3, 7, and 28 d is shown in Table 5. The hydration degrees of C_3S and C_2S (as a percentage of their own mass) are calculated and presented in Table 6.

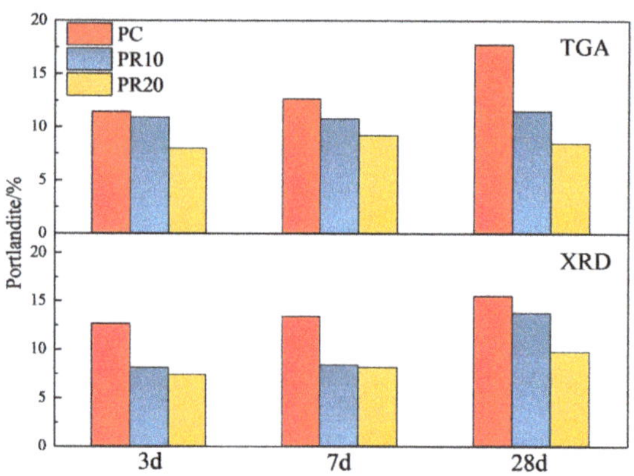

Figure 17. The contrast of $Ca(OH)_2$ content in samples measured by XRD and TGA.

Table 5. The consumption amount of C_3S and C_2S in samples (%).

Samples	Initial Content		Consumption Amount			
			3 d	7 d	28 d	
	C_3S	C_2S	C_3S	C_3S	C_3S	C_2S
PC	56.58	21.76	45.2	48.08	51.45	5.51
PR10	50.92	19.58	40.23	41.25	44.30	7.45
PR20	45.26	17.41	33.04	35.76	41.64	6.09

Table 6. The hydration degree of C_3S and C_2S in samples.

Samples	Consumption Amount			
	3 d	7 d	28 d	
	C_3S	C_3S	C_3S	C_2S
PC	0.80	0.85	0.91	0.25
PR10	0.79	0.81	0.87	0.38
PR20	0.73	0.79	0.92	0.35

It can be seen from Table 6 that the higher the content of RHA is added, the more $Ca(OH)_2$ is consumed. The hydration degree of C_3S in samples blended with RHA is lower than that of PC at 3 d and 7 d, and the hydration degree of the cement single mine in PR10 is higher than that of PR20. This indicates that a higher RHA content affects the hydration of cement single mine in the early stage, which is consistent with the previous analysis. The increase of RHA enhances early water absorption, reducing the water available for cement hydration and thereby lowering the hydration degree of single mines. Additionally, the pozzolanic reaction consumes OH^- and Ca^{2+} released from cement hydration, and $H_3SiO_4^-$ produced by the hydrolysis of RHA particles will use the surface of cement particles as the nucleation position. This results in pozzolanic reaction products covering the cement particles, causing the cement to enter the diffusion-controlled reaction stage prematurely. The slowing down of cement leads to a decrease in the hydration degree of C_3S. The Rietveld quantitative calculation was carried out on the cement single mines, and the cement hydration degree was calculated using Equation (2), as shown in Figure 18.

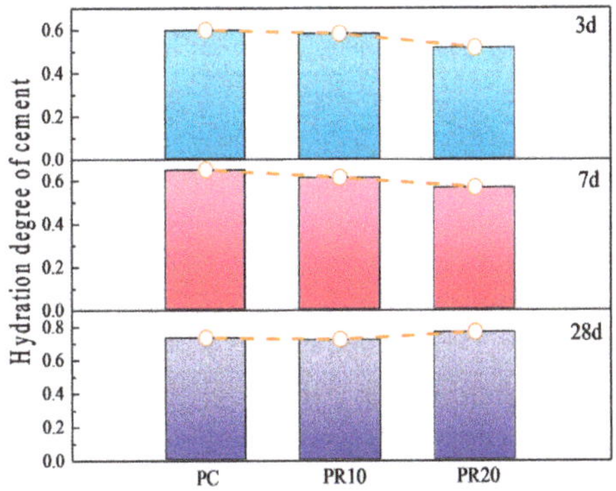

Figure 18. The hydration degree of cement in cementitious material samples.

Figure 18 shows that the cement hydration degree in PR20 significantly increases at 28 d, which is higher than that of other samples. This indicates that more RHA produces water-release at 7–28 d, which facilitates more cement hydration. The $Ca(OH)_2$ content obtained by XRD quantitative analysis in Figure 17 is also significantly lower than that of other samples. This suggests that a higher content of RHA promotes the hydrolysis of unreacted C_3S and C_2S by the internal curing effect in the later stage of hydration. The released Ca^{2+} and OH^- migrate to the internal and external surfaces of RHA particles through the pore solution, further stimulating the pozzolanic reactions on their surfaces, thereby improving the hydration degree of the composite cementitious system, which is consistent with the results from Figure 15. The analysis of Figure 11 reveals that RHA produces a water-release around 24 h, and another internal curing effect occurs again at 7–28 d, as shown in Figure 19. The water-release is primarily related to the RH gradient around the particles. With the consumption of free water in the hydration reaction and the evaporation of some free water, a water-release effect is immediately observed when the environmental humidity decreases [37]. It can be seen that the RH in the cement–

RHA system remains consistently higher than in the control group after 2 d. This is due to the water-release and moisture-retaining effect of RHA. When the hydration process approaches 7 d, a noticeable abrupt change in RH occurs in samples with 20% RHA, with RH significantly higher than in other groups, effectively alleviating the decline of RH within the cement–RHA system. The pore size distribution of RHA particles ranges from 2–20 nm (Figure 5). Based on Kelvin's equation, the corresponding humidity range is calculated to be 75–96% [41], and the theoretical humidity is between 96–99% for some larger pores (Figure 6), which means that when the environmental humidity falls below 99%, the water in the RHA pores will be released to compensate for the decrease of humidity. Different environmental humidities will correspond to different degrees of water-release from the pores. In summary, the internal curing effect of RHA exhibits a certain periodicity in the cement–RHA cementitious system, and the higher the RHA content, the more significant the water-release in the later stage.

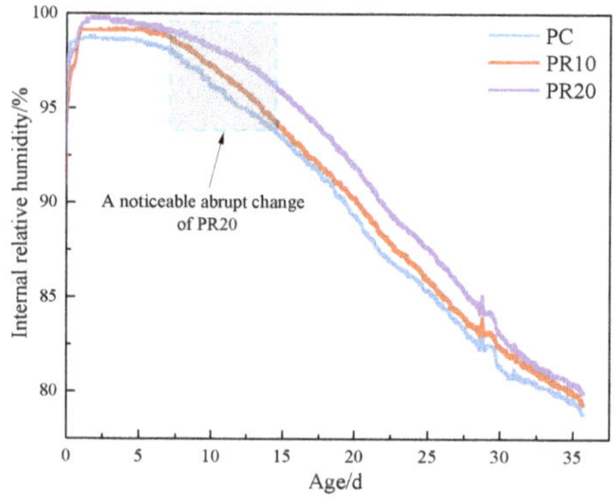

Figure 19. The internal RH of cement–RHA system during hydration.

4. Conclusions

(1) With the increase of RHA, the main exothermic peak of the cement–RHA composite paste shifts forward, and the peak value increases. The induction period is prolonged, and the acceleration period is shortened, 15% RHA extends the induction period to 3 h before entering the acceleration period. The water-release time by RHA at an early age and the impact on the exothermic rate are related to the RHA content.

(2) Around 24 h of hydration, RHA produces an internal curing effect, with the water-release occurring earlier as the increase of RHA. The higher the content, the more water-release will be released, and the more significant the promotion of cement hydration and RHA reaction.

(3) The hydration heat of cement composite pastes is mainly divided into three stages. The first stage (0–18 h) is the superposition of the RHA nucleation effect and chemical effect, and the heat increases with RHA. The second stage (18–51 h) is the superposition of the dilution effect and internal curing effect, that is, the reduced heat caused by the decrease of cement and the promotion of hydration from the water-release by the internal curing effect. The third stage (51–72 h) is the internal curing effect with the water-release.

(4) A moderate content of RHA primarily enhances the hydration degree by the pozzolanic reaction in the early stage. The excessive content of RHA mainly contributes to the internal curing effect in the later stage, and also promotes the hydration of the cementitious system by pozzolanic reactions. The water-release of RHA exhibits a certain periodicity, and the higher the content, the more significant the water-release effect in the later stage.

Author Contributions: Conceptualization, J.W.; Data curation, J.W.; Formal analysis, J.W.; Funding acquisition, F.J.; Investigation, X.H.; Methodology, F.J. and H.C.; Project administration, F.J. and X.H.; Writing–original draft, J.W.; Writing—review & editing, X.H. All authors have read and agreed to the published version of the manuscript.

Funding: This research was funded by Chongqing Municipal Education Commission Science and Technology Research Project (KJQN202301910) and the Science and Technology Plan Project of Suzhou (SNG2022015).

Data Availability Statement: The original contributions presented in the study are included in the article, further inquiries can be directed to the corresponding author.

Conflicts of Interest: The authors declare no conflicts of any commercial or associative interest in connection with the work submitted.

References

1. Wang, Z.; Li, B. Research Progress on Application of Rice Husk Ash in Cement and Concrete. *Mater. Rep.* **2020**, *34*, 9003–9011.
2. Fernandes, I.J.; Moraes, C.A.; Egea, J.R.; Sousa, V.C. Production and characterization of silica materials from rice husk ash by different combustion processes. *Powder Technol.* **2024**, *436*, 119743. [CrossRef]
3. Chen, P. Effect of Pretreatment on Properties of Rice Husk Ash. Master's Thesis, Harbin Institute of Technology, Harbin, China, 2014.
4. Xu, L.; Wu, K.; Wang, P.; Ludwig, H.M. Influence of rice husk ash on early hydration of calcium aluminate cement. *J. Chin. Ceram. Soc.* **2016**, *44*, 1098–1103.
5. Pu, B.-C.; Liu, B.; Li, L.; Jiang, L.; Zhou, J.; Ding, P. Using rice husk ash in alkali-activated ultra-high-performance concrete: Flowability, early age strength and elasticity modulus. *Constr. Build. Mater.* **2024**, *443*, 137771. [CrossRef]
6. Zareei, S.A.; Ameri, F.; Dorostkar, F.; Ahmadi, M. Rice husk ash as a partial replacement of cement in high strength concrete containing micro silica: Evaluating durability and mechanical properties. *Case Stud. Constr. Mater.* **2017**, *7*, 73–81. [CrossRef]
7. Su, Q.; Xu, J.M. Mechanical properties of rice husk ash and glass powder concrete: Experimental and mesoscopic studies. *J. Build. Eng.* **2024**, *95*, 110278. [CrossRef]
8. Amin, M.N.; Khan, K.; Arab, A.M.A.; Farooq, F.; Eldin, S.M.; Javed, M.F. Prediction of sustainable concrete utilizing rice husk ash (RHA) as supplementary cementitious material (SCM): Optimization and hyper-tuning. *J. Mater. Res. Technol.* **2023**, *25*, 1495–1536. [CrossRef]
9. Lo, F.C.; Lee, M.G.; Lo, S.L. Effect of coal ash and rice husk ash partial replacement in ordinary Portland cement on pervious concrete. *Constr. Build. Mater.* **2021**, *286*, 122947. [CrossRef]
10. Prasara-A, J.; Gheewala, S.H. Sustainable utilization of rice husk ash from power plants: A review. *J. Clean. Prod.* **2017**, *167*, 1020–1028. [CrossRef]
11. Thomas, B.S. Green concrete partially comprised of rice husk ash as a supplementary cementitious material–A comprehensive review. *Renew. Sustain. Energy Rev.* **2018**, *82*, 3913–3923. [CrossRef]
12. Christopher, F.; Bolatito, A.; Ahmed, S. Structure and properties of mortar and concrete with rice husk ash as partial replacement of ordinary Portland cement—A review. *Int. J. Sustain. Built Environ.* **2017**, *6*, 675–692.
13. Bui, D.D. Rice Husk Ash as a Mineral Admixture for High Performance Concrete. Ph.D. Thesis, Delft University of Technology, Delft, The Netherlands, 2001.
14. Park, K.B.; Kwon, S.J.; Wang, X.Y. Analysis of the effects of rice husk ash on the hydration of cementitious materials. *Constr. Build. Mater.* **2016**, *105*, 196–205. [CrossRef]
15. Nazari, A.; Toufigh, V. Effects of elevated temperatures and re-curing on concrete containing rice husk ash. *Constr. Build. Mater.* **2024**, *439*, 137277. [CrossRef]
16. Rêgo, J.H.S.; Nepomuceno, A.A.; Figueiredo, E.P.; Hasparyk, N.P.; Borges, L.D. Effect of particle size of residual rice-husk ash in consumption of $Ca(OH)_2$. *J. Mater. Civ. Eng.* **2015**, *27*, 04014178. [CrossRef]
17. Kang, S.H.; Hong, S.G.; Moon, J. The use of rice husk ash as reactive filler in ultra-high performance concrete. *Cem. Concr. Res.* **2019**, *115*, 389–400. [CrossRef]
18. Feng, Q.; Yamamichi, H.; Shoya, M.; Sugita, S. Study on the pozzolanic properties of rice husk ash by hydrochloric acid pretreatment. *Cem. Concr. Res.* **2004**, *34*, 521–526. [CrossRef]
19. Nguyen, V.T. Rice Husk Ash as a Mineral Admixture for Ultra-High Performance Concrete. Ph.D. Thesis, Delft University of Technology, Delft, The Netherlands, 2011.
20. Tuan, N.V.; Ye, G.; Breugel, K.V. Hydration and microstructure of ultra high performance concrete incorporating rice husk ash. *Cem. Concr. Res.* **2011**, *41*, 1104–1111. [CrossRef]
21. Lu, S. Effects of Pretreatment and Calcination Conditions of Rice Husk on the Activity of Rice Husk Ash. Master's Thesis, Shenyang University of Chemical Technology, Shenyang, China, 2019.

22. Van Breugel, K.; Van Tuan, N. Ultra high performance concrete made with rice husk ash for reduced autogenous Shrinkage. In Proceedings of the 4th International FIB Congress: Improving Performance of Concrete Structures, Mumbai, India, 10–14 February 2014.
23. Scrivener, K.; Snellings, R.; Lothenbach, B. *A Practical Guide to Microstructureal Analysis of Cementitious Materials*; CRC Press: Boca Raton, FL, USA, 2016.
24. Rößler, C.; Bui, D.D.; Ludwig, H.M. Rice husk ash as both pozzolanic admixture and internal curing agent in ultra-high performance concrete. *Cem. Concr. Compos.* **2014**, *53*, 270–278.
25. Luxán, M.P.; Madruga, F.; Saavedra, J. Rapid evaluation of pozzolanic activity of natural products by conductivity measurement. *Cem. Concr. Res.* **1989**, *19*, 63–68. [CrossRef]
26. Wang, J.; Xiao, J.; Zhang, Z.; Han, K.; Hu, X.; Jiang, F. Action mechanism of rice husk ash and the effect on main performances of cement-based materials: A review. *Constr. Build. Mater.* **2021**, *288*, 123068. [CrossRef]
27. Frugier, P.; Gin, S.; Minet, Y. SON68 nuclear glass dissolution kinetics: Current state of knowledge and basis of the new GRAAL model. *J. Nucl. Mater.* **2008**, *380*, 8–21. [CrossRef]
28. Sjöberg, S. Silica in aqueous environments. *J. Non-Cryst. Solids* **1996**, *196*, 51–57. [CrossRef]
29. Maraghechi, H.; Rajabipour, F.; Pantano, C.G.; Burgos, W.D. Effect of calciumon dissolution and precipitation reactions of amorphous silica at high alkalinity. *Cem. Concr. Res.* **2016**, *87*, 1–13. [CrossRef]
30. Fertani-Gmati, M.; Jemal, M. Thermochemistry and kinetics of silica dissolution in NaOH aqueous solution. *Thermochim. Acta* **2011**, *513*, 43–48. [CrossRef]
31. Rößler, C.; Bui, D.D.; Ludwig, H.M. Mesoporous structure and pozzolanic reactivity of rice husk ash in cementitious system. *Constr. Build. Mater.* **2013**, *43*, 208–216.
32. Habeeb, G.A.; Mahmud, H.B. Study on properties of rice husk ash and its use as cement replacement material. *Mater. Res.* **2010**, *13*, 185–190. [CrossRef]
33. Marchon, D.; Juilland, P.; Gallucci, E.; Frunz, L.; Flatt, R.J. Molecular and submolecular scale effects of comb-copolymers on tri-calcium silicate reactivity: Toward molecular design. *J. Am. Ceram. Soc.* **2017**, *100*, 817–841. [CrossRef]
34. Zunino, F.; Scrivener, K.L. The influence of the filler effect on the sulfate requirement of blended cements. *Cem. Concr. Res.* **2019**, *126*, 105918. [CrossRef]
35. Bazzoni, A. *Study of Early Hydration Mechanisms of Cement by Means of Electron Microscopy*; ACM: New York, NY, USA, 2014.
36. Juilland, P.; Nicoleau, L.; Arvidson, R.S.; Gallucci, E. Advances in dissolution understanding and their implications for cement hydration. *RILEM Tech. Lett.* **2017**, *2*, 90. [CrossRef]
37. Sensale, G.R.; Ribeiro, A.B.; Goncalves, A. Effects of RHA on autogenous shrinkage of Portland cement pastes. *Cem. Concr. Compos.* **2008**, *30*, 892–897. [CrossRef]
38. Deschner, F.; Lothenbach, B.; Winnefeld, F.; Neubauer, J. Effect of temperature on the hydration of Portland cement blended with siliceous fly ash. *Cem. Concr. Res.* **2013**, *52*, 169–181. [CrossRef]
39. Durdzinski, P.T. Hydration of multi-component cements containing cement clinker, slag, calcareous fly ash and limestone. *Water Resour. Res.* **2016**, *29*, 3867–3878.
40. Scrivener, K.L.; Juilland, P.; Monteiro, P.J. Advances in understanding hydration of Portland cement. *Cem. Concr. Res.* **2015**, *78*, 38–56. [CrossRef]
41. Lura, P.; Jensen, O.M.; Breugel, K.V. Autogenous shrinkage in high-performance cement paste: An evaluation of basic mechanisms. *Cem. Concr. Res.* **2003**, *33*, 223–232. [CrossRef]

materials MDPI

Article

Mechanical Properties of Recycled Concrete Incorporated with Super-Absorbent Polymer and Machine-Made Stone Powder under the Freeze-Thaw Cycle Environment

Lingling Zhang [1], Ronggui Liu [1,2,*] and Feifei Jiang [1,*]

[1] School of Civil Engineering, Nantong Institute of Technology, Nantong 226000, China; zll18662854351@163.com
[2] Faculty of Civil Engineering and Mechanics, Jiangsu University, Zhenjiang 212013, China
* Correspondence: liurg@ujs.edu.cn (R.L.); 999620140019@just.edu.cn (F.J.)

Abstract: Recycled concrete incorporating additional super-absorbent polymer (SAP) and machine-made stone powder (MSP) was prepared using a two-factor, four-level orthogonal test. To enhance the frost resistance of recycled concrete and improve its mechanical properties, such as compressive and flexural strength, the prepared concrete underwent 200 freeze–thaw cycles. Before freeze–thaw cycles, the amount of SAP has a predominant influence on the mechanical properties of recycled concrete in comparison with MSP. After 200 cycles of freeze–thaw, the influence of MSP became more significant than that of SAP. Typically, the compressive strength and flexural strength exhibited a trend of initially increasing and then decreasing as the contents of SAP and MSP increased. The optimized recycled concrete was identified as S16M6, containing 0.16% SAP and 6% MSP, as demonstrated by the minimal strength loss after freeze–thaw cycles. This study also proposed a linear regression model for predicting the mechanical properties which offered valuable guidance for the engineering application of recycled concrete mixed with SAP under the freeze–thaw cycle environment.

Keywords: freeze–thaw cycle; combined incorporation of SAP and MSP; mechanical strength; linear regression

1. Introduction

Concrete is extensively utilized in construction, including bridges, tunnels, buildings, embankments, and harbors, due to its cost-effectiveness and versatility [1–3]. Its indispensable role in urban construction has led to a high demand for concrete production. However, this production has significantly impacted resources, energy, and the environment, including issues related to the supply of raw materials and the recycling of waste concrete [4–6].

Among the components that constitute concrete, aggregate has a significant impact on the quality of concrete due to its high proportion (ca. 60–70%) [7–9]. Currently, there is a shortage in the supply of sandstone. Although these materials are not rare, over-exploitation has led to shortage of sandstone [10,11]. In order to cope with the problem of resource shortage, recycled concrete produced by recycling sandstone from construction waste has been proposed [12–16]. This approach not only addresses raw material shortages but also effectively manages waste concrete. Recycled concrete predominantly uses machine-made sand and recycled aggregate to replace natural sand and aggregate in common concrete [17,18]. Machine-made sand, with particles smaller than 4.75 mm, is produced from various rocks and construction wastes through mechanical processing [19,20]. The strength grade of common concrete is generally determined by the dust content in natural sandstone. The strength of recycled concrete is determined by the content of crusher dust in the machine-made sand. The crusher dust of machine-made sand basic ingredients is machine-made stone powder (MSP), with particles smaller than 0.075 mm [21–24].

The production of recycled concrete from tunnel slag, particularly tunnel slag generated from tunnel excavation, has been recognized as a sustainable approach for slag treatment and slag recycling [25–28]. Slag aggregate-based concrete (SABC), made from tunnel slag, has been successfully used in tunneling projects [25,29,30] and has improved seismic performance and reduced collapse [31,32]. However, to enhance the application of such concrete in cold regions, further study on the effects of MSP content on mechanical properties and frost resistance is necessary [33,34]. Under the coupling effect of external loads and freeze–thaw environment, concrete structures suffer from cracking, resulting in reduced stability of the structure. Therefore, several studies have focused on the effect of MSP on the strength of recycled concrete. However, the effect of MSP on strength and frost resistance of recycled concrete is not uniform [35–38]. Generally, the strength of recycled concrete first decreases and then increases as the MSP content increases, until a good strength is obtained at 10% [39]. However, some studies indicate that MSP contents of 5% and 8% are beneficial to the frost resistance of concrete [29]. Tan et al. [40] believed that mixing an appropriate amount of stone powder could improve the frost resistance of concrete. Most researchers have concluded that incorporating an appropriate amount of MSP can improve the mechanical properties and frost resistance of recycled concrete [25,41,42]. However, the optimal dosage has not been specifically provided. Although this represents a breakthrough in the research on recycled concrete, the solidity and fluidity of machine-made sand are more defective than that of natural sand. Furthermore, excessive incorporation of MSP leads to shrinkage cracking, strength reduction, and impaired frost resistance in the recycled concrete structure [35,43,44]. To overcome these shortcomings, superabsorbent polymer (SAP) is added to ordinary concrete to deal with issues such as low water-cement ratios and significant shrinkage during drying, enhancing concrete strength, shrinkage cracking, structural density, and frost resistance [45–47]. Therefore, this study innovatively proposes to add admixtures to recycled concrete, such as adding SAP to address issues like a low water-cement ratio, poor workability, and high drying shrinkage. However, the incorporation of SAP reduces mechanical properties and frost resistance of the structure through leaving internal cavities [48–51]. When the SAP proportion was 0.2%, the compressive strength of the structure was reduced to 10–12% [52]. In addition, the strength of the structure decreased rather than increased when SAP was incorporated at proportions of 0.1%, 0.2%, 0.3%, 0.4%, and 0.5% [53,54]. Some researchers have suggested that the SAP release of water allows the structure to be porous, leading to a reduction in structural strength [48,55,56]. Therefore, in order to ensure the better mechanical properties and frost resistance of recycled concrete, the content of MSP and SAP added in recycled concrete are worthy of further study.

This study aims to balance the strength and frost resistance of recycled concrete by adjusting MSP (0%, 3%, 6%, 9%) and SAP (0%, 0.08%, 0.16%, 0.24%) dosages. Using the orthogonal test method, the effects on compressive strength, flexural strength, and frost resistance were analyzed to determine the optimal combination of MSP and SAP. The findings offer valuable insights into enhancing the mechanical properties and frost resistance of recycled concrete.

2. Orthogonal Experiments

2.1. Materials

PO 42.5 ordinary silicate cement was produced by the Crane Forest brand, Zhenjiang City, Jiangsu Province, China. The cement exhibited a density of 3180 kg/m^3 and a specific surface area of 368.6 m^2/kg, and the mineral composition is shown in Table 1 as follows. The superabsorbent polymer (SAP), provided by a chemical plant in Jiangsu Province, China, had the characteristics detailed in Table 2. The chemical composition of SAP is shown in Figure 1a. The recycled coarse aggregate, with size ranging from 5 to 20 mm, was produced by waste concrete. The fine aggregate was divided into river sand and machine-made sand. The river sand, categorized as medium sand, was sourced from a building materials factory in Jiangsu Province, China. The machine-made sand was made from natural rocks generated during the excavation of tunnels by Puyan Expressway Co., Ltd., Sanming, China, and its grading curve is depicted in Figure 2. The grading interval of machine-made sand was in Zone I (coarse sand zone), the cumulative percentage was 87%, and the rest was the stone powder (MSP) less than 0.075 mm which was generated during the production of machine-made sand (Figure 3). The types and composition of minerals in the MSP are shown in Table 3. Polycarboxylate superplasticizer produced by Shanghai Chenqi Chemical Technology Co., Ltd. in Shanghai, China. The polycarboxylate superplasticizer had a water reduction rate less than 16%, and the chemical composition of the polycarboxylate superplasticizer is shown in Figure 1b. The dosage of polycarboxylate superplasticizer, which acted as the water-reducing agent, was set to 0.2% of the total weight of the cement.

Table 1. Mineral composition of the cement (wt%).

Mineral Composition	C_3S	C_2S	C_3A	C_4AF	Gypsum	Quartz	Periclase	K_2SO_4	Free Lime
Content	60.2	14.8	3.8	12.7	3.5	0.1	0.1	0.9	1.2

Table 2. Characteristics of SAP.

	Water Absorption Ratio (g/g)	Brine Absorption Ratio (g/g)	Pressure Water Absorption Ratio (g/g)	Packing Density (g/mL)	pH
Numeric value	400–500	60	26	0.65–0.85	6.2

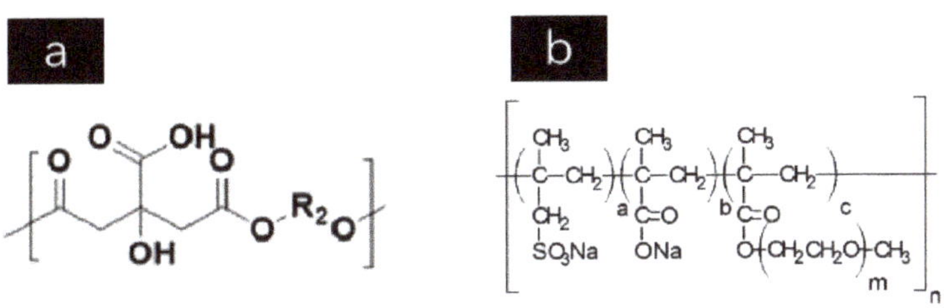

Figure 1. The chemical composition. (**a**) SAP. (**b**) Polycarboxylate superplasticizer.

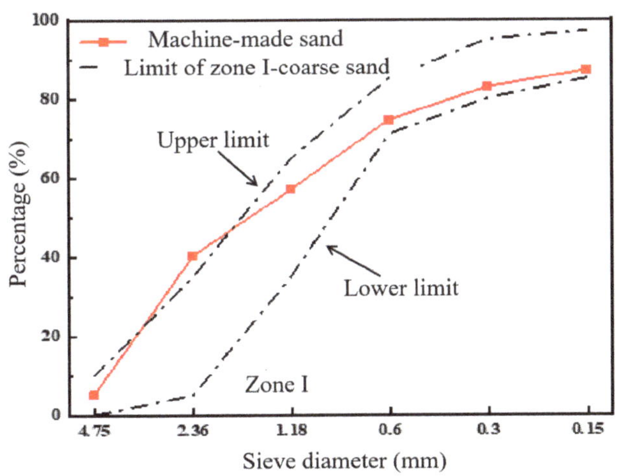

Figure 2. Machine-made sand grading curve.

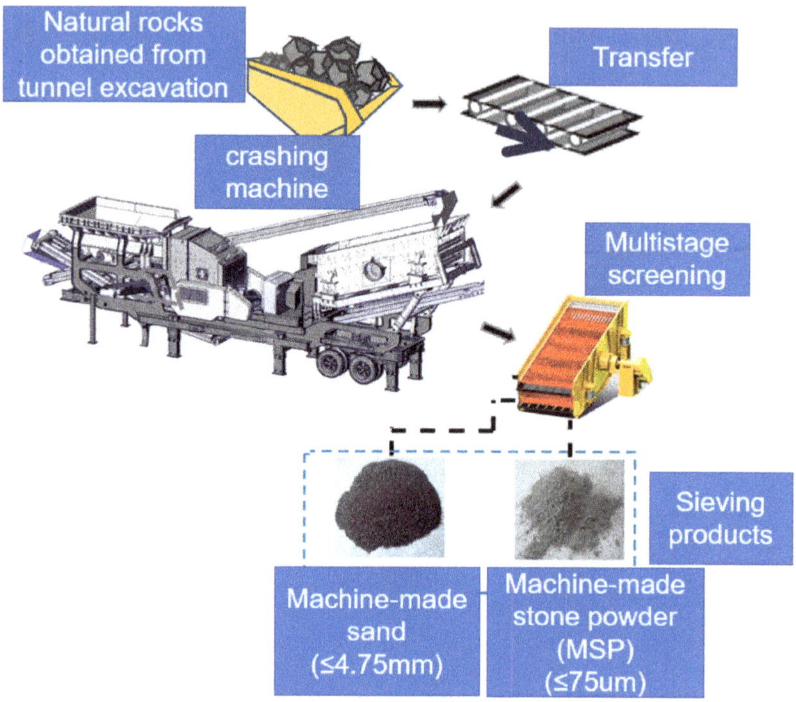

Figure 3. The preparation process of machine-made sand and MSP.

Table 3. Mineral composition of the MSP (wt%).

	Quartz	Potassium Feldspar	Plagioclase	Calcite	Dolomite
MSP	3.2	0.3	0.4	80.1	15.1

2.2. *Orthogonal Design for the Mix Proportions of Recycled Concrete*

Recycled concrete with mixture of SAP and MSP was tested by the orthogonal test method of two factors and four levels. The SAP (0%, 0.08%, 0.16%, 0.24%) and MSP (0%, 3%, 6%, 9%) content was based on the cement mass percentage substituted. Referring to the common concrete mix proportion design code, a total of 16 sets of mix proportions were designed (Table 4). The sixteen samples were labeled as 'SxMy', where x and y represent the amount of SAP and MSP, respectively. There are three identical specimens with each sample.

Table 4. Mixing ratio of concrete (kg/m^3).

Sample	SAP (%)	MSP (%)	Cement	River Sand	Machine-Made Sand	Coarse Aggregate	Water	Polycarboxylate Superplasticizer
S0M0	0.00	0	455.0	141	564.2	1057.8	182.00	0.91
S8M0	0.08	0	454.7	141	564.2	1057.8	191.09	0.91
S16M0	0.16	0	454.3	141	564.2	1057.8	191.09	0.91
S24M0	0.16	0	453.9	141	564.2	1057.8	191.08	0.91
S0M3	0.00	3	441.4	141	564.2	1057.8	182.00	0.91
S8M3	0.08	3	441.0	141	564.2	1057.8	190.82	0.91

Table 4. *Cont.*

Sample	SAP (%)	MSP (%)	Cement	River Sand	Machine-Made Sand	Coarse Aggregate	Water	Polycarboxylate Superplasticizer
S16M3	0.16	3	440.6	141	564.2	1057.8	190.81	0.91
S24M3	0.24	3	440.1	141	564.2	1057.8	190.80	0.91
S0M6	0.00	6	427.7	141	564.2	1057.8	182.00	0.91
S8M6	0.08	6	427.3	141	564.2	1057.8	190.55	0.91
S16M6	0.16	6	427.0	141	564.2	1057.8	190.54	0.91
S24M6	0.24	6	426.6	141	564.2	1057.8	190.53	0.91
S0M9	0.00	9	414.1	141	564.2	1057.8	182.00	0.91
S8M9	0.08	9	413.7	141	564.2	1057.8	190.27	0.91
S16M9	0.16	9	413.3	141	564.2	1057.8	190.27	0.91
S24M9	0.24	9	413.0	141	564.2	1057.8	190.26	0.91

2.3. Test Methodology

This study focused on investigating the effect of the freeze–thaw cycle environment on the mechanical properties of recycled concrete. According to the standard for test methods of concrete physical and mechanical properties, the specimens with corresponding sizes were prepared, as shown in Table 5; three identical specimens were taken from each sample. The mechanical test was performed after 28 days of curing in a standard SHBY-40B curing box (Zhejiang Chenxin Test Machine Manufacturing Co., Ltd., Shaoxing, China). Mechanical property tests involved the measurement of the compressive strength and flexural strength of concrete cubes as plotted in Figure 4a,b. Placing the test block pending test into the SYE-2000 type pressure test (Hebei Sanyu Test Machine Manufacturing Co., Ltd., Cangzhou, China), we started the pressure tester and loaded it with 6 kN/s speed. After the specimen was damaged, we promptly recorded the compressive test data and turned off the pressure tester, then calculated the average of the three specimens. Placing the test specimen into WE-300B digital universal material testing machine (Wuxi Lushitong Test Machine Manufacturing Co., Ltd., Wuxi, China), the upper pressure plate did not touch the specimen; we opened the oil feed valve, the oil cylinder rose slowly, we pressed the "Zero" key to remove the tare, and then pressed a certain loading speed until the test specimen ruptures. We closed the oil feed valve, opened the oil return valve unloading, and recorded the flexural test data in time, taking the average of the three specimens. A total of 200 freeze–thaw cycles were applied in the CABR-HDK9 rapid freezing–thawing equipment (Beijing Jianyan Kunlun Technology Co., Ltd., Beijing, China) as shown in Figure 4c. The minimum temperature during the cycle was set to -18 ± 2 °C and the maximum temperature was set to 5 ± 2 °C. The temperature was controlled by the three temperature sensors in the freeze–thaw box, as shown in Figure 4d. The range of the temperature sensor indicates were controlled within 2 °C.

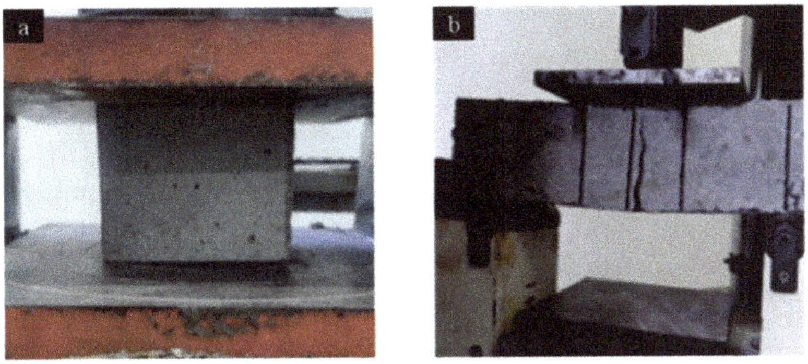

Figure 4. *Cont.*

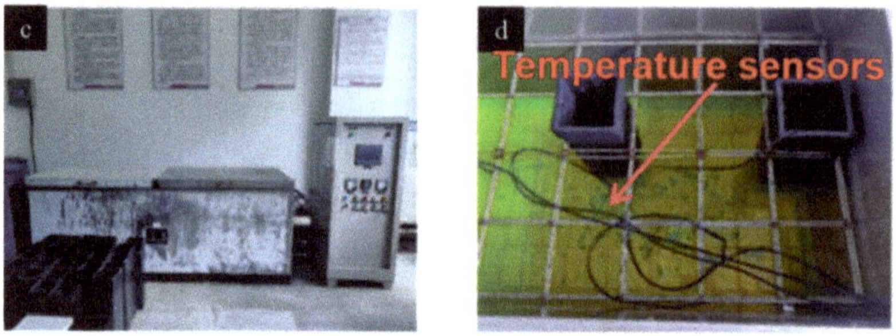

Figure 4. Mechanical test: (**a**) Compressive strength test. (**b**) Flexural compressive strength test. (**c**) CABR-HDK9 rapid freezing–thawing test machine of concrete. (**d**) Layout of temperature sensors.

Table 5. Specimen specifications.

No.	Test Content	Specimen Size/mm
1	Compressive strength	100 × 100 × 100
2	Flexural strength	100 × 100 × 400

2.4. Orthogonal Experimental Range Analysis

The orthogonal experimental range analysis was designed for evaluation of the influence of SAP and MSP in mechanical properties, e.g., compressive strength, flexural strength. The value of range R (Equation (2)) was the difference between the maximum value and the minimum value of the test result $\overline{k_{ij}}$ (Equation (1)) at each factor j. Larger values of R represented the larger influence of the factor on the mechanical properties.

$$\overline{k_{ij}} = \frac{1}{n}k_{ij} \tag{1}$$

$$R = \max\left\{\overline{k_{ij}}\right\} - \min\left\{\overline{k_{ij}}\right\} \tag{2}$$

where K_{ij} is the sum of the mechanical property test results of factor j with i level in the orthogonal table, and n is the number of parameter factors.

2.5. Strength Loss

In order to study the frost resistance of concrete under freeze–thaw cycling, the macroscopic mechanical properties such as compressive strength and flexural strength can be analyzed. The lower the strength loss before and after the freeze–thaw cycle, the better the freezing resistance of the concrete structure. The strength loss of each sample was calculated as Equation (3).

$$f = \frac{f1 - f2}{f1} \tag{3}$$

where f_1 is the strength before freeze–thaw (MPa), and f_2 is the strength after freeze–thaw (MPa).

3. Analysis and Discussion

3.1. Analysis of Compressive Strength Test Results

Table 6 described the compressive strength of 16 recycled concrete samples before and after freeze–thaw cycles and their strength loss. Table 7 presented the orthogonal experimental range analysis of compressive strength. Before freeze–thaw, the R value of SAP (6.28) is greater than that of MSP (4.45). The value R (6.58) of SAP after freeze–thaw cycles is lower than that of MSP (8.10). It could be noted that the influence of SAP on compressive strength is larger than that of MSP before freeze–thaw cycles, But after freeze–thaw cycles, the situation is the opposite, and the MSP content has a greater impact.

Table 6. Compressive strength and strength loss before and after freeze–thaw cycles (Same MSP).

Sample	Factor		Compressive Strength (MPa)		Strength Loss (%)
	SAP (%)	MSP (%)	f_1	f_2	
S0M0	0.00	0	54.95	38.38	30
S8M0	0.08	0	59.61	50.35	16
S16M0	0.16	0	55.26	48.80	12
S24M0	0.24	0	52.47	44.41	15
S0M3	0.00	3	55.27	44.94	19
S8M3	0.08	3	61.45	55.10	10
S16M3	0.16	3	58.73	54.44	7
S24M3	0.24	3	53.37	47.35	11
S0M6	0.00	6	57.62	52.25	9
S8M6	0.08	6	62.19	56.34	9
S16M6	0.16	6	59.23	55.40	6
S24M6	0.24	6	57.39	50.33	12
S0M9	0.00	9	58.71	50.16	15
S8M9	0.08	9	62.88	50.18	20
S16M9	0.16	9	60.71	52.60	13
S24M9	0.24	9	57.78	45.94	20

Table 7. Analysis on the range of compressive strength.

	SAP (%)	MSP (%)
$\overline{k_{1j}}$-1	56.64	55.57
$\overline{k_{2j}}$-1	61.53	57.21
$\overline{k_{3j}}$-1	58.48	59.11
$\overline{k_{4j}}$-1	55.25	60.02
$\overline{k_{1j}}$-2	46.43	45.49
$\overline{k_{2j}}$-2	52.99	50.46
$\overline{k_{3j}}$-2	52.81	53.58
$\overline{k_{4j}}$-2	47.01	49.72
R-1	6.28	4.45
R-2	6.56	8.10

(Ps: $\overline{k_{ij}}$-1, R-1 represented the degree of influence of SAP and MSP on recycled concrete before freeze–thaw cycles; $\overline{k_{ij}}$-2, R-2 indicated the degree of influence after 200 freeze–thaw cycles).

From the Table 7, among the four levels of SAP (0%, 0.08%, 0.16%, 0.24%), the optimal level of compressive strength before and after freeze–thaw cycles is 0.08% due to the larger $\overline{k_{ij}}$ value. Among the four levels of MSP (0%, 3%, 6%, 9%), the optimal level before freeze–thawing is 9%, and after freeze–thaw cycles is 6%. This shows that the more MSP affects the compressive strength of the structure in a freeze–thaw environment, destroys its internal structure, leads to a greater reduction in mechanical properties, and a large difference in R value before and after freeze–thaw.

Table 6 and Figure 5 show the compressive strength of recycled concrete before and after freeze–thaw cycles. Generally, freeze–thaw cycles may cause concrete to lose moisture and increase structural voids [57]. Therefore, there is a loss of compressive strength after freeze–thaw cycles, and the minimum strength loss all is content of SAP 0.16% or MSP 6%. From Table 6, when the MSP content is the same, with the increase in SAP content, compressive strength increases first and then decreases before and after freeze–thaw cycles, which is consistent with the Babatunde et al. [53]. As can be seen from Figure 5, when SAP content is the same, with the increase in MSP content, compressive strength always increases before freeze–thaw, but after freeze–thaw it increases first and then decreases, which is related to the addition of excessive MSP. Therefore, there is an optimal content of SAP and MSP.

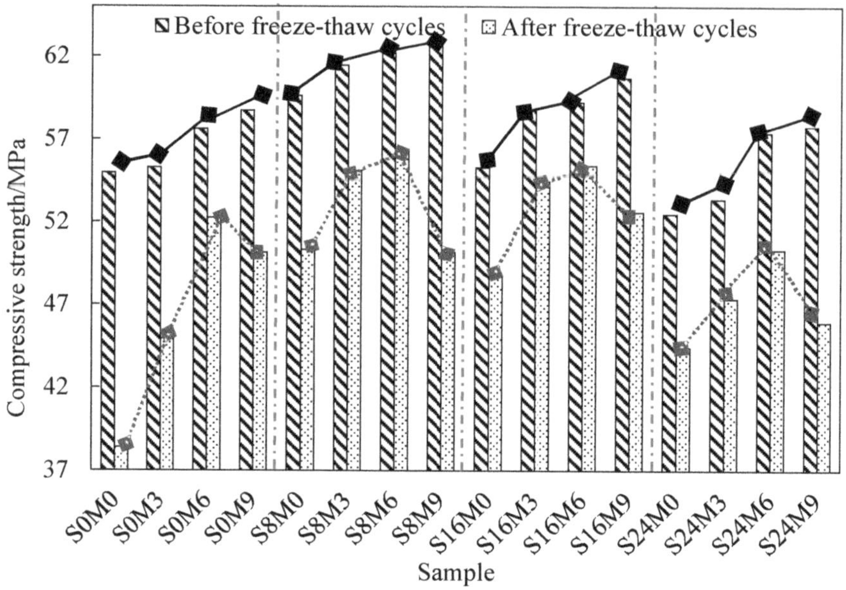

Figure 5. Compressive strength before and after freeze–thaw cycles (same SAP).

The maximum compressive strength before freeze–thaw is 62.88 MPa (Sample: S8M9), the maximum compressive strength after freeze–thaw is 56.34 MPa (Sample: S8M6), and the minimum compressive strength loss before and after freeze–thaw is 6% (Sample: S16M6). The compressive strength loss of S8M9 before and after freeze–thaw is 20%, which may have contributed to the excessive MSP hindering the hydration reaction and causing an increase in internal defects in the structure. Compared with S8M6 and S16M6, the compressive strength of S16M6 decreases by 4.76% before freeze–thaw and 1.69% after freeze–thaw, and the compressive strength of S16M6 after freeze–thaw is second only to S8M6. This is mainly because the increase in SAP content reduces the porosity of recycled concrete, increases the compactness, and improves the frost resistance [58]. As SAP content continues to increase, the compressive strength of S24M6 after freezing and thawing decreases by 9.15% compared with S16M6. This is because excessive SAP will remain in the concrete and develop into a larger void, which affects the freeze resistance of concrete [58]. Therefore, it could be considered that recycled concrete with best frost resistance is the sample S16M6 (SAP 0.16%, MSP 6%), and its compressive strength is 44% higher than S0M0.

3.2. Analysis of Flexural Strength Test Results

Table 8 describes the flexural strength of 16 recycled concrete samples before and after freeze–thaw cycles and their strength loss. Table 9 shows the orthogonal experimental range analysis of flexural strength. The R-value analysis of flexural strength is consistent with that of compressive strength. The R-value is SAP > MSP before freeze–thaw and SAP < MSP after freeze–thaw. This may be due to the strong ability of SAP to absorb water and retain water. Before the freeze–thaw cycle, SAP can absorb water to form a solid gel and increase the mechanical properties of the structure. However, after the freeze–thaw cycle, the strength of the structure will be affected because of the formation of ice by water [59]. The micro-aggregate effect of MSP enhances the density of concrete, reduces the porosity of concrete, and better hinders the entry of external water. Therefore, the more MSP is mixed in the freeze–thaw environment, the greater the improvement of structural strength [25]. Therefore, of particular note is that the amount of MSP of recycled concrete in the freeze–thaw cycle environment.

Table 8. Flexural strength and strength loss before and after freeze–thaw cycles (Same MSP).

Sample	Factor		Flexural Strength (MPa)		Strength Loss (%)
	SAP (%)	MSP (%)	f_1	f_2	
S0M0	0.00	0	5.5	1.6	71
S8M0	0.08	0	5.4	3.6	33
S16M0	0.16	0	5.6	3.8	32
S24M0	0.24	0	4.7	2.2	53
S0M3	0.00	3	5.9	2.7	54
S8M3	0.08	3	5.7	4.2	26
S16M3	0.16	3	6.0	4.5	25
S24M3	0.24	3	4.8	2.8	42
S0M6	0.00	6	6.5	3.9	40
S8M6	0.08	6	6.8	5.0	26
S16M6	0.16	6	7.4	6.5	12
S24M6	0.24	6	6.1	4.9	20
S0M9	0.00	9	6.1	3.1	41
S8M9	0.08	9	6.4	3.8	49
S16M9	0.16	9	6.5	3.4	48
S24M9	0.24	9	4.9	2.6	47

Table 9. Analysis on the range of flexural strength.

	SAP (%)	MSP (%)
$\overline{k_{1j}}$-1	6.08	5.30
$\overline{k_{2j}}$-1	6.00	5.60
$\overline{k_{3j}}$-1	6.38	6.70
$\overline{k_{4j}}$-1	5.13	5.98
$\overline{k_{1j}}$-2	3.00	2.80
$\overline{k_{2j}}$-2	3.98	3.55
$\overline{k_{3j}}$-2	4.55	5.08
$\overline{k_{4j}}$-2	3.13	3.23
R-1	1.25	1.40
R-2	1.55	2.28

(Note: $\overline{k_{ij}}$-1, R-1, $\overline{k_{ij}}$-2, R-2 are consistent with Table 4).

From the Table 9, among the four levels of SAP (0%, 0.08%, 0.16%, 0.24%), the optimal level of flexural strength before and after freeze–thaw cycles is 0.16% due to the larger $\overline{k_{ij}}$ value. Among the four levels of MSP (0%, 3%, 6%, 9%), the optimal level before and after freeze–thaw cycles is 6%.

As shown in Table 8 and Figure 6, the flexural strength of recycled concrete is evaluated both before and after freeze–thaw cycles. When the MSP content is kept constant (Table 8), an increase in SAP leads to an initial increase in flexural strength, followed by a decrease, with the highest strength observed at the third level of SAP (0.16%). Similarly, when the SAP content is held constant (Figure 6), increasing the MSP content also results in an initial increase in flexural strength, which then decreases, with the maximum strength occurring at the third level of MSP (6%).

The maximum flexural strength before freeze–thaw is 7.4 MPa (Sample: S16M6), the maximum compressive strength after freeze–thaw is 6.5 MPa (Sample: S16M6), and the minimum compressive strength loss before and after freeze–thaw is 12% (Sample: S16M6). The analysis of flexural strength further proves that S16M6 (SAP 0.16%, MSP 6%) is the best admixture ratio of recycled concrete in freeze–thaw environment.

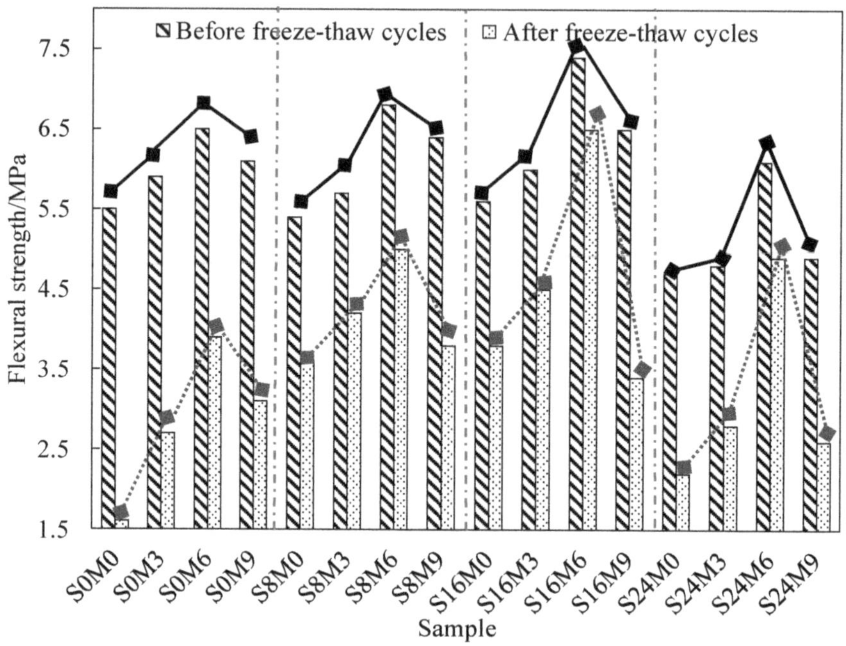

Figure 6. Flexural strength values of concrete before and after freeze–thaw cycles (Same SAP).

3.3. Regression Analysis

In order to further analyze the relationship between compressive strength and flexural strength before and after 200 freeze–thaw cycles, three sets of strength values under the optimal mix proportion (S16M6: 0.16% SAP and 6% MSP) are selected to establish the linear regression model:

$$y = 0.9433x - 0.4736 \; R^2 = 0.99$$

where x is the strength value before freeze–thaw cycles, and y is the strength model predicted value after freeze–thaw cycles.

The strength value of S16M6 before and after 200 freeze–thaw cycles and the model prediction of value are shown in Table 10. The standard deviation of regression analysis is 0.1. This regression model is helpful to predict the compressive and flexural strength of recycled concrete mixed with SAP after freezing and thawing.

Table 10. Strength value of S16M6.

Strength Value	Before 200 Freeze–Thaw Cycles (MPa)	Experimental Values after 200 Freeze–Thaw Cycles (MPa)	Model Predicted Value after 200 Freeze–Thaw Cycles (MPa)
Compressive strength	59.230	55.400	55.398
Flexural strength	7.400	6.500	6.507

4. Conclusions

This study investigated the mechanical properties and frost resistance of recycled concrete with the combined incorporation of SAP (0%, 0.08%, 0.16%, and 0.24%), and MSP (0%, 3%, 6%, and 9%). The optimal mix proportions for performance under freeze–thaw conditions were determined, and a linear regression model was established to predict the mechanical strength before and after freeze–thaw cycles. The detailed conclusions of this study were as follows:

1. Based on the extreme value analysis of the orthogonal test, it was found that, for both compressive strength and flexural strength, the R-value of SAP was greater than that of MSP before freeze–thaw cycles, while after freeze–thaw cycles, the R-value of MSP exceeded that of SAP. We proposed that the influence of SAP on the mechanical properties was dominant in the normal environment (before freeze–thaw) due to its higher water absorbency and water retention performance. However, after freeze–thaw, the outer water was blocked by MSP through the micro-aggregate effect resulting in the improved mechanical properties of recycled concrete. This indicated that the amount of MSP should be given special attention when designing recycled concrete for freeze–thaw environments. According to the larger $\overline{k_{ij}}$ value, the optimal levels of the $\overline{k_{ij}}$ value for compressive strength were SAP at 0.08% and MSP at 6% (before freeze–thaw) or 9% (after freeze–thaw), while the optimal levels of the $\overline{k_{ij}}$ value for flexural strength were SAP at 0.16% and MSP at 6%.
2. From the perspective of compressive strength, as the SAP content increased, compressive strength first increased and then decreased, both before and after freeze–thaw cycles. With increasing MSP content, compressive strength consistently increased before freeze–thaw cycles, and first increased and then decreased after freeze–thaw cycles. The maximum compressive strength before freeze–thaw was observed in S8M9, and after freeze–thaw in S8M6. The minimum strength loss before and after freeze–thaw was found in S16M6. After freeze–thaw cycles, the excessive MSP in S8M9 resulted in increasing defects within the structure, caused a large loss of structural strength, and reduced frost resistance. As the SAP content increased, the density of the structure increased, and the frost resistance improved. However, excessive SAP formed voids in the concrete and reduced the frost resistance of the structure. The mix with the best frost resistance was S16M6 (SAP 0.16%, MSP 6%).
3. From the perspective of flexural strength, flexural strength also followed a trend of first increasing and then decreasing with the increase in both SAP and MSP content, both before and after freeze–thaw cycles. The S16M6 mix consistently showed the highest flexural strength and the smallest strength loss, both before and after freeze–thaw cycles.
4. A linear regression model was established for the compressive and flexural strengths of recycled concrete with the optimal mix ratio of SAP 0.16% and MSP 6%, before and after freeze–thaw cycles. This model provides a valuable reference for predicting the behavior of recycled concrete mixed with SAP in freeze–thaw environments.

Author Contributions: Methodology, L.Z. and F.J.; Formal analysis, L.Z.; Resources, R.L.; Data curation, R.L. and F.J.; Writing—original draft, L.Z.; Writing—review & editing, R.L. and F.J. All authors have read and agreed to the published version of the manuscript.

Funding: The project was funded by the Provincial Science and Technology Service Platform Cultivation Project (XQPT202102); Nantong Institute of Technology doctoral research start-up fund: 2023XK(B)04; Basic Science (Natural Science) Research Project of Higher Education Institutions in Jiangsu Province: 24KJD560004; Jiangsu University 'Blue Project': 2024.

Institutional Review Board Statement: Not applicable.

Informed Consent Statement: Not applicable.

Data Availability Statement: All data generated or analyzed in this research were included in this published article. Additionally, readers can access all data used to support the conclusions of the current study from the corresponding author upon request.

Conflicts of Interest: The authors declare no conflicts of interest.

References

1. Kim, G. New advances in cement and concrete research. *Materials* **2023**, *16*, 4162. [CrossRef]
2. Jonny, N. Smart materials and technologies for sustainable concrete construction. *Dev. Built Environ.* **2023**, *15*, 100177.

3. Cheng, H.; Zhou, Q.; Lou, S.; Zhang, G.; Liu, Y.; Lie, Z. Simulation of the working behavior of Shibahe reservoir rock-filled concrete gravity dam during construction. *J. Tsinghua Univ. (Sci. Technol.)* **2022**, *62*, 1408–1416.
4. Habert, G.; Miller, S.; John, V.; Provis, J.; Favier, A. Environmental impacts and decarbonization strategies in the cement and concrete industries. *Nat. Rev. Earth Environ.* **2020**, *1*, 559–573. [CrossRef]
5. Li, C.; Li, J.; Ren, Q.; Zheng, Q.; Jiang, Z. Durability of concrete coupled with life cycle assessment: Review and perspective. *Cem. Concr. Compos.* **2023**, *139*, 105041. [CrossRef]
6. Olsson, J.; Miller, S.; Kneifel, J. A review of current practice for life cycle assessment of cement and concrete. *Resour. Conserv. Recycl.* **2024**, *206*, 107619. [CrossRef]
7. Cao, Y.; Wang, Y.; Zhang, Z.; Wang, H. Recycled sand from sandstone waste: A new source of high-quality fine aggregate. *Resour. Conserv. Recycl.* **2022**, *179*, 106116. [CrossRef]
8. Kumar, S.; Skariah, B.; Gupta, V. Sandstone wastes as aggregate and its usefulness in cement concrete—A comprehensive review. *Renew. Sustain. Energy Rev.* **2018**, *81*, 1147–1153. [CrossRef]
9. Banerjee, D.; Nanda, R. Waste materials as concrete aggregate—A systematic review. *IOP Conf. Ser. Earth Environ. Sci.* **2024**, *1326*, 012149. [CrossRef]
10. Singh, P.; Giri, C. Effect of Fines in Sandstone Manufactured Sand Concrete. *J. Futur. Sci. Appl.* **2021**, *2721*, 252840316.
11. Rathore, K.; Agrwal, V.; Nagar, R. Green concrete: Using quarry waste of sandstone as fine aggregate with high levels of microfines. *Mater. Today Proc.* **2020**, *32*, 728–733. [CrossRef]
12. Sharma, P.; Parashar, A.; Sharma, N. The influence of fines in sandstones on the performance of concrete. *Mater. Today Proc.* **2022**, *62*, 4126–4129. [CrossRef]
13. Mundra, S.; Agarwal, V.; Nagar, R. Sandstone cutting waste as partial replacement of fine aggregates in concrete: A mechanical strength perspective. *J. Build. Eng.* **2020**, *32*, 101534. [CrossRef]
14. Wu, B.; Xiong, Y. Influence of sandstone sand manufactured by excavated waste on properties of cement mortar. *Case Stud. Constr. Mater.* **2024**, *20*, e02777. [CrossRef]
15. Sharma, B.; Singh, A.; Joshi, S. Utilization of sandstone waste in cement mortar for sustainable production of building materials through biomineralization. *J. Sustain. Cem. Based Mater.* **2023**, *8*, 712–720. [CrossRef]
16. Chen, J. Experimental analysis of influence factors on pavement concrete performance: Different content of stone powder in manufactured sand. *Mater. Des. Process. Commun.* **2021**, *2*, 1–8. [CrossRef]
17. Gerges, N.; Issa, C.; Khalil, N. Effects of recycled waste on the modulus of elasticity of structural concrete. *Sci. Rep.* **2024**, *14*, 16189. [CrossRef]
18. Oftadeh, H.; Manthouri, M. Application of deep boltzmann machine in diagnosis processes of hepatitis types b & c. *Azerbaijan J. High Perform. Comput.* **2022**, *5*, 252914373.
19. Wu, H.; Gao, Q. Study on Freeze—Thaw Resistance of Cement Concrete with Manufactured Sand Based on BP Neural Network. *Buildings* **2024**, *14*, 2952. [CrossRef]
20. Li, Y.; Mu, J.; Hao, J.; Liu, Y.; Jiang, X.; Luo, X. Research on the particle characteristics of manufactured sands affecting the flow ability of fresh mortar. *Constr. Build. Mater.* **2023**, *287*, 382131. [CrossRef]
21. Fu, T.; Liang, J.; Lan, Y.; Xiao, Y.; Wei, J. Effect of Stone Powder Content on Mechanical Properties of Manufactured Sand Concrete. *Iran. J. Sci. Technol. Trans. Civ. Eng.* **2022**, *46*, 4255–4264. [CrossRef]
22. Wang, J.; Xue, C.; Zhang, Y. Study of Early-Age Hydration, Mechanical Properties Development, and Microstructure Evolution of Manufactured Sand Concrete Mixed with Granite Stone Powder. *Materials* **2023**, *16*, 4857. [CrossRef]
23. Zhou, A.; Miao, G.; Zhang, Y.; Qian, R.; Zhang, Y.; Shi, J.; Yang, Y. Effects of stone powder on microstructure of manufactured-sand concrete and its gas permeability. *J. Build. Eng.* **2024**, *91*, 109539. [CrossRef]
24. Liu, T.; Zhu, L.; Yu, B.; Li, T. Experimental Studies to Investigate Efficacies of Granite Porphyry Powder as Homologous Manufactured Sand Substitute. *J. Adv. Concr. Technol.* **2023**, *21*, 307–321.
25. Wang, F.; Meng, F.; Feng, T. Effect of stone powder content on the mechanical properties and microstructure of tunnel slag aggregate-based concrete. *Constr. Build. Mater.* **2023**, *388*, 131692. [CrossRef]
26. Hang, M.; Wang, J.; Zhou, X. Design and Study of Physical and Mechanical Properties of Concrete Based on Ferrochrome Slag and Its Mechanism Analysis. *Buildings* **2023**, *13*, 54. [CrossRef]
27. Islam, M.; Sohel, K.; Al-Jabri, K. Properties of Concrete with Ferrochrome Slag as a Fine Aggregate at Elevated Temperatures. *Case Stud. Constr. Mater.* **2021**, *15*, e00599. [CrossRef]
28. Kambole, C.; Paige-Green, P.; Kupolati, W. Ferrochrome slag aggregate: Physical-chemical characteristics and potential use in lightly cemented road bases. *World J. Eng.* **2023**, *20*, 93–102. [CrossRef]
29. Tang, Y.; Qiu, W.; Liu, D. Experimental Study on the Properties of Mortar and Concrete Made with Tunnel Slag Machine-Made Sand. *Materials* **2022**, *15*, 4817. [CrossRef]
30. Amran, M.; Murali, G.; Khalid, N. Slag uses in making an ecofriendly and sustainable concrete: A review. *Constr. Build. Mater.* **2020**, *272*, 121942. [CrossRef]
31. Zhang, J.; Liu, X.; Liu, J. Seismic performance and reparability assessment of recycled aggregate concrete columns with ultra-high-strength steel bars. *Eng. Struct.* **2023**, *277*, 115426. [CrossRef]
32. Son, P.; Nam, N.; Tran, N. Steel slag aggregate low-cement concrete: Engineering performance, microstructure and sustainability. *Constr. Build. Mater.* **2024**, *436*, 136827.

33. Hu, A.; Xue, G.; Shang, Z. Study on pore structure and the mechanical properties of sandstone-concrete binary under freeze–thaw environment. *Sci. Rep.* **2023**, *13*, 18280. [CrossRef]
34. Li, J.; Zhu, L.; Zhou, K. Experimental Investigation on the Effects of Ambient Freeze–Thaw Cycling on Creep Mechanical Properties of Sandstone Under Step Loading. *IEEE Access* **2019**, *7*, 108513–108520. [CrossRef]
35. Xu, H.; Chen, X.; Dang, W.; Guo, X.; Guan, J.; Li, J.; Xiong, Q. Technical Application of Machine-made Sand in the Production of the Commercial Concrete. *IOP Conf. Ser. Earth Environ. Sci.* **2021**, *647*, 012105. [CrossRef]
36. Yan, J.; Zhong, S.; Chen, S.; Yang, L.; Peng, G.; Deng, A. Study on the Application of Sediment-Based Embankment Building and Ultra-High-Performance Concrete (UHPC) Preparation in the Resource Utilization of Yellow River Sediment. *Materials* **2022**, *15*, 5668. [CrossRef]
37. Liu, H.; Ma, Y.; Ma, J. Frost Resistance of Desert Sand Concrete. *Adv. Civ. Eng.* **2021**, *2021*, 6620058. [CrossRef]
38. Kang, J.; Chen, X.; Yu, Z. Effect of polypropylene fiber on frost resistance and carbonation resistance of manufactured sand concrete. *Structures* **2023**, *56*, 104896. [CrossRef]
39. Liu, D.; Zhang, W.; Tang, Y.; Jian, Y.; Lai, Y. Orthogonal Experimental Study on Concrete Properties of Machine-Made Tuff Sand. *Materials* **2022**, *15*, 3516. [CrossRef]
40. Tan, Y.; Long, J.; Xiong, W.; Chen, X.; Zhao, B. Effects of Polypropylene Fibers on the Frost Resistance of Natural Sand Concrete and Machine-Made Sand Concrete. *Polymers* **2022**, *14*, 4054. [CrossRef]
41. Li, J.; Fan, J. Mechanical properties and microstructure of artificial sand concrete based on rock-powder content. *Desalination Water Treat.* **2024**, *319*, 100497. [CrossRef]
42. Che, D.; Wang, J.; Ding, Z.; Ma, B.; Zhang, Y. Research on the Performance of Manufactured Sand Concrete with Different Stone Powder Content. *EDP Sci.* **2021**, *283*, 01044.
43. Huang, J.; Xu, C.; Chen, S. Enhancing Mechanical Properties and Microstructures of Mass-Manufactured Sand Concrete by Incorporating Granite Powder. *Materials* **2024**, *17*, 2234. [CrossRef] [PubMed]
44. Xu, F.; Li, Z.; Li, T.; Wang, S. The Mechanical Properties and Microstructure of Tailing Recycled Aggregate Concrete. *Materials* **2024**, *17*, 1058. [CrossRef] [PubMed]
45. Ma, S.; Huang, C.; Baah, P. The influence of water-to-cement ratio and superabsorbent polymers (SAPs) on solid-like behaviors of fresh cement pastes. *Constr. Build. Mater.* **2021**, *275*, 122160. [CrossRef]
46. Niu, X.; Zhang, Y.; Elakneswaran, Y.; Sasaki, M.; Kawai, H. Effect of Superabsorbent Polymer (SAP) Size on Microstructure and Compressive Strength of Concrete. *Polymers* **2024**, *16*, 197. [CrossRef]
47. Tang, C.; Dong, R.; Tang, Z. Effect of SAP on the properties and microstructure of cement-based materials in the low humidity environment. *Case Stud. Constr. Mater.* **2024**, *20*, e03001. [CrossRef]
48. Yao, Y.; Zhu, Y.; Yang, Y. Incorporation superabsorbent polymer (SAP) particles as controlling pre-existing flaws to improve the performance of engineered cementitious composites (ECC). *Constr. Build. Mater.* **2012**, *28*, 139–145. [CrossRef]
49. Sun, B.; Wu, H.; Song, W.; Li, Z.; Yu, J. Design methodology and mechanical properties of Superabsorbent Polymer (SAP) cement-based materials. *Constr. Build. Mater.* **2019**, *204*, 440–449. [CrossRef]
50. Liu, J.; Farzadnia, N.; Shi, C. Effects of superabsorbent polymer on interfacial transition zone and mechanical properties of ultra-high performance concrete-ScienceDirect. *Constr. Build. Mater.* **2020**, *23*, 117142. [CrossRef]
51. Huang, X.; Liu, X.; Rong, H.; Yang, X.; Duan, Y.; Ren, T. Effect of Super-Absorbent Polymer (SAP) Incorporation Method on Mechanical and Shrinkage Properties of Internally Cured Concrete. *Materials* **2022**, *15*, 7854. [CrossRef] [PubMed]
52. Ding, H.; Zhang, L.; Zhang, P. Factors Influencing Strength of Super Absorbent Polymer (SAP) Concrete. *Trans. Tianjin Univ.* **2017**, *23*, 245–257. [CrossRef]
53. Babatunde, J.; Adewumi, J.; William, P. Early-age and long-term strength development of high-performance concrete with SAP-ScienceDirect. *Constr. Build. Mater.* **2021**, *267*, 121798.
54. Zhang, X.; Zhang, R.; Ning, G.; Duan, Y.; Dou, X. Experimental study on the effect of SAP content on concrete strength. *Build. Sci.* **2021**, *11*, 79–85. (In Chinese)
55. Yang, H.; Liu, M.; Biao, H.; Wu, L.; Liu, J.; Wang, W. The influence of SAP particle size and extra water on the dynamic mechanical behavior of SAP-modified cement-based composites. *Constr. Build. Mater.* **2022**, *347*, 128622. [CrossRef]
56. Qin, X.; Lin, Y.; Mao, J.; Sun, X.; Xie, Z.; Huang, Q. Research of Water Absorption and Release Mechanism of Superabsorbent Polymer in Cement Paste. *Materials* **2023**, *15*, 3062. [CrossRef]
57. Luo, S.; Bai, T.; Guo, M.; Wei, Y.; Ma, W. Impact of Freeze-Thaw Cycles on the Long-Term Performance of Concrete Pavement and Related Improvement Measures: A Review. *Materials* **2022**, *15*, 4568. [CrossRef]
58. Viktor, M.; Wyrzykowski, M.; Schröfl, C. Application of super absorbent polymers (SAP) in concrete construction—Update of RILEM state-of-the-art report. *Mater. Struct.* **2021**, *54*, 1–20.
59. Mechtcherine, V.; Schrfl, C.; Wyrzykowski, M. Effect of superabsorbent polymers (SAP) on the freeze-thaw resistance of concrete: Results of a RILEM interlaboratory study. *Mater. Struct.* **2017**, *50*, 14. [CrossRef]

materials

MDPI

Article

Stress-Strain Behavior and Strength Development of High-Amount Phosphogypsum-Based Sustainable Cementitious Materials

Ying Shi, Yue Li, Hongwei Wang *, Yixuan Ma and Xinyue Lu

School of Resources and Safety Engineering, Central South University, Changsha 410083, China
* Correspondence: hongwei.wang@csu.edu.cn

Abstract: Phosphogypsum is a common industrial solid waste that faces the challenges of high stockpiling and low utilization rates. This study focuses on the mechanical properties and internal characteristics of cementitious materials with a high phosphogypsum content. Specifically, we examined the effects of varying amounts of ground granulated blast furnace slag (5–28%), fly ash (5–20%), and hydrated lime (0.5–2%) on the stress–strain curve, unconfined uniaxial compressive strength, and elastic modulus (E_{50}) of these materials. The test results indicate that increasing the ground granulated blast furnace slag content can significantly enhance the mechanical properties of phosphogypsum-based cementitious materials. Additionally, increasing the fly ash content can have a similar beneficial effect with an appropriate amount of hydrated lime. Furthermore, microscopic analysis of the cementitious materials using a scanning electron microscope revealed that the high sulfate content in phosphogypsum leads to the formation of calcium aluminate as the main product. Concurrently, a continuous reaction of the raw materials contributes to the strength development of the cementitious materials over time. The results could provide a novel method for improving the reusing phosphogypsum amount in civil engineering materials.

Keywords: solid waste; cementitious material; stress–strain behavior; compressive strength

1. Introduction

Ordinary Portland Cement (OPC), as a common cementitious material, is in high demand in the road and construction industries. However, the cost of OPC production and the accompanying environmental pollution have been unresolved issues for OPC as a cementitious material. The cement industry generally requires 1.2 tonnes of hydrated limestone and 0.3 tonnes of clay to produce one tonne of cement clinker [1]. In the process of producing one tonne of OPC, 700 kg of carbon dioxide is emitted into the atmosphere [2–6], as well as other toxic and harmful gasses [7–10]. Therefore, many scholars are committed to researching and developing low-cost green cementitious materials that can replace OPC [11–13]. Industrial production generates a large amount of solid waste [14–17]. These solid wastes, at present, are mainly disposed of at landfills with low comprehensive utilization [18–22]. Therefore, preparing low-cost green cementitious materials using various types of solid wastes is a pathway with great potential for development [23–28].

Phosphogypsum (PG), as a typical industrial solid waste, is generally produced as a by-product from the production of phosphoric acid. It is primarily composed of calcium sulfate dihydrate, along with impurities such as soluble phosphorus, fluorine, and heavy metals [29,30]. It was reported that approximately five tonnes of PG are generated in the production of one tonne of phosphoric acid. The global production of PG is about 300 million tonnes per year [31,32]. Its mass utilization would be 10–15% under ideal conditions [33]. The use of PG to prepare cementitious materials has been studied [34–36]. Min et al. [37] used PG, FA, lime, and cement to prepare cementitious materials. They found that relying on the interaction between the raw materials, they could obtain a cementitious material with excellent comprehensive performance. Xu et al. [38] mixed phosphonyls with iron tailings to prepare cementitious materials. They concluded that phosphonyls could effectively promote the generation of calcium alumina in the cementitious system, fill the large pores in the system, and significantly improve the compressive strength of cementitious materials. Liu et al. [39] aimed to address the issues of long setting time and low early strength in gelling agents. They explored a novel approach by adding polymerized aluminum compounds to regulate the nucleation and growth of calcite. This research developed a low-carbon gelling agent based on waste gypsum and metallurgical slag. Zhang et al. [40] prepared a low-carbon calcium sulphoaluminate cement by replacing natural gypsum with PG. They discussed the mechanical properties and hydration mechanism of this new cement.

Previous studies show that although PG can provide a large amount of sulfate, it lacks the silicon and aluminum elements needed for hydration reactions [41]. Therefore, the synergistic reaction with substances containing more silicon and aluminum elements mixed with PG [42] can effectively enhance the mechanical properties of PG-based cementitious materials. However, at present, most studies use low-volume (<45%) PG to partially replace OPC as composite cementitious materials [43–46]. There is still a knowledge gap in using high-volume (50–90%) PG to fabricate cementitious material without OPC. Therefore, this study planned to prepare low-cost cementitious materials using high-volume PG without OPC. Additionally, other solid wastes containing some amount of silicon and aluminum elements, i.e., ground granulated blast furnace slag (GGBS), and fly ash (FA), are required. GGBS effectively enhances the strength growth of the cementitious material in the early stage, while FA ensures the strength increase in the later stage of the cementitious material. At the same time, PG can raise sufficient amounts of sulfate to facilitate the volcanic ash reaction process during this period.

The properties of the relevant cementitious materials in different aspects are represented by a number of other parameters [47–49], while the mechanical properties are generally characterized by parameters such as peak strength, stress–strain curves, E_{50}, etc. [50,51]. Referring to previous scholars' research on cementitious materials [52–54]. The following exploration of test results was carried out in this paper: (1) analyzing the effect of changes in the components of the cementitious materials on the mechanical properties of paste; (2) studying the relationship between compressive strength and E_{50} of paste; (3) determining the microstructure of paste based on the scanning electron microscope micrographs.

2. Experimental Program

2.1. Raw Materials

The PG used in this study was sourced from Xifeng, Guizhou, China. Prior to use, it underwent water washing to remove attached pollutants. Based on pre-experimental research, the mass ratio of PG to water was 1:3, with stirring conducted for a total of half an hour. The GGBS, with S95-grade activity, was procured from the Baifeng Mineral Products Processing Plant, Shijiazhuang, China. F-grade FA was obtained from Longjiang FA Development and Utilisation Co. (Heilongjiang, China). Hydrated lime (HL) from Shanghai McLean Biochemical Technology Co. (Shanghai, China) was added to the reaction system to enhance the volcanic ash reaction. The particle size analysis of PG, GGBS, and FA was conducted using a Mastersizer 3000 laser diffraction particle size analyzer, with the results illustrated in Figure 1. The d50 values for PG, GGBS, FA, and HL were 50.5 μm, 73.7 μm, 15.2 μm, and 4.9 μm, respectively.

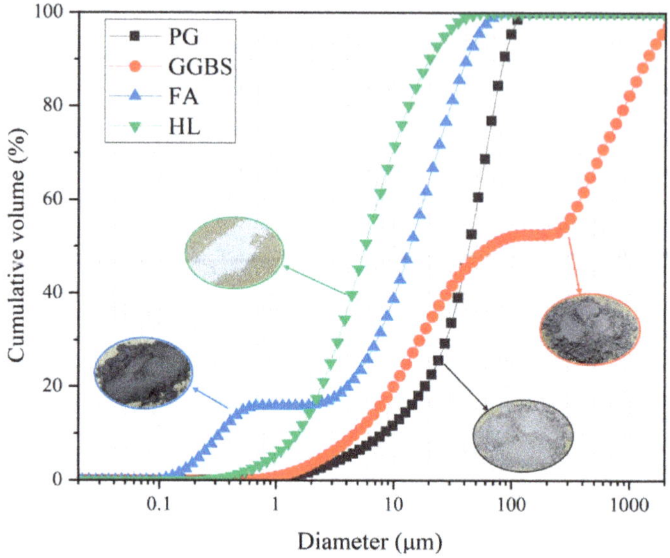

Figure 1. Particle size analysis of raw materials.

In this study, an X-ray fluorescence spectrometer (XRF) was used to analyze the main chemical constituents of solid waste. The results are shown in Figure 2. The main components of PG were 41.9% CaO and 48.6% SO_3. GGBS was mainly composed of 34.3% CaO, 20.6% SiO_2, and 11.4% Al_2O_3. In contrast, FA was composed of 44.6% SiO_2 and 26.7% Al_2O_3.

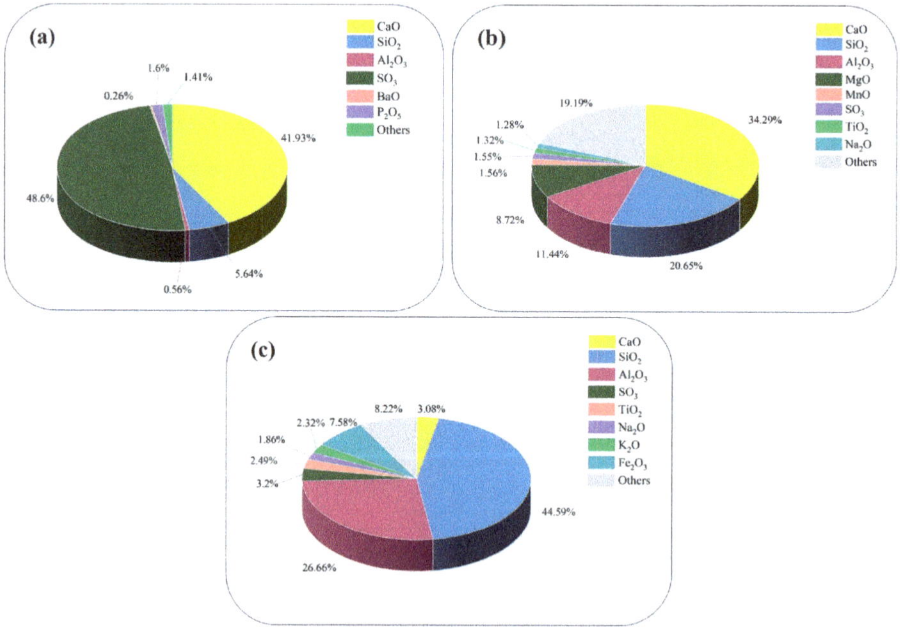

Figure 2. XRF analysis of raw materials: (**a**) PG; (**b**) GGBS; (**c**) FA.

2.2. Test Method

To examine the effects of three raw materials on the mechanical properties of PG-based paste specimens, the experimental design involved fixing the content of two raw materials while varying the content of the third. The raw material mixing for the cementitious materials was designed based on findings from previous laboratory studies [55]. Using the two groups G-1 and G-2 from Table 1 as examples, this study explores the impact of varying GGBS content on the properties of the cementitious materials, while maintaining constant levels of FA and HL. Given that PG, at a dosage greater than 50%, can serve as both a calcium source and a provider of sulfate ions, its individual variation is not considered in this research [55]. Tables 1–3 provide specific content information for each group. Based on previous laboratory findings, the water-to-binder ratio of paste was fixed at 0.5.

Table 1. The ratio of experimental groups with different content of GGBS.

No.	GGBS (%)	FA (%)	HL (%)	PG (%)	Total (%)
G-1	5	5	1.25	88.75	100
G-2	28	5	1.25	65.75	100
G-3	5	20	1.25	73.75	100
G-4	28	20	1.25	50.75	100
G-5	5	12.5	0.5	82.00	100
G-6	28	12.5	0.5	59.00	100

Table 2. Ratio of experimental groups with different content of FA.

No.	GGBS (%)	FA (%)	HL (%)	PG (%)	Total (%)
F-1	5	5	1.25	88.75	100
F-2	5	20	1.25	73.75	100
F-3	16.5	5	0.5	78	100
F-4	16.5	20	0.5	63	100
F-5	16.5	5	2	76.5	100
F-6	16.5	20	2	61.5	100
F-7	28	5	1.25	65.75	100
F-8	28	20	1.25	50.75	100

Table 3. Ratio of experimental groups with different content of HL.

No.	GGBS (%)	FA (%)	HL (%)	PG (%)	Total (%)
C-1	5	12.5	0.5	82	100
C-2	5	12.5	2	80.5	100
C-3	16.5	5	0.5	78	100
C-4	16.5	5	2	76.5	100
C-5	16.5	20	0.5	63	100
C-6	16.5	20	2	61.5	100

Figure 3 illustrates the specimen preparation process and testing apparatus. Firstly, dry PG, GGBS, FA, and HL were mixed for two minutes using a planetary mixer to obtain a homogeneous mixture. A corresponding mass of water was then added and mixed for another two minutes. Subsequently, the mixture was poured into 40 mm × 40 mm × 160 mm molds, and the poured molds were placed on a shaking table to eliminate air bubbles. Afterward, cling film was attached to the surface of the molds and maintained for three days before demolding. The specimens were further wrapped with cling film and cured in a constant temperature and humidity box (SHBY-60B, Cangzhou Yixuan Testing Instrument, Cangzhou, China) at 20 °C and 95% humidity.

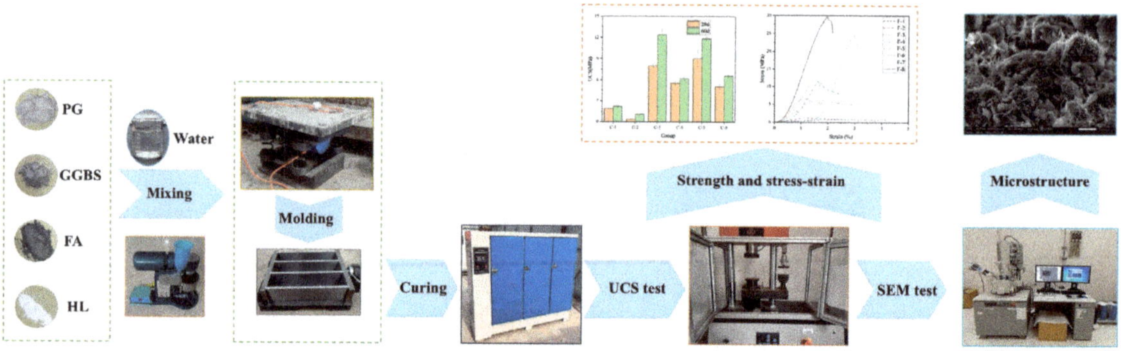

Figure 3. Specimen preparation process and test.

The specimens' unconfined compressive strength (UCS) at 28 days and 60 days was determined using a WHY-200/10 (Cangzhou Yixuan Testing Instrument) microcomputer-controlled compression tester. The corresponding data were analyzed to obtain the stress–strain curves. The UCS tests in this study refer to the Chinese standard GB/T17671 [56]. The uniaxial compressive strength result was the average of the test results of three specimens.

After the compressive strength test, specimens with a volume of 5 mm^3, featuring natural fractures and flat surfaces, were selected. These specimens were dried in a constant-temperature oven at 60 °C until they reached a constant weight. Prior to testing, the samples were coated with gold foil to enhance conductivity. Scanning electron microscope images of these samples were obtained using a Phenom ProX (Thermo Fisher Scientific, Waltham, MA, USA) scanning electron microscope with an accelerating voltage of 10 KV.

3. Result and Discussion

3.1. Stress and Strain

The corresponding stress–strain curve data for each group were exported and fitted to the curve when using the WHY-200/10 microcomputer-controlled compression tester to test the uniaxial compressive strength. The horizontal axis represents the axial strain of the cementitious material, while the vertical axis represents the axial stress of the cementitious material. Based on the change in stress rate, the stress–strain curve of the specimen is divided into four stages: compaction stage, elastic deformation stage, plastic deformation stage, and post-peak damage stage. In the compaction stage, the specimen closes the pores due to external pressure, resulting in a slow increase in stress with strain [57]. The elastic deformation stage is entered after the pores are closed, characterized by an approximately linear change in stress with strain. Subsequently, as the specimen continues to deform and the internal pores and cracks expand, it transitions into the plastic deformation stage under high pressure [58,59]. At this time, the stress growth rate of the specimen will continue to decline, reducing to zero when the peak stress is reached. Continuing to apply pressure to the specimen at this point will cause it to enter the post-peak damage stage. In this stage, the bearing capacity of the specimen will be reduced, and obvious cracks will appear on the surface [60].

Figure 4 illustrates the stress–strain curves for each group in Table 1 at 60 days. From the figure, it is evident that in the group with higher FA content, a higher GGBS content shortens the compaction stage of the specimen. This is because GGBS contains a certain amount of cement clinker and a significant amount of reactive silica-alumina, both of which fully participate in the hydration reaction [61]. This suggests that a significant amount of hydration products is generated inside the specimen to fill numerous pores. When the content of FA and HL is fixed, increasing the GGBS content in cementitious materials markedly enhances the slopes of the stress–strain curves in both the elastic and plastic deformation stages. This indicates a notable positive effect of increasing GGBS content on improving the deformation resistance of cementitious materials. Additionally, specimens in the group with a higher GGBS content ratio exhibited damage at higher strain values and demonstrated more ductile behavior.

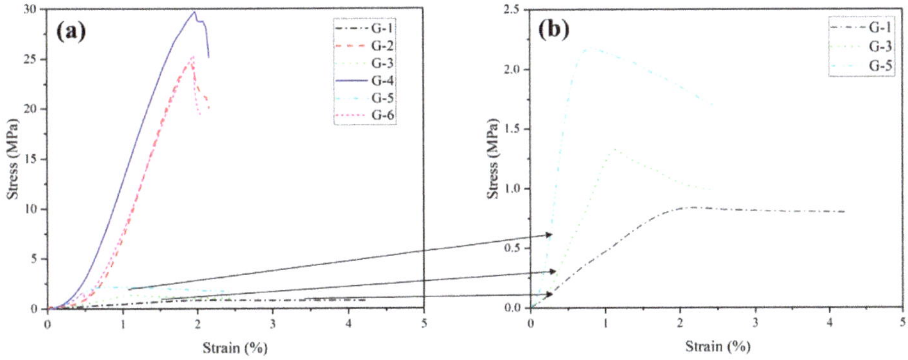

Figure 4. (**a**) Stress–strain relationship for different GGBS content levels; (**b**) is an enlargement of the typical curves in (**a**).

Figure 5 illustrates the stress–strain curves for each group in Table 2 at 60 days. At higher GGBS and HL dosages, increasing the FA dosage significantly shortens the compaction stage of the cementitious materials. This indicates that sufficient amounts of GGBS and HL greatly enhance the participation of FA in the reaction, leading to the generation of more hydration products that fill internal pores. In contrast, with low GGBS and HL dosages, increasing the FA dosage prolongs the compaction stage of the specimens, suggesting that FA requires adequate alkalinity to dissolve the vitreous surface, thereby enabling the internal reactive silica-alumina to participate in the reaction [62]. After entering the ascending branch of the elastic deformation stage and the plastic deformation stage, increasing the FA content was able to increase the slope, indicating that FA participated in the reaction system in all groups and improved the deformation resistance of the specimens. However, the elevation of FA content decreases the failure strain values of the specimens, indicating a gradual transition from ductile to brittle specimens.

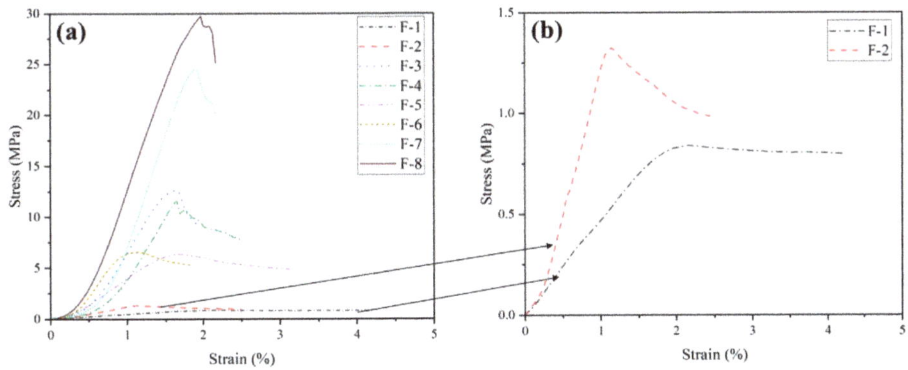

Figure 5. (**a**) Stress–strain relationship for different FA content levels; (**b**) is an enlargement of the typical curves in (**a**).

Figure 6 illustrates the stress–strain curves at 60 days for the groups in Table 3. From the figure, it can be seen that the compaction stages of the group with high HL content are all lower than those of the group with low HL content, indicating that HL can effectively neutralize the residual acid of PG and promote the participation of GGBS and FA in the reaction [63] so that there are enough hydration products inside the specimen to fill the pores. If the GGBS and FA content is low, the slopes of the group with high HL content will be lower than that of the group with low HL content in both the elastic deformation stage and plastic deformation stage, indicating that too much HL will be harmful to the deformation resistance of the specimen. However, if the content of GGBS and FA is increased, the HL will be fully utilized, increasing the specimens' deformation resistance. The effect of HL content on the failure strain of the specimen is related to the content of other raw materials. Based on the high content of GGBS and FA, increasing the content of HL will make the specimen reach the failure strain value earlier and show higher brittleness. However, if the HL content is increased based on low GGBS and FA content, the specimen reaches the failure strain value later and shows higher ductility.

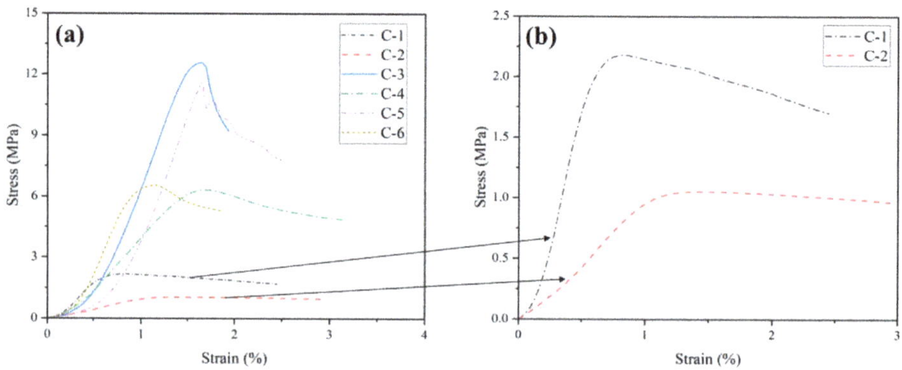

Figure 6. (**a**) Stress–strain relationship for different HL content levels; (**b**) is an enlargement of the typical curves in (**a**).

3.2. Compressive Strength

Figure 7a demonstrates the UCS at 28 and 60 days for each group of Table 1. Figure 7b demonstrates the growth rate of each group's strength at 60 days relative to 28 days for different GGBS content, calculated as shown in Equation (1). It can be observed that the strength of specimens with high GGBS content is significantly higher than that of specimens with low GGBS content, regardless of the variation in FA and HL content. This is attributed to the fact that GGBS contains a certain amount of cement clinker and a large amount of active silica-alumina, effectively enhancing the strength of cementitious materials [61]. The presence of high GGBS content can effectively ensure the strength growth value of the specimens from 28 days of maintenance to 60 days. The strength growth rate of the high GGBS content group was higher than that of the low GGBS content group, except for G-4. The strength growth rate in the G-3 group was 74%, significantly higher than the 18% observed in the G-4 group. This difference is attributed to the much higher initial strength value of the G-3 group at 28 days. The G-4 group formulations outperform the others in this study.

$$\text{Growth rate}(\%) = \frac{UCS_{60d} - UCS_{28d}}{UCS_{28d}} \quad (1)$$

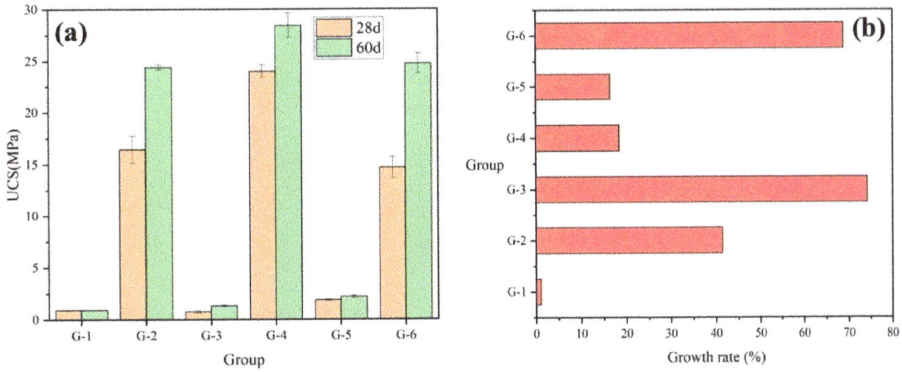

Figure 7. Effect of GGBS content on (**a**) UCS and (**b**) growth rate.

Figure 8a illustrates the UCS of each group of Table 2 at 28 and 60 days. Figure 8b illustrates the growth rate of each group's strength at 60 days relative to 28 days for different FA contents. As shown in the figure, while keeping the low HL content constant, the strength of the high FA content group (F-4) is lower than that of the low FA content group (F-5). Conversely, the strength of the high FA content group is higher than that of the low FA content group for all other ratio groups, indicating that to promote FA's participation in the reaction, it is necessary to ensure a certain level of alkalinity [64]. Thus, due to high alkalinity, the strength growth rate of 30% in the high FA content group (F-6) was significantly higher than the 13% strength growth rate in the low FA content group (F-5).

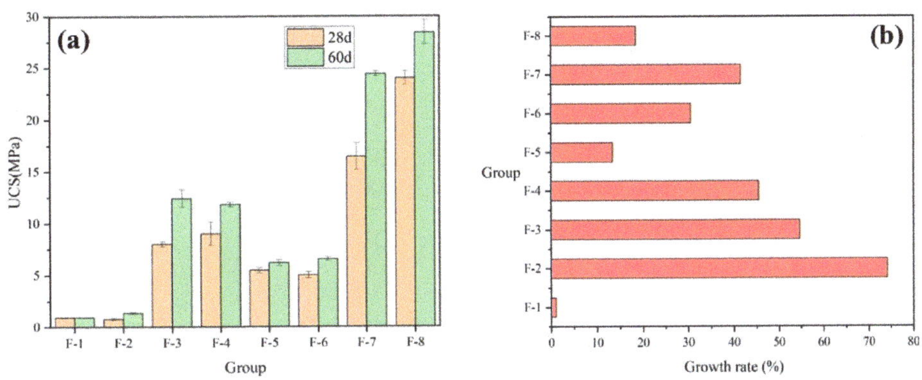

Figure 8. Effect of FA content on (**a**) UCS and (**b**) growth rate.

Figure 9a illustrates the UCS of each group of Table 3 at 28 and 60 days. Figure 9b illustrates the growth rate of each group at 60 days strength relative to 28 days strength for different HL contents. It can be seen from the figure that the high HL admixture has a negative effect on the hydration reaction inside the specimen [65]. The strengths of the specimens in the high HL content group are much lower than those in the low HL content group. While keeping the low content of GGBS and FA unchanged, the strength growth rate of the high HL content group (C-2) was 176%, which was 16% higher than the strength growth rate of the low HL content group (C-1). Conversely, other groups exhibit the opposite trend: the C-4 strength growth rate of 13% is lower than the C-3 strength growth rate of 54%, and the C-6 strength growth rate of 30% is lower than the C-5 strength growth rate of 45%. This suggests that the low lime content group still requires a certain amount of GGBS to undergo hydrolysis reaction [66], generating HL crystals to maintain the alkalinity of the reaction system.

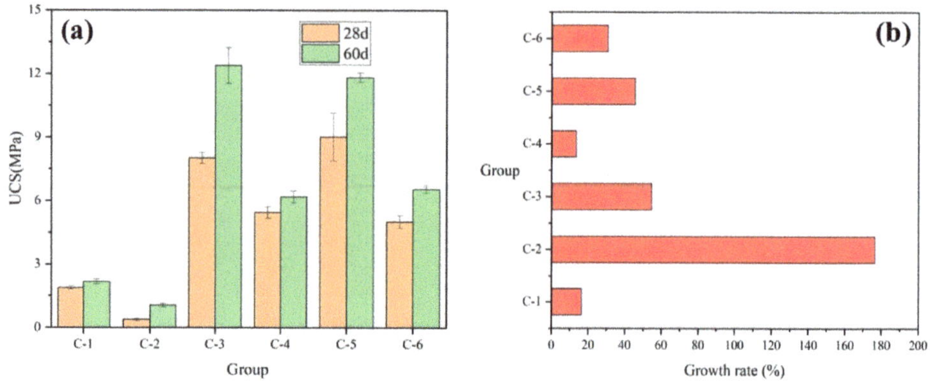

Figure 9. Effect of HL content on (**a**) UCS and (**b**) growth rate.

3.3. Prediction Model

As shown in Table 4 below, the different dosages of raw materials used in this study were grouped into three level values. A reasonable prediction model was constructed using UCS at a curing time of 60 days as an example.

Table 4. Range of variation in prediction model factors.

Variables	Coded	Levels		
		−1	0	1
GGBS/wt.%	A	5	16.5	28
FA/wt.%	B	5	12.5	20
HL/wt.%	C	0.5	1.25	2

The corresponding simplified mathematical model is shown in Equation (2). Table 5 displays the statistical indicators of the constructed model, including essential metrics such as the F-value, *p*-value, and R^2.

$$\begin{aligned} UCS(60d) = 9.89 + 10.88\ A + 0.63\ B - 2.63\ C + 1.26\ AB - 1.82\ AC + 0.23\ BC \\ +2.75\ A^2 + 0.91\ B^2 - 1.57\ C^2 \end{aligned} \quad (2)$$

Table 5. The model fitting parameters.

Curing Time	Mean	F-Value	*p*-Value	R^2
60 d	11.01	28.17	0.0009	0.9807

The F-value for the model in Table 5 is 28.17, indicating that the model is significant. The model has a probability of 0.09%, and the observed large F-value is not attributed to noise. In this study, the *p*-value for the model is less than 0.05, affirming the significance of the model. Meanwhile, the R^2 value of the model is 0.9807, indicating that the model accurately represents the actual situation [67,68].

The comparison of experimental and predicted UCS is shown in Figure 10. Using the model fitted to the test data, the UCS values for a curing time of 60 days can be predicted. The predicted values are generally comparable to the experimental values, with a root mean square error (RMSE) of only 1.268 MPa. Consequently, the reliability of the fitted model is considered high.

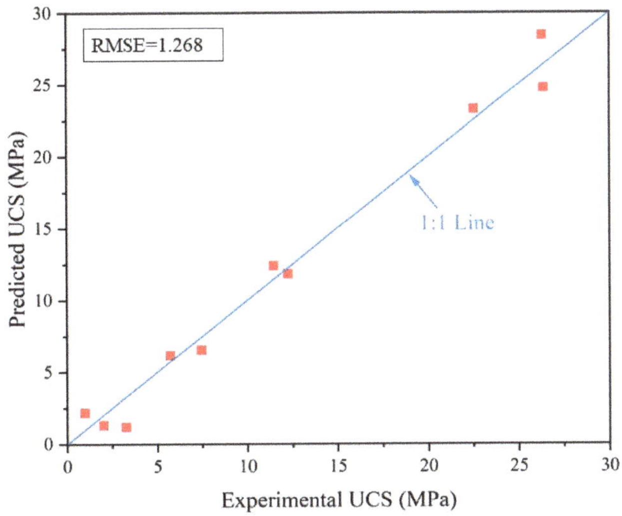

Figure 10. Comparison of experimental and predicted UCS.

Based on the cementitious materials prepared in this study, two experimental groups with different ratios were created. A compressive strength test was conducted after 60 days of curing. The results were used to verify the accuracy of the previously constructed model, as shown in Table 6. The findings demonstrate that the model possesses a notable degree of generalizability.

Table 6. Model validation.

GGBS (%)	FA (%)	HL (%)	PG (%)	Predicted UCS (MPa)	Experimental UCS (MPa)
28.00	12.50	2.00	57.50	17.50	16.33
16.50	12.50	1.25	69.75	9.89	10.13

3.4. Elastic Modulus

The E_{50} is a constant of proportionality of material over the range of elastic deformation and is a parameter that clearly reflects the mechanical properties of the prepared cementitious material [69,70]. In this study, the constant of proportionality of the prepared cementitious material at 50% peak stress is taken and calculated as shown below:

$$E_{50} = \frac{\sigma_{50}}{\varepsilon_{50}} \tag{3}$$

where E_{50} is the constant of proportionality of the cementitious material at 50% peak stress, MPa. The σ_{50} is the 50% peak stress of the cementitious material, MPa. The ε_{50} is the the strain value at the 50% peak stress of the cementitious material.

The following Figure 11 shows the E_{50} for different GGBS content groups. Elevated GGBS content significantly increases the E_{50} of the specimens, with the highest group, G-4, reaching an E_{50} of 2000 MPa, nearly 12 times higher compared to the corresponding low GGBS content group G-3.

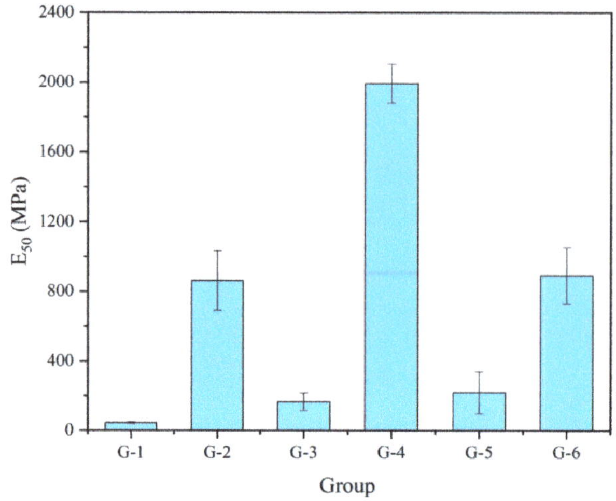

Figure 11. E_{50} for different GGBS content groups.

The following Figure 12 shows the E_{50} for different FA content groups. Based on low HL content, increasing the FA content decreases the E_{50} of the specimens (F-3 vs. F-4). However, increasing the FA content in the remaining groups significantly increases the E_{50}, which is consistent with the change in the strength of the specimens. The highest of these groups, F-8, improved nearly three times compared to F-7.

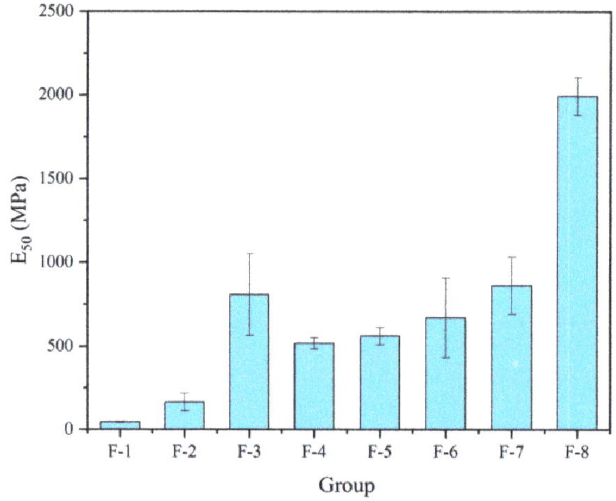

Figure 12. E_{50} for different FA content groups.

The following Figure 13 shows the E_{50} for different HL content groups. Based on high FA content, increasing the HL content enhances the E_{50} to some extent (C-5 vs. C-6), which is related to the need for a certain alkalinity of the FA. Whereas, increasing the HL content for the other groups significantly reduces the E_{50}.

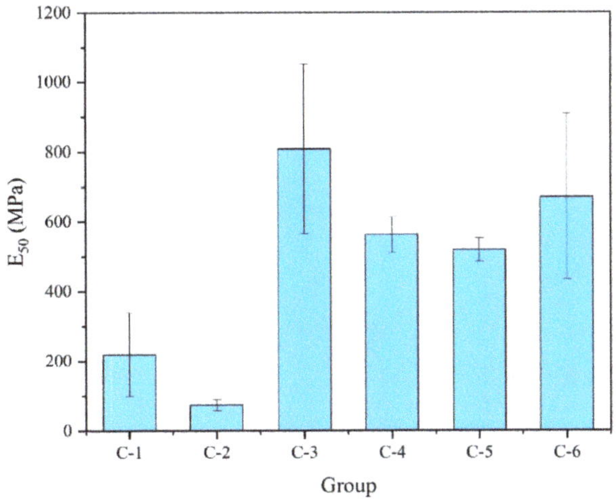

Figure 13. E_{50} for different HL content groups.

Based on the UCS and E_{50} discussed above, the corresponding data for each group were curve-fitted to build the model equations, as shown in Figure 14 (data points identified with dashed boxes in the figure are discrete points). The E_{50} of each group has a gradually increasing trend with UCS, and the higher the UCS, the larger the E_{50}. The corresponding linear correlation equation is E_{50} = 64.11UCS, and the correlation coefficient of the equation is R^2 = 0.95, which is highly reliable.

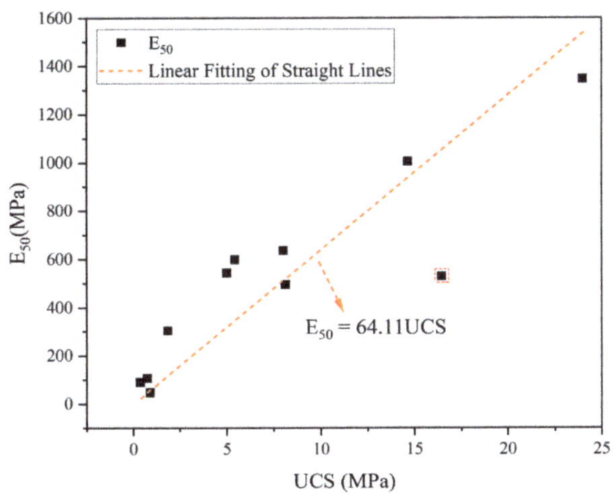

Figure 14. Relationship between E_{50} and UCS.

3.5. SEM

In this study, the typical microstructure of this system was analyzed using the optimized proportioning group based on the strength results as a representative specimen. The results are shown in Figure 15. From Figure 15d, it can be seen that the main product of the reaction of the cementitious materials is ettringite, which is associated with the high sulfate content provided by PG [25]. As shown in Figure 15b,c, a certain amount of PG, GGBS, and FA is still present within the collodion material, indicating that the hydration reaction has not been fully completed in a short period. In the later stages of the hydration reaction, the glassy layer on the surface of the FA gradually dissolves. This dissolution releases reactive silica-aluminum compounds [71,72]. These reactive substances react with the abundant sulfates in the system, producing adequate amounts of ettringite to fill the pores within the cementitious material. Consequently, the cementitious material is expected to exhibit significant strength growth in the later period (60 days).

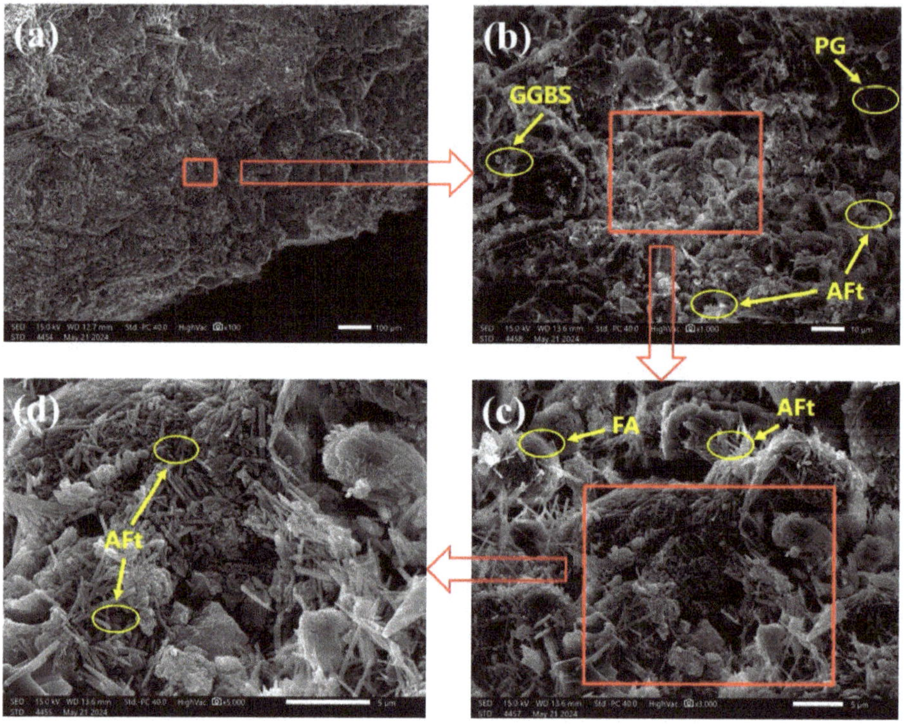

Figure 15. SEM images of typical specimens. (**a**) specimen (×100); (**b**) is a local enlargement of (**a**) (×1000); (**c**) is a local enlargement of (**b**) (×3000); (**d**) is a local enlargement of (**c**) (×5000).

4. Conclusions

This study analyzes the mechanical properties and microstructure of a PG-based cementitious material developed using various solid wastes as raw materials to improve the reusing of PG in civil engineering materials. Different raw materials' effects on the cementitious material's properties are clarified, and relevant fitting models are constructed. The main conclusions are as follows:

(1) A high GGBS content (28%) can compensate for the pores inside the cementitious material and improve deformation resistance. Similarly, a high FA content (20%) can achieve the same function with sufficient alkalinity. However, excessive alkalinity negatively affects the material's performance.
(2) High GGBS content significantly improves the UCS of cementitious materials. Moreover, increasing FA content under appropriate HL conditions also enhances UCS.
(3) A prediction model was constructed based on the UCS of cementitious materials. The statistical indices of the constructed model indicate a certain degree of reliability, and the model's predicted data are generally consistent with the test data.
(4) The pattern of change in E_{50} and UCS of cementitious materials tends to be the same, so the relationship between the two is modeled as a linear equation, represented as $E_{50} = 64.11\text{UCS}$.

(5) SEM results indicate that ettringite primarily fills the pores within the studied cementitious material. As the curing age increases, the unreacted materials in the system gradually participate in the hydration reaction, continuously producing ettringite to further fill the pores of the cementitious material.

Author Contributions: Conceptualization, Formal analysis, Investigation, Writing—original draft preparation, Y.S., H.W. and Y.L.; Formal analysis, Software, Y.S., H.W., Y.L. and Y.M.; Conceptualization, Writing—review and editing, Y.S., H.W., Y.L. and X.L. All authors have read and agreed to the published version of the manuscript.

Funding: This research was supported by the National Natural Science Foundation of China (grant no. 42177160, Grant No. 52304114) and the Natural Science Foundation of Hunan Province (grant no. 2024JJ6510).

Institutional Review Board Statement: Not applicable.

Informed Consent Statement: Not applicable.

Data Availability Statement: The authors confirm that the data supporting the findings of this study are available within the article.

Conflicts of Interest: The authors declare no conflicts of interest.

References

1. Kumar, S.; Kumar, R.; Bandopadhyay, A. Innovative methodologies for the utilisation of wastes from metallurgical and allied industries. *Resour. Conserv. Recycl.* **2006**, *48*, 301–314. [CrossRef]
2. Hasanbeigi, A.; Price, L.; Lu, H.; Lan, W. Analysis of energy-efficiency opportunities for the cement industry in Shandong Province, China: A case study of 16 cement plants. *Energy* **2010**, *35*, 3461–3473. [CrossRef]
3. Naqi, A.; Jang, J. Recent Progress in Green Cement Technology Utilizing Low-Carbon Emission Fuels and Raw Materials: A Review. *Sustainability* **2019**, *11*, 537. [CrossRef]
4. Rashad, A.M.; Zeedan, S.R. The effect of activator concentration on the residual strength of alkali-activated fly ash pastes subjected to thermal load. *Constr. Build. Mater.* **2011**, *25*, 3098–3107. [CrossRef]
5. Poudyal, L.; Adhikari, K. Environmental sustainability in cement industry: An integrated approach for green and economical cement production. *Resour. Environ. Sustain.* **2021**, *4*, 100024. [CrossRef]
6. Zhi, X.; An, X.P. Low carbon technology roadmap of China cement industry. *J. Sustain. Cem.-Based Mater.* **2023**, *12*, 771–774. [CrossRef]
7. Rashad, A.M. Metakaolin as cementitious material: History, scours, production and composition—A comprehensive overview. *Constr. Build. Mater.* **2013**, *41*, 303–318. [CrossRef]
8. Valipour, M.; Yekkalar, M.; Shekarchi, M.; Panahi, S. Environmental assessment of green concrete containing natural zeolite on the global warming index in marine environments. *J. Clean. Prod.* **2014**, *65*, 418–423. [CrossRef]
9. Liu, C.; Tian, J.; Cheng, K.; Xu, X.; Wang, Y.; Liu, X.; Liu, Z.; Bian, R.; Zhang, X.; Xia, S.; et al. Topsoil microbial biomass carbon pool and the microbial quotient under distinct land-use types across China: A data synthesis. *Soil Sci. Environ.* **2023**, *2*, 5. [CrossRef]
10. Yue, Q.; Cheng, K.; Sheng, J.; Wang, L.; Ji, C.; Zhang, Y.; Zheng, J. Greenhouse gas emissions from managed freshwater wetlands under intensified aquaculture. *Soil Sci. Environ.* **2023**, *2*, 3. [CrossRef]
11. Ahmad, J.; Kontoleon, K.J.; Al-Mulali, M.Z.; Shaik, S.; El Ouni, M.H.; El-Shorbagy, M.A. Partial Substitution of Binding Material by Bentonite Clay (BC) in Concrete: A Review. *Buildings* **2022**, *12*, 634. [CrossRef]
12. Wang, H.; Zentar, R.; Wang, D. Rheological Characterization of Fine-Grained Sediments under Steady and Dynamic Conditions. *Int. J. Geomech.* **2022**, *22*, 04021260. [CrossRef]
13. Wang, H.; Zentar, R.; Wang, D. Predicting the compaction parameters of solidified dredged fine sediments with statistical approach. *Mar.Geores.Geotechnol.* **2023**, *41*, 195–210. [CrossRef]
14. Ren, F.-R.; Chen, K.-J.; Tian, Z.; Zhang, Y. The investment and treatment efficiencies of industrial solid waste in China's Yangtze and non-Yangtze River Economic Belts. *J. Mater. Cycles Waste Manag.* **2022**, *24*, 900–916. [CrossRef]
15. Wu, C. Investigation of foundation theory of safety & security complexity. *J. Saf. Sustain.* **2024**, *1*, 14–25.
16. Zhang, Y.; Qi, H.; Li, C.; Zhou, J. Enhancing safety, sustainability, and economics in mining through innovative pillar design: A state-of-the-art review. *J. Saf. Sustain.* **2024**, *1*, 53–73. [CrossRef]
17. Zhao, S.; Wang, F.Y.; Tan, D.Y.; Yang, A.W. A deep learning informed-mesoscale cohesive numerical model for investigating the mechanical behavior of shield tunnels with crack damage. *Structures* **2024**, *66*, 106902. [CrossRef]
18. Zhou, B.Y.; Sun, C.X.; Yi, H.T. Solid Waste Disposal in Chinese Cities: An Evaluation of Local Performance. *Sustainability* **2017**, *9*, 2234. [CrossRef]
19. Ramanayaka, S.; Tsang, D.C.W.; Hou, D.Y.; Ok, Y.S.; Vithanage, M. Green synthesis of graphitic nanobiochar for the removal of emerging contaminants in aqueous media. *Sci. Total Environ.* **2020**, *706*, 135725. [CrossRef]

20. Wang, L.; Chen, L.; Tsang, D.C.W.; Li, J.S.; Yeung, T.L.Y.; Ding, S.M.; Poon, C.S. Green remediation of contaminated sediment by stabilization/solidification with industrial by-products and CO_2 utilization. *Sci. Total Environ.* **2018**, *631–632*, 1321–1327. [CrossRef]
21. Adesina, A. Performance and sustainability overview of sodium carbonate activated slag materials cured at ambient temperature. *Resour. Environ. Sustain.* **2021**, *3*, 100016. [CrossRef]
22. Hu, J.; Wang, Y.; Zhang, S.; Chang, T.; Sun, L. Preparation and classification, hydration mechanism and durability of magnesium cementitious materials. *J. Min. Sci. Technol.* **2023**, *8*, 856–867. [CrossRef]
23. Yang, P.; Liu, L.; Suo, Y.; Qu, H.; Xie, G.; Zhang, C.; Deng, S.; Lv, Y. Investigating the synergistic effects of magnesia-coal slag based solid waste cementitious materials and its basic characteristics as a backfill material. *Sci. Total Environ.* **2023**, *880*, 163209. [CrossRef] [PubMed]
24. Wang, H.; Zentar, R.; Wang, D.; Ouendi, F. New Applications of Ordinary Portland and Calcium Sulfoaluminate Composite Binder for Recycling Dredged Marine Sediments as Road Materials. *Int. J. Geomech.* **2022**, *22*, 04022068. [CrossRef]
25. Zentar, R.; Wang, H.; Wang, D. Comparative study of stabilization/solidification of dredged sediments with ordinary Portland cement and calcium sulfo-aluminate cement in the framework of valorization in road construction material. *Constr. Build. Mater.* **2021**, *279*, 122447. [CrossRef]
26. Zhu, M.; Xie, G.; Liu, L.; Wang, R.; Ruan, S.; Yang, P.; Fang, Z. Strengthening mechanism of granulated blast-furnace slag on the uniaxial compressive strength of modified magnesium slag-based cemented backfilling material. *Process Saf. Environ. Protect.* **2023**, *174*, 722–733. [CrossRef]
27. Liu, J.; Tan, X.; He, Y.; Zheng, k.; Wang, H.; He, Y.; Xie, N.; Wang, D. Semi-flexible pavement application of red mud-based cementitious materials. *J. Chang. Univ. (Nat. Sci. Ed.)* **2022**, *42*, 21–32. [CrossRef]
28. Zhang, D.; Wang, A. A Review of the Properties and Engineering Applications of Geopolymer Cementitious Materials. *J. Archit. Civ. Eng.* **2020**, *37*, 13–38. [CrossRef]
29. Calderón-Morales, B.R.S.; García-Martínez, A.; Pineda, P.; García-Tenório, R. Valorization of phosphogypsum in cement-based materials: Limits and potential in eco-efficient construction. *J. Build. Eng.* **2021**, *44*, 102506. [CrossRef]
30. Cui, Y.; Bai, J.; Chang, I.S.; Wu, J. A systematic review of phosphogypsum recycling industry based on the survey data in China–applications, drivers, obstacles, and solutions. *Environ. Impact Assess. Rev.* **2024**, *105*, 107405. [CrossRef]
31. Rashad, A.M. Phosphogypsum as a construction material. *J. Clean. Prod.* **2017**, *166*, 732–743. [CrossRef]
32. Bilal, E.; Bellefqih, H.; Bourgier, V.; Mazouz, H.; Dumitraş, D.-G.; Bard, F.; Laborde, M.; Caspar, J.P.; Guilhot, B.; Iatan, E.-L.; et al. Phosphogypsum circular economy considerations: A critical review from more than 65 storage sites worldwide. *J. Clean. Prod.* **2023**, *414*, 137561. [CrossRef]
33. Chernysh, Y.; Yakhnenko, O.; Chubur, V.; Roubík, H. Phosphogypsum Recycling: A Review of Environmental Issues, Current Trends, and Prospects. *Appl. Sci.* **2021**, *11*, 1575. [CrossRef]
34. Li, X.; Zhou, Y.; Zhu, Q.; Zhou, S.; Min, C.; Shi, Y. Slurry Preparation Effects on the Cemented Phosphogypsum Backfill through an Orthogonal Experiment. *Minerals* **2019**, *9*, 31. [CrossRef]
35. Min, C.; Shi, Y.; Liu, Z. Properties of cemented phosphogypsum (PG) backfill in case of partially substitution of composite Portland cement by ground granulated blast furnace slag. *Constr. Build. Mater.* **2021**, *305*, 124786. [CrossRef]
36. Zhou, S.; Li, X.; Zhou, Y.; Min, C.; Shi, Y. Effect of phosphorus on the properties of phosphogypsum-based cemented backfill. *J. Hazard. Mater.* **2020**, *399*, 122993. [CrossRef]
37. Chen, M.; Liu, P.; Kong, D.; Li, Y.; Chen, Y.; Cui, G.; Wang, J.; Yu, K.; Wu, N. Performance Study and Multi-Index Synergistic Effect Analysis of Phosphogypsum-Based Composite Cementitious Material. *Coatings* **2022**, *12*, 1918. [CrossRef]
38. Xu, J.Q.; Chen, P.; Zhang, C.Y.; Yang, Y.H. Influence of phosphogypsum on mechanical properties and microstructure of iron tailings cementitious material. *Arch. Civ. Mech. Eng.* **2023**, *24*, 18. [CrossRef]
39. Liu, X.; Tang, P.; Chen, W. Development of an ettringite-based low carbon binder by promoting the nucleation and Ostwald ripening process. *Constr. Build. Mater.* **2024**, *427*, 136282. [CrossRef]
40. Zhang, J.; Cui, K.; Yang, Y.; Chang, J. Investigation on the preparation of low carbon cement materials from industrial solid waste phosphogypsum: Clinker preparation, cement properties, and hydration mechanism. *J. Clean Prod.* **2024**, *452*, 142203. [CrossRef]
41. Min, C.; Li, X.; He, S.; Zhou, S.; Zhou, Y.; Yang, S.; Shi, Y. Effect of mixing time on the properties of phosphogypsum-based cemented backfill. *Constr. Build. Mater.* **2019**, *210*, 564–573. [CrossRef]
42. Ma, F.; Chen, L.; Lin, Z.; Liu, Z.; Zhang, W.; Guo, R. Microstructure and Key Properties of Phosphogypsum-Red Mud-Slag Composite Cementitious Materials. *Materials* **2022**, *15*, 6069. [CrossRef] [PubMed]
43. Dong, S.; Zhuo, Q.; Chen, L.; Wu, F.; Xie, L. Reuse of Pretreated Red Mud and Phosphogypsum as Supplementary Cementitious Material. *Sustainability* **2023**, *15*, 2856. [CrossRef]
44. Huang, Y.; Lin, Z. Investigation on phosphogypsum-steel slag-granulated blast-furnace slag-limestone cement. *Constr. Build. Mater.* **2010**, *24*, 1296–1301. [CrossRef]
45. Yang, X.; Jia, Y.; Yang, C.; He, X.; Jia, Y.; Xu, L. Research on formulation optimization and hydration mechanism of phosphogypsum-based filling cementitious materials. *Front. Environ. Sci.* **2022**, *10*, 1012057. [CrossRef]
46. Zeng, L.-L.; Bian, X.; Zhao, L.; Wang, Y.-J.; Hong, Z.-S. Effect of phosphogypsum on physiochemical and mechanical behaviour of cement stabilized dredged soil from Fuzhou, China. *Geomech. Energy Environ.* **2021**, *25*, 100195. [CrossRef]

47. Zhu, J.; Qu, Z.; Huang, Y.; Song, L.; Liu, S.; Min, H.; Li, Z. Investigating Mechanical Properties of Alkali-Activated Slag Cementitious Material for Load-Bearing Layer of Sandwich Panels. *Materials* **2023**, *16*, 6398. [CrossRef]
48. Xu, Z.; Han, X.; Feng, K.; Zhang, W.; Ma, K. Improvement of Mechanical Properties and Microstructure of Cementitious Materials in a Hot-Dry Environment. *Crystals* **2022**, *12*, 981. [CrossRef]
49. Jongpradist, P.; Jamsawang, P.; Kongkitkul, W. Equivalent void ratio controlling the mechanical properties of cementitious material-clay mixtures with high water content. *Mar. Georesources Geotechnol.* **2019**, *37*, 1151–1162. [CrossRef]
50. Deng, Y.; Yan, C.; Zhang, J.; Yin, L.; Liu, S.; Yan, Y. Preparation and mechanical characterization of engineered cementitious composites with high-volume fly ash and waste glass powder. *J. Clean. Prod.* **2022**, *333*, 130222. [CrossRef]
51. Sun, L.; Hao, Q.; Zhao, J.; Wu, D.; Yang, F. Stress strain behavior of hybrid steel-PVA fiber reinforced cementitious composites under uniaxial compression. *Constr. Build. Mater.* **2018**, *188*, 349–360. [CrossRef]
52. Lu, D.; Du, X.; Ma, C.; Wang, Y. Study on static and dynamic stress–strain relationship of cement mortar material. *Mater. Res. Innov.* **2014**, *18*, S2-799–S2-805. [CrossRef]
53. Kohees, M.; Rajeev, P.; Sanjayan, J.G. Stress-strain relationship of cement paste under triaxial compression. *Proc. Inst. Civ. Eng.-Constr. Mater.* **2022**, *175*, 233–241. [CrossRef]
54. Zhou, J.-k.; Ge, L.-m. Effect of strain rate and water-to-cement ratio on compressive mechanical behavior of cement mortar. *J. Cent. South Univ.* **2015**, *22*, 1087–1095. [CrossRef]
55. Shi, Y.; Li, Y.; Wang, H. Eco-friendly solid waste-based cementitious material containing a large amount of phosphogypsum: Performance optimization, micro-mechanisms, and environmental properties. *J. Clean. Prod.* **2024**, *471*, 143335. [CrossRef]
56. *GB/T 17671-2021*; Test Method of Cement Mortar Strength (ISO Method). Standardization Administration of the People's Republic of China: Beijing, China, 2021; p. 24.
57. Huang, K.; Ma, Q.; Ma, D.; Bosco, M. Effect of Basalt Fiber on Static and Dynamic Mechanical Properties of Metakaolin-Based Cement Clay. *Adv. Civ. Eng.* **2020**, *2020*, 1359163. [CrossRef]
58. Zhang, J.; Deng, A.; Jaksa, M. Enhancing mechanical behavior of micaceous soil with jute fibers and lime additives. *J. Rock Mech. Geotech. Eng.* **2021**, *13*, 1093–1100. [CrossRef]
59. Mirzababaei, M.; Miraftab, M.; Mohamed, M.; McMahon, P. Unconfined Compression Strength of Reinforced Clays with Carpet Waste Fibers. *J. Geotech. Geoenviron. Eng.* **2013**, *139*, 483–493. [CrossRef]
60. Wang, D.; Wang, H.; Di, S. Mechanical properties and microstructure of magnesia–fly ash pastes. *Road Mater. Pavement Des.* **2018**, *20*, 1243–1254. [CrossRef]
61. Zhang, F.; Li, Y.; Zhang, J.; Gui, X.; Zhu, X.; Zhao, C. Effects of slag-based cementitious material on the mechanical behavior and heavy metal immobilization of mine tailings based cemented paste backfill. *Heliyon* **2022**, *8*, e10695. [CrossRef]
62. Ma, W.; Brown, P.W. Hydrothermal reactions of fly ash with Ca(OH)2 and CaSO4·2H2O. *Cem. Concr. Res.* **1997**, *27*, 1237–1248. [CrossRef]
63. Mashifana, T.; Okonta, F.N.; Ntuli, F. Geotechnical Properties and Application of Lime Modified Phosphogypsum Waste. *Mater. Sci.* **2018**, *24*, 312–318. [CrossRef]
64. Zhang, S.; Li, V.C.; Ye, G. Micromechanics-guided development of a slag/fly ash-based strain-hardening geopolymer composite. *Cem. Concr. Compos.* **2020**, *109*, 103510. [CrossRef]
65. Seo, J.H.; Park, S.M.; Yang, B.J.; Jang, J.G. Calcined Oyster Shell Powder as an Expansive Additive in Cement Mortar. *Materials* **2019**, *12*, 1322. [CrossRef] [PubMed]
66. Sevim, O.; Sengul, C.G.; Kartal, S.; Toklu, K.; Caglar, Y. Effect of Supplementary Cementitious Materials with Similar Specific Surface Area on Cementitious Composite Systems. *J. Test. Eval.* **2023**, *51*, 2971–2985. [CrossRef]
67. Lin, T.; Wang, C.; Yousfani, S.; Wang, W.; Qian, X.; Xun, Z. Study on extraction technology of bamboo fiber by response surface. *Acta Mater Compos. Sin* **2018**, *35*, 876–884.
68. Xu, X.H.; He, M.Z. *Experimental Design and Application of Design-Expert and SPSS*; Science Press: Beijing, China, 2010. (In Chinese)
69. He, Z.-H.; Zhan, P.-M.; Du, S.-G.; Liu, B.-J.; Yuan, W.-B. Creep behavior of concrete containing glass powder. *Compos. Part B-Eng.* **2019**, *166*, 13–20. [CrossRef]
70. Wang, P.; Xie, M.; Liu, L. Study on Early Shrinkage and Mechanical Properties of Concrete with Various Cementitious Materials. *Buildings* **2022**, *12*, 1543. [CrossRef]
71. Van Roode, M.; Douglas, E.; Hemmings, R.T. X-ray diffraction measurement of glass content in fly ashes and slags. *Cem. Concr. Res.* **1987**, *17*, 183–197. [CrossRef]
72. Diamond, S.; Kilgour, C.L. A Procedure for the Interpretation of Fly Ash Particle Size Data. *Mater. Res. Soc. Symp. Proc.* **1985**, *65*, 105–114. [CrossRef]

materials

MDPI

Article

Predicting the Compressive Strength of Sustainable Portland Cement–Fly Ash Mortar Using Explainable Boosting Machine Learning Techniques

Hongwei Wang [1], Yuanbo Ding [1], Yu Kong [2], Daoyuan Sun [1], Ying Shi [1] and Xin Cai [1,*]

[1] School of Resources and Safety Engineering, Central South University, Changsha 410083, China; hongwei.wang@csu.edu.cn (H.W.)

[2] China Construction Fifth Engineering Division Corp., Ltd., Changsha 410004, China

* Correspondence: xincai@csu.edu.cn

Abstract: Unconfined compressive strength (UCS) is a critical property for assessing the engineering performances of sustainable materials, such as cement–fly ash mortar (CFAM), in the design of construction engineering projects. The experimental determination of UCS is time-consuming and expensive. Therefore, the present study aims to model the UCS of CFAM with boosting machine learning methods. First, an extensive database consisting of 395 experimental data points derived from the literature was developed. Then, three typical boosting machine learning models were employed to model the UCS based on the database, including gradient boosting regressor (GBR), light gradient boosting machine (LGBM), and Ada-Boost regressor (ABR). Additionally, the importance of different input parameters was quantitatively analyzed using the SHapley Additive exPlanations (SHAP) approach. Finally, the best boosting machine learning model's prediction accuracy was compared to ten other commonly used machine learning models. The results indicate that the GBR model outperformed the LGBM and ABR models in predicting the UCS of the CFAM. The GBR model demonstrated significant accuracy, with no significant difference between the measured and predicted UCS values. The SHAP interpretations revealed that the curing time (T) was the most critical feature influencing the UCS values. At the same time, the chemical composition of the fly ash, particularly Al_2O_3, was more influential than the fly-ash dosage (FAD) or water-to-binder ratio (W/B) in determining the UCS values. Overall, this study demonstrates that SHAP boosting machine learning technology can be a useful tool for modeling and predicting UCS values of CFAM with good accuracy. It could also be helpful for CFAM design by saving time and costs on experimental tests.

Keywords: sustainable cement–fly ash mortar; compressive strength; boosting machine learning; SHAP explanation

1. Introduction

Cement is one of the most extensively used construction materials worldwide. In addition to being a building material, it can also be used as a binder in the stabilization/solidification of soil [1–4]. However, the highly energy-intensive and carbon-emitting production of cement has raised concerns [5]. Cement production accounts for around 5% of global CO_2 emissions [6]. Some research focuses on cement production [7], while others focus on reducing cement use. It is suggested that environmentally friendly "supplementary cementing materials" (SCMs) be used to replace part of the cement used to produce mortar. Recently, SCMs such as silica fume, fly ash, powdered granulated blast furnace slag, and natural pozzolans have been employed [8–10].

Fly ash (FA) is a byproduct of coal-fired power stations burning pulverized coal to generate electricity [11,12]. FA has been used as a low-cost adsorbent [13]. A promising approach to improve the utilization of fly ash is to convert it into low-grade zeolites [14], which can be used in agriculture and many other fields [15]. In addition, because of its potential to increase the mechanics, impermeability, and durability of mortar and reduce carbon dioxide emissions, the use of FA in mortar has piqued interest in recent decades. The effect of FA on cement hydration and the pozzolanic reaction mechanism has been the subject of numerous investigations [16]. The unconfined compressive strength (UCS) of cement–fly ash mortar (CFAM) is contributed by the hydration reaction, packing effect, and pozzolanic reaction. The chemical composition of FA is widely acknowledged as one of the most critical parameters influencing CFAM performance [17]. Some research shows a high correlation between the CaO/SiO_2 ratio and the UCS of CFAM, $(CaO + MgO + R_2O)/SiO_2$ ratio, and UCS of CFAM [11]. There are two main types of FA according to the ASTM standard C 618 [18]. If the $SiO_2 + Al_2O_3 + Fe_2O_3$ content is more than 70 wt.%, this FA is classified as F fly ash. Meanwhile, FA is classified as C fly ash if it contains contents of between 50 and 70 wt.% of $SiO_2 + Al_2O_3 + Fe_2O_3$.

One of the most essential performances of CFAM is its UCS [19,20]. The UCS of CFAM is usually obtained through an experiment in the laboratory. However, the experiment is time- and labor-consuming, because it involves lots of compressive tests for each type of FA and different curing times. Therefore, an approach that accurately predicts the UCS of CFAM is preferred. The UCS of CFAM is often predicted using traditional linear or nonlinear regression techniques on experimental data. Cyr et al. [21] proposed an empirical model to predict the UCS of CFAM. This model considered cement hydration, pozzolanic reaction, and the physical filling effect on the UCS, but the curing time was not considered. Qadir et al. [22] used a nonlinear relationship to quantify cement mortar's compressive, split tensile, and flexural strengths as a function of the water-to-binder ratio (W/B), fly ash content, and curing time. However, with these empirical models, it is challenging to derive a precise regression equation. These equations' generalization capabilities are also unsatisfactory, because they frequently only apply to a particular type of FA. More advanced techniques are required to replace traditional regression techniques in the UCS prediction of CFAM.

Machine learning has been used in various fields, like rock blasting and risk assessment [23–26]. In recent years, machine learning models have become the most popular technique for predicting compressive strength and other parameters in civil engineering [27–29]. These machine learning models include multiple linear regression, M5P tree model, support vector machine, decision trees, Random Forest, gene expression programming, and artificial neural networks [30]. Only a few studies predicted the UCS of CFAM using machine learning techniques. Mohammed et al. [31] used linear and nonlinear regression, M5P tree, and artificial neural network technical approaches to predict the UCS of CFAM, with the parameters of the fly ash incorporation ratio, water-to-binder ratio, and curing time. Moreover, as reported in [32,33], developing ensemble machine learning models is useful as it integrates the advantages of different models. The three main ensemble machine learning algorithm categories are bagging, boosting, and stacking [33].

Among them, boosting machine learning models have been widely used in the field of biomedicine [34,35], and they have higher prediction accuracies than other individual machine learning models [36]. Boosting is an ensemble machine learning technique that adjusts the sample weights of the next model's learning data based on the previous model's learning outcomes. It connects numerous weak learners to create strong learners. As a result, the results of the previous learning will affect the next learning, and the weights of the data will also increase with the learning time. However, the use of boosting machine learning models is not extensively reported in the literature related to the civil engineering field [37,38]. In particular, to the best of the authors' knowledges, no studies focus on predicting the UCS of CFAM with boosting machine learning models. Moreover, these studies' developed machine learning models are "black boxes" and not explainable. Thus, although these models have high accuracy, a user cannot understand their working principles, and this problem limits the application of machine learning models in civil engineering practice. Lundberg and Lee [22] recently developed a SHapley Additive exPlanations (SHAP) framework to interpret machine learning models. The SHAP approach could assess the contribution by features to the predictions quantitatively. It could explain how the features affect the machine learning models in both global and local views. Only a few studies have used SHAP to explain a developed machine learning model in civil engineering [39,40].

Therefore, in this study, we focus on developing three typical explainable boosting machine learning models—gradient boosting regressor (GBR), light gradient boosting machine (LGBM), and Ada-Boost regressor (ABR)—to predict the UCS of CFAM. Secondly, we use SHAP to explain the best model in this study, with global and local explanations. Lastly, we check the importance of input parameters, especially the chemical composition of FA, for the models. Ultimately, we compare the prediction accuracy of the best boosting machine learning model and the other commonly used machine learning models.

2. Machine Learning Techniques

The UCS of CFAM was determined using a variety of boosting machine learning techniques, including gradient boosting regressor (GBR), light gradient boosting machine (LGBM), and Ada-Boost regressor (ABR). In the end, the prediction performance of the best boosting machine learning model was compared with other commonly used machine learning models.

2.1. Gradient Boosting Regressor (GBR)

Friedman first proposed the GBR model as a robust and interpretable boosting model used for regression and classification [41]. The GBR is a boosting model that combines several weak learning models to create a robust predictive model. Decision tree models were used to build ensembles, and to correct the predictions produced by previous models, the number of trees was increased one at a time. Ending tree growth as soon as possible avoids the overfitting problem that decision tree learning faces. The GBR model is a powerful machine learning tool that can fit boosted decision trees by achieving a minimal loss gradient. It also offers improved performance and better stability. GBR provides excellent non-linear data prediction capabilities [42]. Figure 1 presents a diagram of the GBDT algorithm.

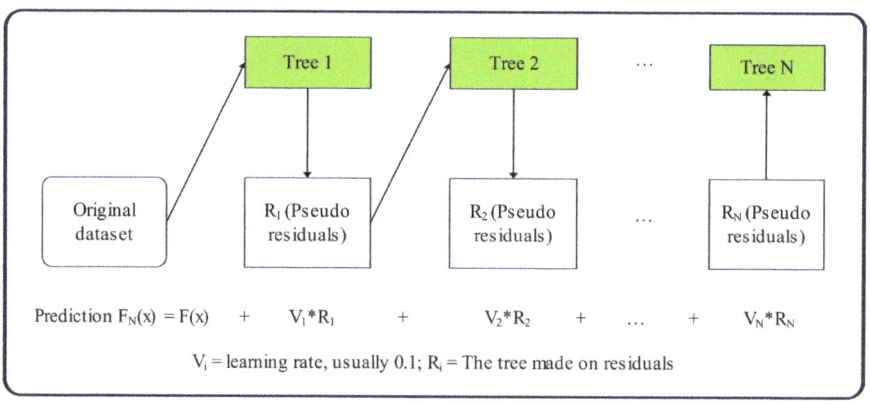

Figure 1. Diagram of the GBDT algorithm.

2.2. Light Gradient Boosting Machine (LGBM)

The LGBM was created by Microsoft in 2017 [43]. It is an ensemble boosting framework based on gradient boosting decision tree (GBDT). However, if the database contains a lot of data, both the prediction accuracy and the forecasting speed of the GBDT dramatically decline. In terms of memory consumption reduction and operating time acceleration while also maintaining high accuracy, LGBM improves the capabilities of GBDT [44]. According to reports [45], LGBM can accelerate GBDT's training procedure up to 25 times while maintaining nearly the same accuracy. As a result, LGBM has become quite popular in machine learning because of its advantages of quick convergence speed and low memory utilization [46]. Figure 2 presents a diagram of the LGBM algorithm.

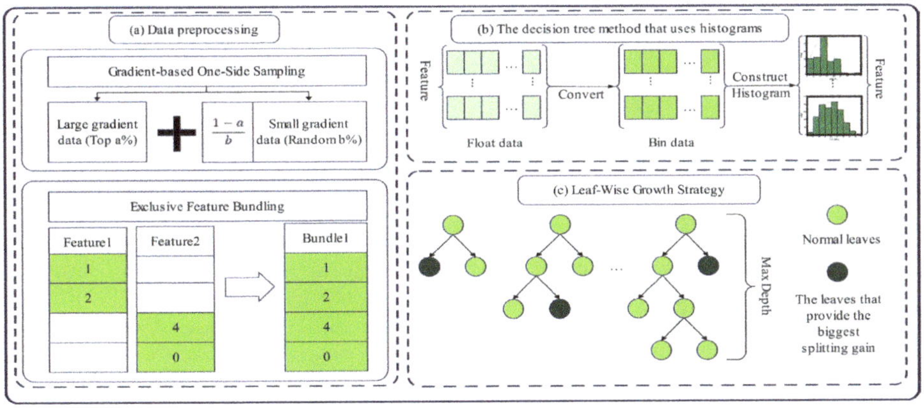

Figure 2. Diagram of the LGBM algorithm.

2.3. Ada-Boost Regressor (ABR)

Another boosting machine learning technique is the ABR, which was created by Yoav Freund and Robert Schapire [47]. It can randomly combine several weak learners from the dataset by weight to produce a boosted strong learner as the last result. The initial weight from the training set is used to train a base regressor, and the weight of the training sample is updated based on the learning error rate. For data points with high error rates to receive greater focus in the following base regression, the weights of the training sample points with high learning error rates in the prior base learner should increase. By changing the weight, the next base regression is trained, and so on, until the total number of base regressors reaches the predetermined number [32,48]. Figure 3 presents a diagram of the ABR algorithm.

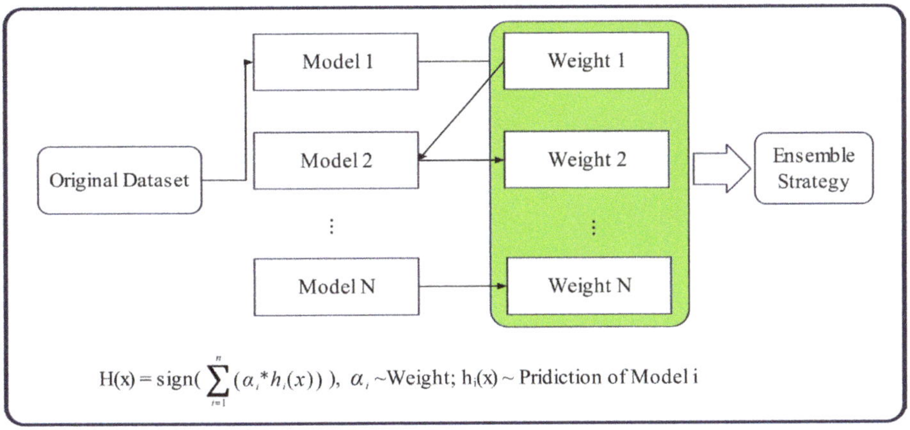

Figure 3. Diagram of the ABR algorithm.

2.4. Other Commonly Used Machine Learning Approaches

In addition, another ten commonly used machine learning approaches are used to compare against the best boosting machine learning model, including linear regression, Random Forest regressor, decision tree regressor, K Neighbors regressor, Bayesian ridge, ridge regression, extra trees regressor, least angle regression, Huber regressor, and lasso regression.

2.5. SHapley Additive exPlanations (SHAP)

Lundberg and Lee developed the SHapley Additive exPlanations (SHAP) in 2017 [49]. The SHAP is a Shapley game-theory-based method, which can be used to explain "black-box" machine learning models. The SHAP aims to assess the influence of each feature on the prediction results, and it can determine whether the influence of the feature is positive or negative. Accordingly, the SHAP can aid in explaining machine learning models from global and local views. In the meantime, the SHAP can conclude the contribution by each feature to each observation. The SHAP can aid in understanding the machine learning models' underlying mechanisms and develop machine learning models that users can trust.

3. Model Development

In this study, we created all of the models using the Python platform (version 3.4). Figure 4 describes the entire proposed workflow for developing the models in this study, including five steps. The first step is data collection to build a database of UCS of CFAM. The second step is data analysis. The third step randomly divides all the data into two parts: 70% as a training dataset and 30% as a testing dataset. The fourth step is model training with the training dataset and hyperparameter tuning. The fifth step is model evaluation and validation with the testing dataset. Then, the SHAP method is used to interpret the best performed model. In the end, the best model is compared with other commonly used machine learning models.

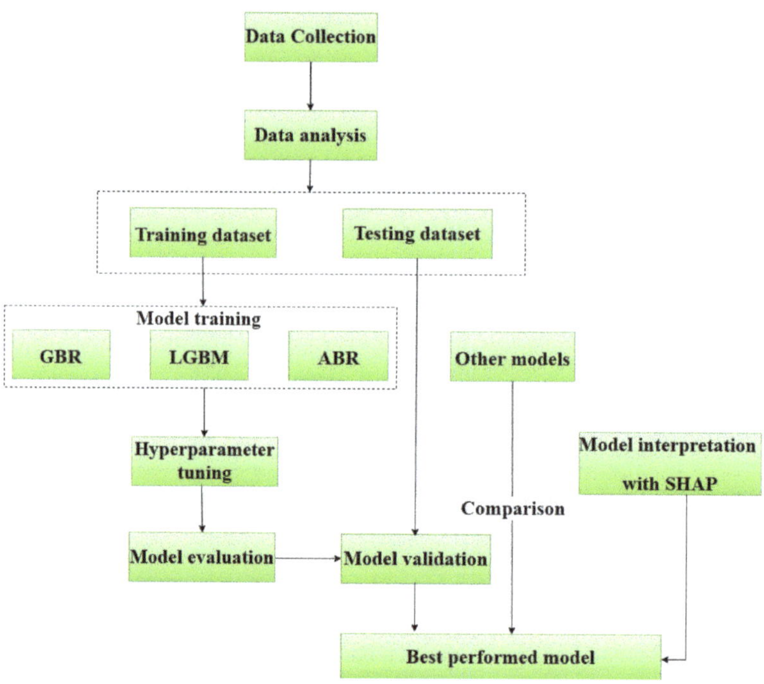

Figure 4. Workflow to develop models in this study.

3.1. Pretreatment of Data

It should be noted that, in this study, all data needed to be normalized with the Z-score normalization method so that the ranges were consistent. The database should be randomly divided into two groups, as follows: 70% of the data used for training the machine learning model and the remaining 30% of the data used for testing the machine learning model.

3.2. Cross-Validation Accuracy

The k-fold cross-validation approach enabled the model to be trained and validated multiple times, leading to a more accurate model with less overfitting [50]. Figure 5 shows a schematic description of the 10-fold cross-validation method. The training dataset was randomly divided into ten groups, of which nine were chosen for the model's training and one was chosen for the model's validation. Moreover, this process was repeated ten times, until each separate group was employed for validation.

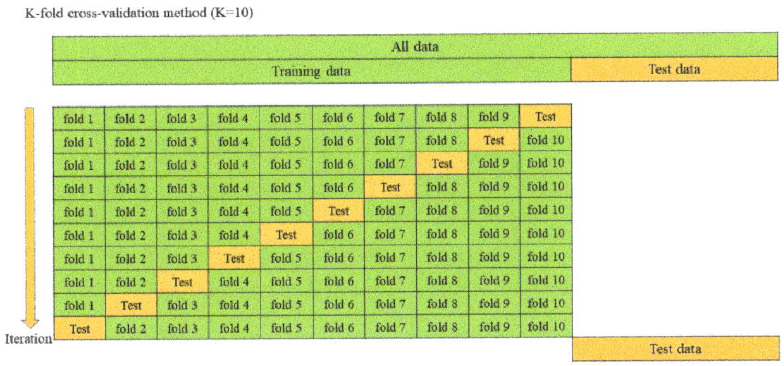

Figure 5. K-fold cross-validation method (K = 10).

3.3. Hyperparameter Tuning

Hyperparameter tuning is important in creating machine learning models, because the optimal hyperparameters can make the model more accurate. Hyperparameter values are usually chosen through a manual process of trial and error. However, manual tuning is frequently time-consuming and produces unsatisfactory results. Some automated approaches to identifying hyperparameters, such as grid search and random search hyperparameter optimization, have been developed to enhance the hyperparameter tuning process. This study used hyperparameter tuning for all machine learning models through the random grid search method.

3.4. Performance Evaluation of the Models

Five important parameters are frequently used to estimate and quantify a machine learning model's performance, as follows: coefficient of determination (R^2), root mean square error ($RMSE$), mean absolute error (MAE), mean squared error (MSE), and mean absolute percent error ($MAPE$). The definitions of these parameters are shown in Equations (1)–(5).

$$R^2 = 1 - \frac{\sum_{k=1}^{n} (y'_k - y_k)^2}{\sum_{k=1}^{n} ((y_k - \overline{y_k})^2} \tag{1}$$

$$RMSE = \sqrt{\frac{\sum_{k=1}^{n} (y'_k - y_k)^2}{n}} \tag{2}$$

$$MSE = \frac{\sum_{k=1}^{n} (y'_k - y_k)^2}{n} \tag{3}$$

$$MAE = \frac{\sum_{k=1}^{n} |y'_k - y_k|}{n} \tag{4}$$

$$MAPE = \frac{\sum_{k=1}^{n} \left| (y'_k - y_k)/y_k \right|}{n} \tag{5}$$

where y_k is the experimental value indexed with k; y'_k is the predicted value indexed with k; $\overline{y_k}$ is the mean of y_k; and n is the total number of data samples.

4. Database

4.1. Data Collection

The database in this study was established with a total of 395 observations from 20 different independent research studies [5,8,11,51–67]. The mortar specimens in the literature are mostly cubes, and the UCS of the cylinder mortars were transformed into those of cube specimens with an empirical formula. Table 1 presents the ranges of this database's input and output parameters. The input parameters consisted of the fly-ash dosage (FAD), water-to-binder ratio (W/B), sand-to-binder ratio (S/B), curing time (T), and main chemical composition of fly ash (CaO, SiO_2, Al_2O_3, and Fe_2O_3). The output parameter is the unconfined compressive strength (UCS) of the Portland cement–fly ash mortar.

Table 1. Ranges of all data from the literature.

No.	FAD	W/B	S/B	T	CaO	SiO_2	Al_2O_3	Fe_2O_3	UCS	Reference
	%			d	%	%	%	%	MPa	
1	0–30	0.50	3.00	3–28	9.37	46.70	19.21	7.55	15.52–53.27	[5]
2	0–35	0.50	3.00	1–140	6.30	50.00	23.90	6.00	12.30–62.30	[8]
3	25	0.50	3.00	28–91	2.54–6.17	52.20–62.60	17.70–23.00	8.85–6.15	41.10–69.50	[11]
4	42–44	0.50	3.00	3–365	3.10–6.49	48.30–62.70	21.20–25.50	6.34–8.5	21.10–71.50	[51]
5	5–25	0.50	3.00	7–180	2.00	58.30	31.70	5.90	10.40–33.00	[52]
6	40–70	0.40	2.75	7–28	1.61	51.80	26.40	13.20	7.50–36.00	[53]
7	10–30	0.50	2.50	3–28	18.10	46.25	46.25	5.60	9.00–41.10	[54]
8	10–50	0.42–0.44	3.00	7–90	0.98	60.02	29.77	6.68	8.31–53.93	[55]
9	0–30	0.50	3.00	1–28	20.17	36.19	19.67	14.96	6.88–49.75	[56]
10	0–40	0.51–0.55	2.75	7–180	13.00	44.40	23.50	10.20	29.00–77.00	[57]
11	0–40	0.30–0.40	3.00	3–90	2.86	53.33	27.65	6.04	21.36–80.38	[58]
12	20	0.49	2.75	3–90	13.20–63.80	20.20–43.20	5.40–21.50	2.90–12.10	15.00–44.50	[59]
13	0–40	0.50	2.75	7–90	14.40–65.40	20.90–41.10	4.80–21.60	3.40–11.30	33.00–63.50	[60]
14	0–30	0.44	3.00	3–28	6.10–62.87	20.21–41.40	4.94–26.20	2.85–16.00	15.91–46.25	[61]
15	0–40	0.40	2.00	1–28	51.29–61.87	18.95–20.65	5.60–7.53	3.82–4.13	11.20–44.90	[62]
16	0–10	0.49	2.75	7–300	39.69–61.00	23.84–32.80	4.20–13.77	3.40–4.78	29.42–66.92	[63]
17	0–20	0.60	3.50	3–180	1.68–66.40	16.70–64.45	3.97–24.83	2.46–4.67	24.37–43.85	[64]
18	0–25	0.40	3.00	1–360	3.87–62.83	21.56–57.60	4.44–21.90	2.70–2.78	7.00–74.95	[65]
19	10–30	0.50	3.00	2–90	8.11	47.26	27.63	4.35	12.70–57.00	[66]
20	15–35	0.50	3.00	2–90	2.99–25.72	36.56–57.39	10.00–23.20	4.20–9.65	10.60–61.10	[67]

4.2. Statistical Analysis

Table 2 shows the descriptive statistics of the parameters in the database used for the model's development and validation, including minimum, maximum, mean, median, sum, and standard deviation values.

Figure 6 presents frequency histograms for each variable of the database. A few of the frequency histograms were discovered to be approximately normally distributed. The database included mortar samples with a wide range of FADs (0–80%) (Figure 6a), with the highest frequency of mortar samples containing 20–30% FA. The majority of the water-to-binder ratios (W/B) were in a range from 0.5 to 0.55 (Figure 6b). As shown in Figure 6c, the majority of the specimens had sand-to-binder ratios (S/B) that ranged from 3.0 to 3.2. In terms of the curing times (T), most values were found to be in the range of 0–50 d (Figure 6d), while the longest curing time was 365 d (Table 2). Regarding the chemical compositions of the FAs, more than 240 mortar samples were made with low-CaO-content (0–10%) FA(Figure 6e). SiO_2 was the main chemical in the compositions of most of the used FAs (Figure 6f). The frequency distributions of the Al_2O_3 and Fe_2O_3 contents were more uniform, with the highest ranges being 20–25% (Figure 6g) and 6-8% (Figure 6h), respectively. The UCS values were normally distributed, varying from 7.5 to 71.5 MPa, with the highest frequency of 40–50 MPa (Figure 6i).

A correlation heat map is a method of perceiving information that displays correlated values along two dimensions using colors. This study presents the correlations between two variables in Figure 7, evaluating their linear dependence. It should be noted that some input variables had high correlations, as follows: CaO with SiO_2 (R = −0.93) and CaO with Fe_2O_3 (R = −0.73). If two variables are correlated, only one variable should be used for modeling [68]. Therefore, in this study, it was decided that CaO should be eliminated. In this way, SiO_2 and Al_2O_3 could be kept and used for modeling.

Table 2. Descriptive statistics for the database.

	Mean	Median	Standard Deviation	Min	Max	Sum
FAD (%)	22.49	20.00	14.39	0.00	70.00	8883.00
W/B	0.48	0.50	0.06	0.30	0.60	189.68
S/B	2.92	3.00	0.28	2.00	3.50	1151.50
T (d)	50.07	28.00	65.97	1.00	365.00	19,776.00
CaO (%)	15.08	6.30	19.10	0.98	66.40	5956.44
SiO_2 (%)	46.81	50.00	12.90	16.70	64.45	18,489.05
Al_2O_3 (%)	21.39	21.90	8.57	3.97	46.25	8449.17
Fe_2O_3 (%)	7.01	6.22	3.02	2.46	16.00	2770.45
UCS (MPa)	39.14	39.30	16.26	6.88	80.38	15,459.61

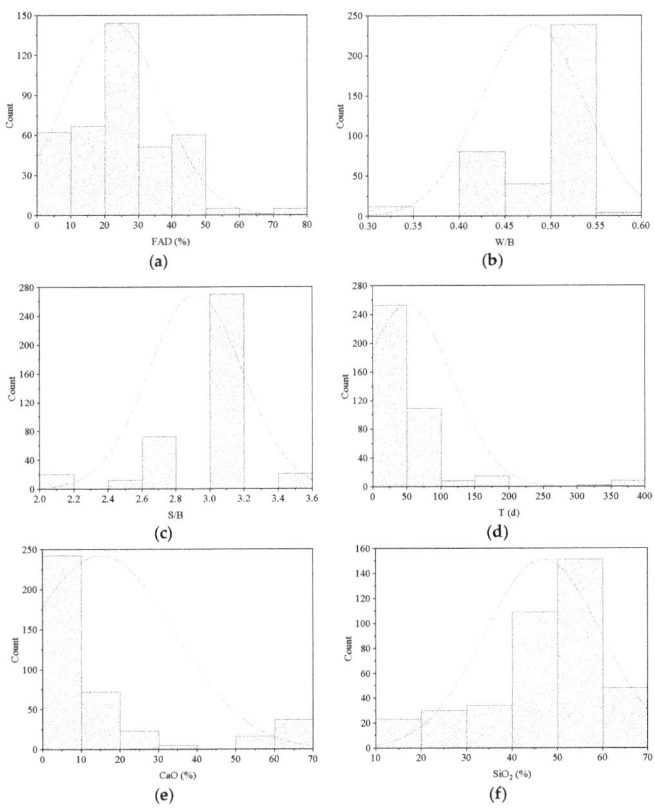

Figure 6. *Cont.*

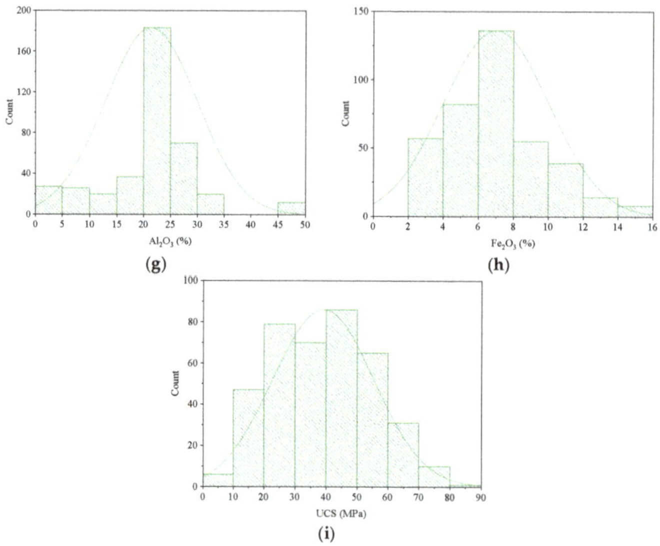

Figure 6. Histograms of the variables: (**a**) fly-ash dosage; (**b**) water–binder ratio; (**c**) sand–binder ratio; (**d**) curing time; (**e**) CaO content of the fly ash; (**f**) SiO_2 content of the fly ash; (**g**) Al_2O_3 content of the fly ash; (**h**) Fe_2O_3 content of the fly ash; (**i**) unconfined compressive strength of Portland cement–fly ash mortar.

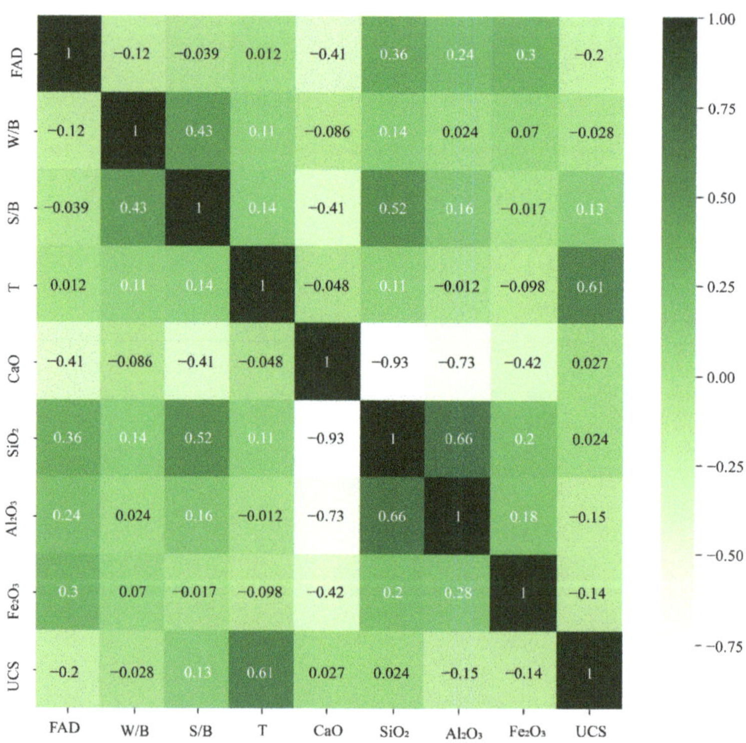

Figure 7. Heatmap of the correlations among the variables.

5. Results and Discussion

5.1. Hyperparameter Tuning and 10-Fold Cross-Validation

First, the random grid search procedure was used to find the optimal hyperparameters for each model. Table 3 lists the optimal hyperparameters and their associated values for each model. The models' performances can be improved on the condition of these optimal hyperparameters.

Table 3. Optimal hyperparameters of the used models.

Hyperparameter	Models		
	GBR	LGBM	ADA
n Estimator	170	120	220
Learning rate	0.05	0.3	0.3
Max depth	11	−1	-
Subsample	0.3	1.0	-

To minimize the bias caused by the random sampling of the training dataset, a typical 10-fold cross-validation approach was used. According to reports [69,70], this approach can provide the models with generalization and reliability properties. Figure 8 illustrates the five error parameters of each fold in the 10-fold cross-validation, including MAE, R^2, $RMSE$, $MAPE$, and MSE. As can be observed, all three models' 10-fold results exhibited fluctuations but still maintained high accuracy. Especially, Table 4 presents the average statistical results of the 10-fold cross-validation for the training dataset. It can be observed that the GBR model had fewer errors and higher R^2 values than the LGBM and ABR models. For instance, the average R^2 value of the GBR model over 10-folds was 0.941, and the maximum and minimum R^2 values were 0.903 and 0.958. As for the LGBM model, the average R^2 value over the 10-folds was 0.921, whereas the maximum and minimum values were 0.944 and 0.850, respectively. Similarly, the ABR model had average, maximum, and minimum R^2 values of 0.796, 0.856, and 0.737, respectively.

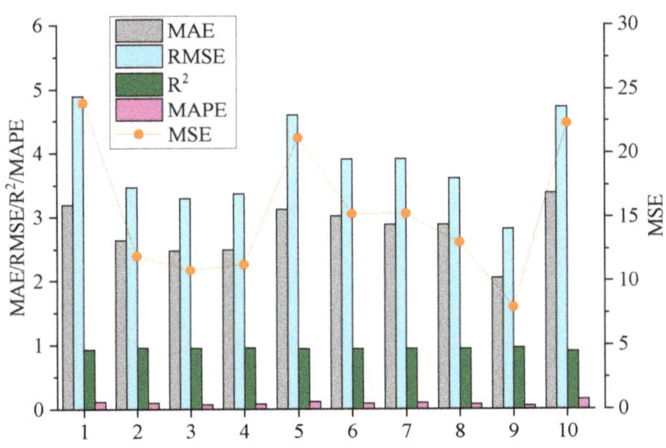

(**a**) GBR

Figure 8. *Cont.*

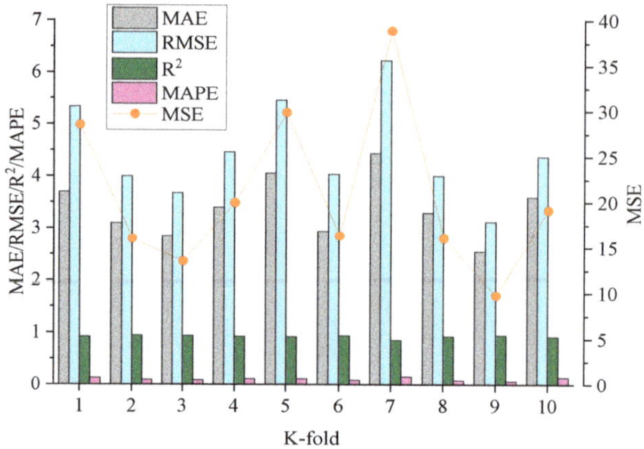

(b) LGBM

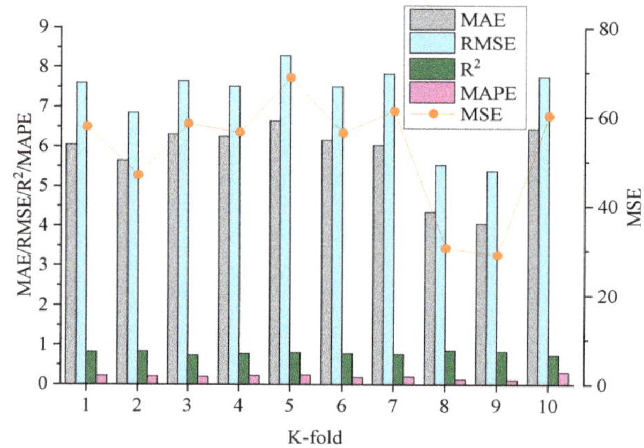

(c) ABR

Figure 8. Results of the 10-fold cross-validation: (**a**) GBR model; (**b**) LGBM model; (**c**) ABR model.

Table 4. Average statistical results of 10-fold cross-validation.

Models		*MAE*	*RMSE*	R^2	*MAPE*	*MSE*
GBR	Mean	2.813	3.857	0.941	0.094	15.303
	SD	0.380	0.652	0.016	0.028	5.144
LGBM	Mean	3.397	4.474	0.921	0.108	20.805
	SD	0.546	0.889	0.026	0.027	8.422
ABR	Mean	5.793	7.192	0.796	0.203	52.588
	SD	0.837	0.929	0.041	0.054	12.466

5.2. Comparison of the Performances of the Different Models

The optimal hyperparameters shown in Table 3 were used to train all models on the training dataset, and the testing dataset was used to evaluate them. Figure 9 compares the models' results on the training and testing datasets with the optimal hyperparameters. It is worth noting that the models' performances on the training dataset was determined with a 10-fold cross-validation. The results demonstrate that the GBR model performed best on the training and testing datasets. It had the highest R^2 value and the lowest *RMSE*, *MAE*, *MAPE*, and *MSE* values.

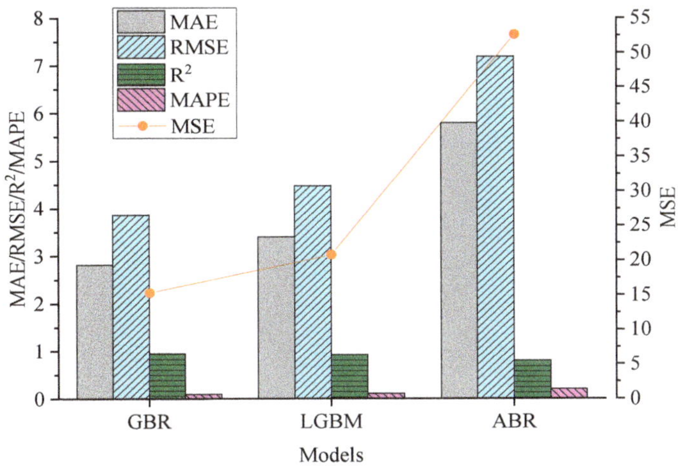

(**a**) Training dataset

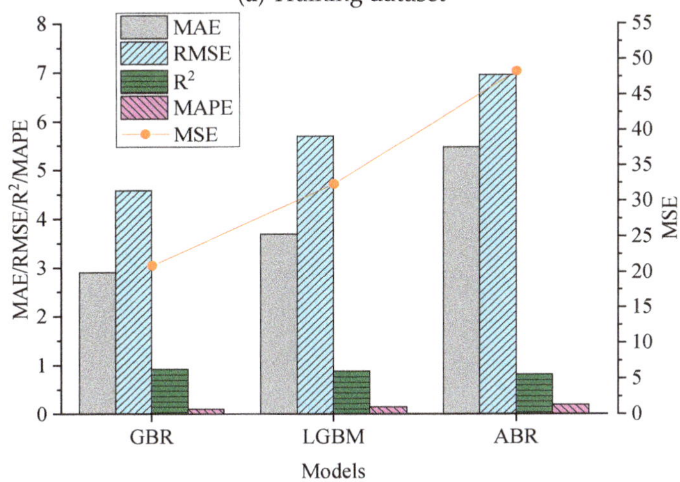

(**b**) Testing dataset

Figure 9. Performances of the optimized models on the (**a**) training dataset and (**b**) testing dataset.

The best GBR, LGBM, and ABR models were created with the training dataset, and the validity of these models must be evaluated with the testing dataset. Figure 10 compares the experimental and predicted UCSs using the developed model for the testing dataset. It can be concluded that most of the predicted UCS values agreed with the experimental UCS values, for the best GBR and LGBM and ABR models. This indicates the great potential of these models for predicting the UCS of CFAM. The best GBR model had the highest prediction accuracy.

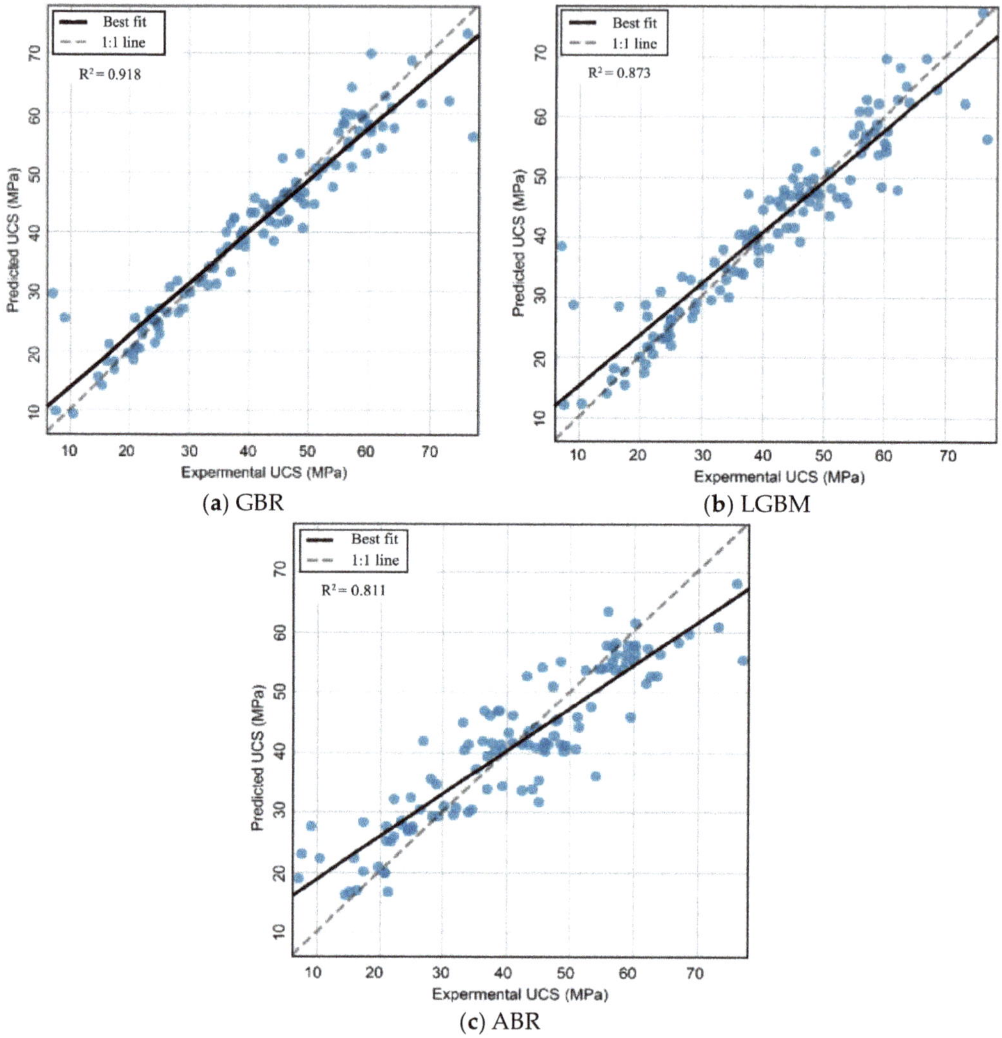

Figure 10. Relationship between the experimental and predicted compressive strengths: (**a**) GBR model; (**b**) LGBM model; (**c**) ABR model.

5.3. Residual Analysis

In order to evaluate the adequacy of the best GBR, LGBM, and ABR models, the residuals for the best GBR, LGBM, and ABR models were statistically analyzed. Figure 11 shows the residuals versus predicted values for the best GBR, LGBM, and ABR models. The residuals seem to exist around the 0 line randomly for all of the best GBR, LGBM, and ABR models. This confirms that this study's best GBR, LGBM, and ABR models have statistical significance. In addition, the best GBR model had the lowest residual values, indicating that the best GBR model is the best prediction model for the UCS of CFAM.

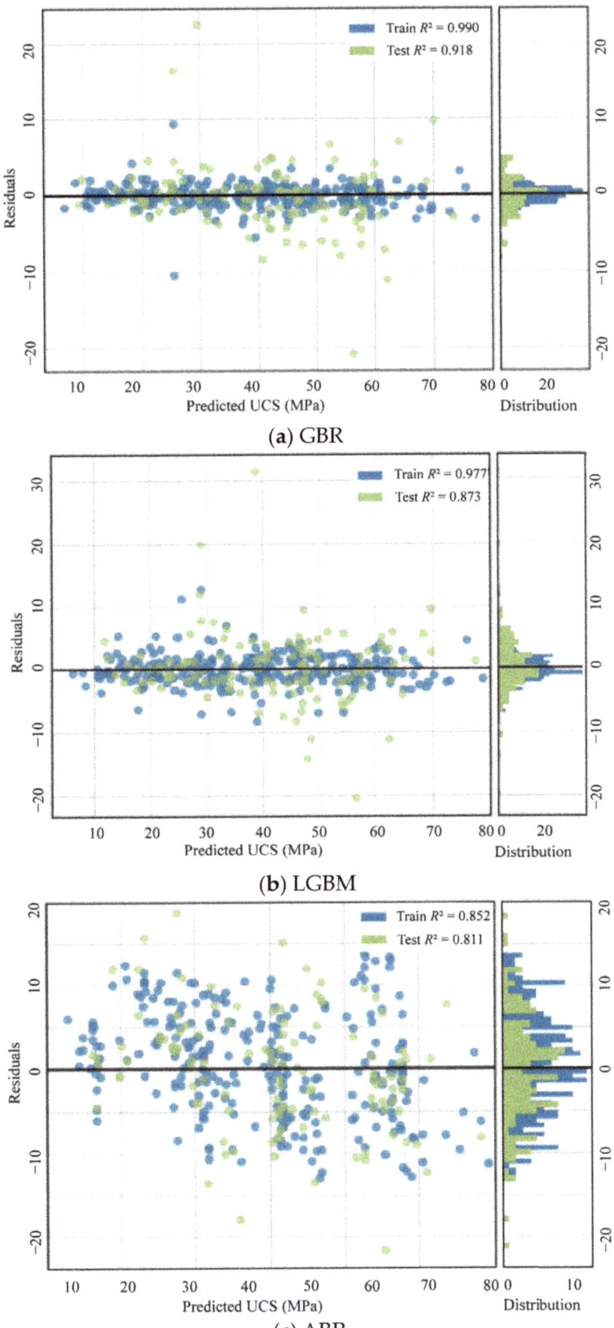

Figure 11. Residual analysis: (**a**) GBR model; (**b**) LGBM model; (**c**) ABR model.

6. Interpretability and Feature Importance Analysis

In this section, the SHAP method is used to interpret the best performing GBR model in order to investigate how the inputs influence the outputs of the developed GBR models. Figure 12a shows the SHAP summary plot. The SHAP value of each feature was represented by a dot in Figure 12a for a particular sample. The detailed SHAP values, which can be either positive or negative, are represented on the x-axis. The positive numbers indicate that the feature will raise the forecasts, while the negative ones suggest lowering them. Moreover, the feature values are coded by color, with blue representing lower feature values and red representing greater feature values. Figure 12b illustrates the feature of global importance factors. These values are the averages of the absolute Shapley values of each feature across the data. The more significant the feature, the higher the importance factor.

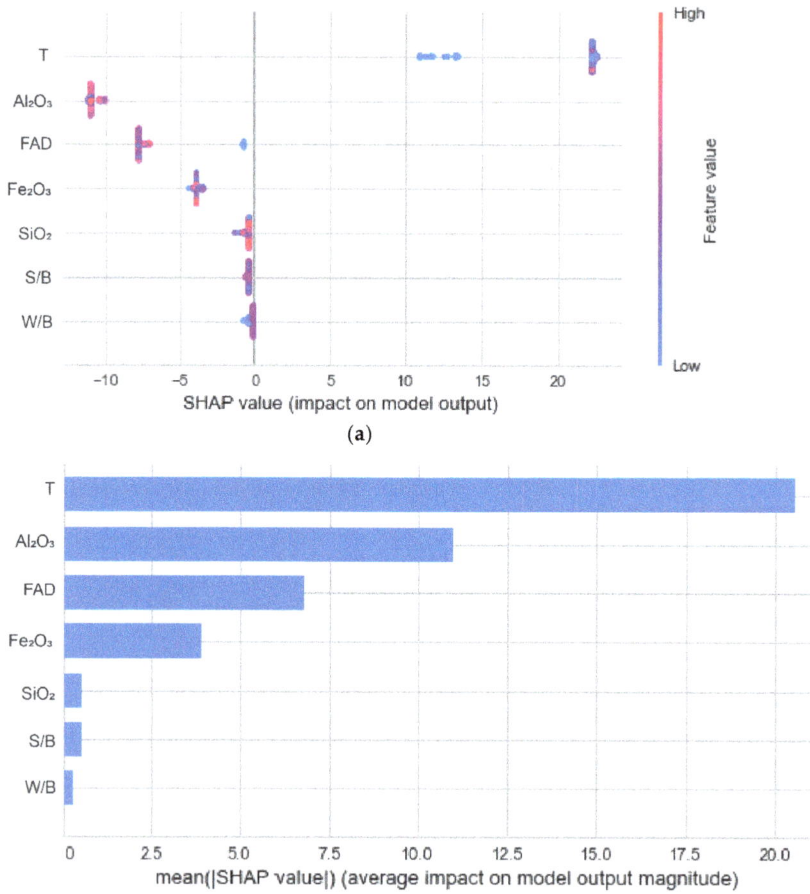

Figure 12. (**a**) SHAP global interpretation plot and (**b**) relative importance of each feature in the GBR model.

It can be found from Figure 12 that an increase in T results in an increase in the UCS, while the increases in Al_2O_3, FAD, and Fe_2O_3 cause the UCS to decrease. The rest of the features, like SiO_2, S/B, and W/B, slightly impact the UCS. Significantly, the sand-to-binder ratio (S/B) is the least important variable in the GBR model. This indicates that changes in the S/B value play a very small role in predicting the UCS of CFAM. The curing time (T) was the most important variable in the GBR model. This indicates that changes in the T value play the most important role in predicting the UCS of CFAM with the GBR model. In addition, this analysis indicates that the chemical composition of FA, especially Al_2O_3, is a more important effect parameter than the fly-ash dosage (FAD) and water-to-binder ratio (W/B) in the prediction of the UCS of CFAM with the GBR model. As a highlight of this study compared with previous studies, this also proves that the chemical composition of FA is an important parameter influencing the UCS of CFAM. It is necessary to consider the chemical composition of FA as an input parameter in the prediction of the UCS of CFAM.

Figure 13 shows the SHAP feature dependency plot, which provides the SHAP value change for two features. The feature values and the corresponding SHAP values are shown on the x- and y-axes in Figure 13. The design variable most dependent on each characteristic, as shown in Figure 13, is represented by a colored bar on the plot's right side. It should be noted that a high SHAP value indicates the model attempting to predict a high UCS value based on the corresponding feature values. In contrast, a low SHAP value indicates the model trying to output a low UCS value. For instance, as shown in Figure 10a, the dependency and interaction effects demonstrate that the features of $7 \leq T \leq 100$ days and $W/B \leq 0.450$ might result in high predicted UCS values.

Figure 14 exhibits the GBR model's local explanation with two selected specimens (No. 50 and No. 187). The base value represents the average experimental UCS value of 39.14 MPa. The bolded value of the (f(x)) represents the predicted UCS value. A red-marked feature (T) means an increase in the predicted UCS value over the base value. At the same time, blue-marked features (Al_2O_3, FAD, Fe_2O_3, SiO_2, etc.) cause predicted UCS values to decrease to below the base value. For instance, for the specimen No. 50, as shown in Figure 11a, the predicted UCS value was 36.62 MPa. Following the order, the most important variables are T, Al_2O_3, FAD, and Fe_2O_3. Among them, a higher T will boost the predicted UCS value, while higher Al_2O_3, FAD, and Fe_2O_3 contents will drag the predicted UCS value down.

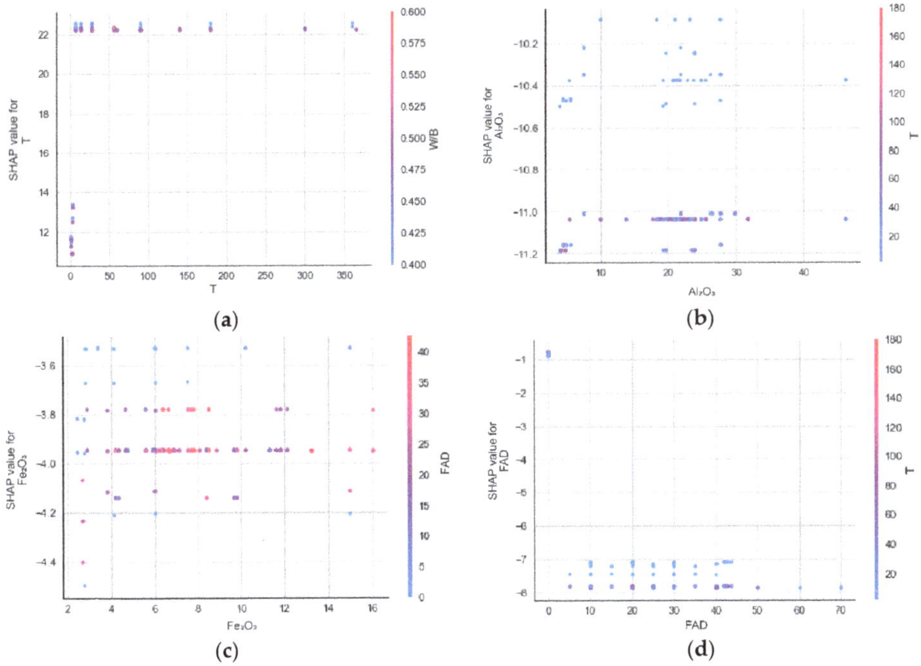

Figure 13. *Cont.*

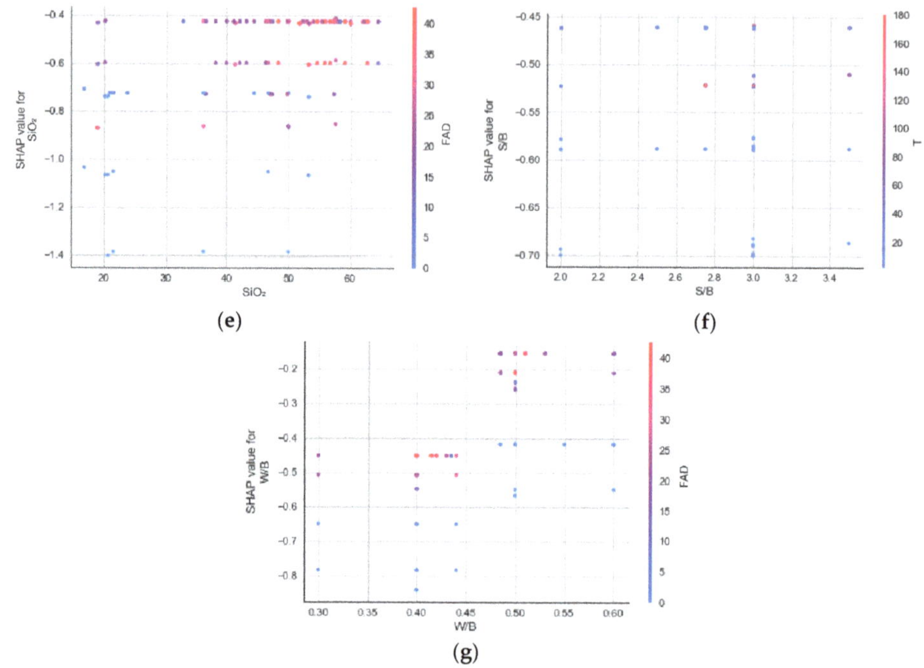

Figure 13. SHAP feature dependence plots: (**a**) T; (**b**) Al_2O_3; (**c**) FAD; (**d**) Fe_2O_3; (**e**) SiO_2; (**f**) S/B; (**g**) W/B.

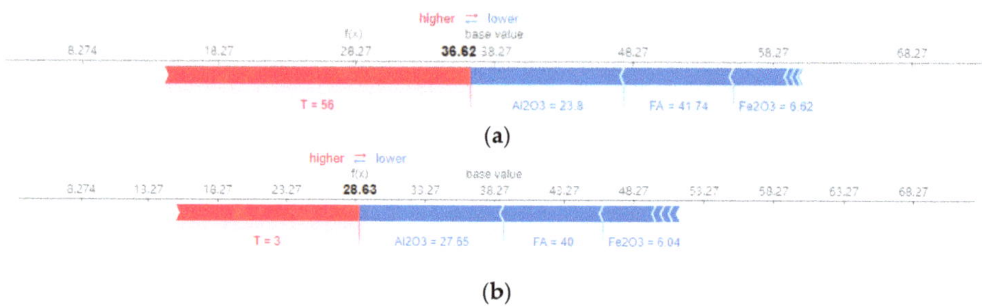

Figure 14. SHAP force plot of the selected two instances for local interpretation: (**a**) specimen No. 50; (**b**) specimen No. 187.

7. Comparison with Other Commonly Used Machine Learning Models

In addition, this study compared ten commonly used machine learning models with the best GBR model. Table 5 lists the performances of these different machine learning models using the training dataset. Similar to the GBR model, these models' parameters are determined using similar approaches. As shown in Table 5, the best GBR model exhibited lower errors than the other ten commonly used machine learning models. This indicates that its predicted values were significantly closer to the experimental values. In comparison, it should be noted that the performance of the extra trees regressor (R^2 = 0.877) and Random Forest regressor (R^2 = 0.868) models can also be accepted and used to predict the UCS of CFAM. However, their errors were higher than those of the best GBR model. The other eight commonly used machine learning models did not provide satisfying accuracies. They had higher errors and lower R^2 values. Therefore, the best GBR model in this study performed best in predicting the UCS of CFAM.

Table 5. Comparison with other various machine learning models.

No.	Model	*MAE*	*RMSE*	R^2	*MAPE*	*MSE*
1	Best GBR	3.423	4.594	0.917	0.113	21.720
2	Extra Trees Regressor	4.118	5.562	0.877	0.126	31.575
3	Random Forest Regressor	4.260	5.759	0.868	0.135	33.667
4	Decision Tree Regressor	5.668	7.802	0.754	0.172	63.521
5	K Neighbors Regressor	7.916	10.011	0.606	0.288	102.230
6	Bayesian Ridge	10.259	12.298	0.404	0.375	153.298
7	Ridge Regression	10.249	12.296	0.404	0.373	153.248
8	Linear Regression	10.248	12.297	0.403	0.373	153.279
9	Least Angle Regression	10.248	12.297	0.403	0.373	153.279
10	Huber Regressor	10.282	12.374	0.394	0.372	155.150
11	Lasso Regression	10.558	12.610	0.374	0.393	161.729

In engineering, the best GBR model can be used to select an appropriate type of FA and determine the approximate dosage range to reduce the time costs of preliminary tests and avoid UCSs below the limit due to inappropriate admixtures.

8. Conclusions

This study investigated and modeled the UCS of CFAM, with the following input parameters: chemical composition of fly ash, fly-ash dosage, water–binder ratio, sand–binder ratio, and curing time. The innovative aspect of this study is the development of boosting machine learning models (GBR, LGBM, and ABR models) to predict the UCS of CFAM. The best model was explained using the SHapley Additive exPlanations (SHAP) approach. Moreover, the best boosting machine learning model was compared with ten other commonly used machine learning models to better understand the performance of the boosting machine learning model. The following conclusions can be drawn:

(1) The GBR model performed better than the LGBM and ABR models and could be the best boosting machine learning model for predicting the UCS of CFAM.

(2) Compared to the other ten commonly used machine learning models, the GBR model exhibited significant accuracy in predicting the UCS of CFAM.

(3) The SHAP interpretations indicate that curing time (T) is the most important feature. The chemical composition of fly ash, especially Al_2O_3, is a more important effect parameter than fly-ash dosage (FAD) and water-to-binder ratio (W/B). The increase in T results in an increase in the UCS, while increases in Al_2O_3, FAD, and Fe_2O_3 cause the UCS to decrease.

In conclusion, this study shows the strong power of boosting machine learning models for predicting the UCS of CFAM. The better prediction performances of the boosting machine learning models indicate their importance in civil engineering. The best GBR model in this study could help engineers or researchers predict the UCS of CFAM under the conditions of different curing times, different types of fly ash, different fly ash contents, and different water-to-binder ratios.

Author Contributions: Conceptualization, H.W. and Y.D.; methodology, H.W. and Y.D.; validation, Y.K. and D.S.; investigation, X.C.; data curation, H.W.; writing—original draft preparation, H.W.; writing—review and editing, Y.S., D.S. and X.C.; visualization, Y.D. and Y.K.; project administration, H.W. and X.C.; funding acquisition, H.W. All authors have read and agreed to the published version of the manuscript.

Funding: This research was funded by the National Natural Science Foundation of China (grant no. 52304114) and the Natural Science Foundation of Hunan Province (grant no. 2024JJ6510).

Institutional Review Board Statement: Not applicable.

Informed Consent Statement: Not applicable.

Data Availability Statement: The authors confirm that the data supporting the findings of this study are available within the article.

Conflicts of Interest: Author Yu Kong was employed by the company China Construction Fifth Engineering Division Corp. Ltd. The remaining authors declare that the research was conducted in the absence of any commercial or financial relationships that could be construed as a potential conflict of interest.

References

1. Zentar, R.; Wang, H.; Wang, D. Comparative study of stabilization/solidification of dredged sediments with ordinary Portland cement and calcium sulfo-aluminate cement in the framework of valorization in road construction material. *Constr. Build. Mater.* **2021**, *279*, 122447. [CrossRef]
2. Wang, H.; Zentar, R.; Wang, D.; Ouendi, F. New Applications of Ordinary Portland and Calcium Sulfoaluminate Composite Binder for Recycling Dredged Marine Sediments as Road Materials. *Int. J. Geomech.* **2022**, *22*, 04022068. [CrossRef]
3. Wang, D.; Wang, H.; Larsson, S.; Benzerzour, M.; Maherzi, W.; Amar, M. Effect of basalt fiber inclusion on the mechanical properties and microstructure of cement-solidified kaolinite. *Constr. Build. Mater.* **2020**, *241*, 118085. [CrossRef]
4. Wang, D.; Wang, Z.; Wang, H. Feasibility and performance assessment of novel framework for soil stabilization using multiple industrial wastes. *Constr. Build. Mater.* **2024**, *449*, 138228. [CrossRef]
5. Jiang, D.B.; Li, X.G.; Lv, Y.; Zhou, M.K.; He, C.H.; Jiang, W.G.; Liu, Z.L.; Li, C.J. Utilization of limestone powder and fly ash in blended cement: Rheology, strength and hydration characteristics. *Constr. Build. Mater.* **2020**, *232*, 117228. [CrossRef]
6. Damtoft, J.S.; Lukasik, J.; Herfort, D.; Sorrentino, D.; Gartner, E.M. Sustainable development and climate change initiatives. *Cem. Concr. Res.* **2008**, *38*, 115–127. [CrossRef]
7. Poudyal, L.; Adhikari, K. Environmental sustainability in cement industry: An integrated approach for green and economical cement production. *Resour. Environ. Sustain.* **2021**, *4*, 100024. [CrossRef]
8. De Weerdt, K.; Kjellsen, K.O.; Sellevold, E.; Justnes, H. Synergy between fly ash and limestone powder in ternary cements. *Cem. Concr. Compos.* **2011**, *33*, 30–38. [CrossRef]
9. Shi, Y.; Li, Y.; Wang, H. Eco-friendly solid waste-based cementitious material containing a large amount of phosphogypsum: Performance optimization, micro-mechanisms, and environmental properties. *J. Clean. Prod.* **2024**, *471*, 143335. [CrossRef]
10. Adesina, A. Performance and sustainability overview of sodium carbonate activated slag materials cured at ambient temperature. *Resour. Environ. Sustain.* **2021**, *3*, 100016. [CrossRef]
11. Cho, Y.K.; Jung, S.H.; Choi, Y.C. Effects of chemical composition of fly ash on compressive strength of fly ash cement mortar. *Constr. Build. Mater.* **2019**, *204*, 255–264. [CrossRef]
12. Wang, A.Q.; Zhang, C.Z.; Sun, W. Fly ash effects—II. The active effect of fly ash. *Cem. Concr. Res.* **2004**, *34*, 2057–2060. [CrossRef]
13. Hijazi, D.A.; BiBi, A.; Al-Ghouti, M.A. Sustainable waste utilization: Geopolymeric fly ash waste as an effective phenol adsorbent for environmental remediation. *Resour. Environ. Sustain.* **2024**, *15*, 100142. [CrossRef]
14. Ahmaruzzaman, M. A review on the utilization of fly ash. *Prog. Energy Combust. Sci.* **2010**, *36*, 327–363. [CrossRef]
15. Sotiriou, V.; Michas, G.; Xiong, L.; Drosos, M.; Vlachostergios, D.; Papadaki, M.; Mihalakakou, G.; Kargiotidou, A.; Tziouvalekas, M.; Salachas, G.; et al. Effects of heavy metal ions on white clover (*Trifolium repens* L.) growth in Cd, Pb and Zn contaminated soils using zeolite. *Soil. Sci. Environ.* **2023**, *2*, 4. [CrossRef]
16. Sakai, E.; Miyahara, S.; Ohsawa, S.; Lee, S.H.; Daimon, M. Hydration of fly ash cement. *Cem. Concr. Res.* **2005**, *35*, 1135–1140. [CrossRef]
17. Antiohos, S.; Tsimas, S. Investigating the role of reactive silica in the hydration mechanisms of high-calcium fly ash/cement systems. *Cem. Concr. Compos.* **2005**, *27*, 171–181. [CrossRef]
18. *ASTM C618*; Standard Specification for Coal Ash and Raw or Calcined Natural Pozzolan for Use in Concrete. ASTM International: West Conshohocken, PA, USA, 2023.
19. Sengul, O.; Tasdemir, C.; Tasdemir, M.A. Mechanical properties and rapid chloride permeability of concretes with ground fly ash. *ACI Mater. J.* **2005**, *102*, 414–421.
20. Ogawa, Y.; Uji, K.; Ueno, A.; Kawai, K. Contribution of fly ash to the strength development of mortars cured at different temperatures. *Constr. Build. Mater.* **2021**, *276*, 117228. [CrossRef]
21. Cyr, M.; Lawrence, P.; Ringot, E. Efficiency of mineral admixtures in mortars: Quantification of the physical and chemical effects of fine admixtures in relation with compressive strength. *Cem. Concr. Res.* **2006**, *36*, 264–277. [CrossRef]
22. Qadir, W.; Ghafor, K.; Mohammed, A. Characterizing and Modeling the Mechanical Properties of the Cement Mortar Modified with Fly Ash for Various Water-to-Cement Ratios and Curing Times. *Adv. Civ. Eng.* **2019**, *2019*, 7013908. [CrossRef]
23. Kahraman, E.; Hosseini, S.; Taiwo, B.O.; Fissha, Y.; Jebutu, V.A.; Akinlabi, A.A.; Adachi, T. Fostering sustainable mining practices in rock blasting: Assessment of blast toe volume prediction using comparative analysis of hybrid ensemble machine learning techniques. *J. Saf. Sustain.* **2024**, *1*, 75–88. [CrossRef]
24. Hu, P.; Tanchak, R.; Wang, Q. Developing risk assessment framework for wildfire in the United States—A deep learning approach to safety and sustainability. *J. Saf. Sustain.* **2024**, *1*, 26–41. [CrossRef]

25. Dong, L.; Wang, J. Intelligent Safety Ergonomics: A Cleaner Research Direction for Ergonomics in the Era of Big Data. *Int. J. Environ. Res. Public Health* **2023**, *20*, 423. [CrossRef]
26. Dong, L.-J.; Wang, J.; Wang, J.-C.; Wang, H.-W. Safe and intelligent mining: Some explorations and challenges in the era of big data. *J. Cent. South Univ.* **2023**, *30*, 1900–1914. [CrossRef]
27. Munir, M.J.; Kazmi, S.M.S.; Wu, Y.F.; Lin, X.S.; Ahmad, M.R. Development of novel design strength model for sustainable concrete columns: A new machine learning-based approach. *J. Clean. Prod.* **2022**, *357*, 131988. [CrossRef]
28. Azimi-Pour, M.; Eskandari-Naddaf, H. ANN and GEP prediction for simultaneous effect of nano and micro silica on the compressive and flexural strength of cement mortar. *Constr. Build. Mater.* **2018**, *189*, 978–992. [CrossRef]
29. Nguyen, H.; Vu, T.; Vo, T.P.; Thai, H.T. Efficient machine learning models for prediction of concrete strengths. *Constr. Build. Mater.* **2021**, *266*, 120950. [CrossRef]
30. Salehi, H.; Burgueno, R. Emerging artificial intelligence methods in structural engineering. *Eng. Struct.* **2018**, *171*, 170–189. [CrossRef]
31. Mohammed, A.; Rafiq, S.; Sihag, P.; Kurda, R.; Mahmood, W.; Ghafor, K.; Sarwar, W. ANN, M5P-tree and nonlinear regression approaches with statistical evaluations to predict the compressive strength of cement-based mortar modified with fly ash. *J. Mater. Res. Technol.* **2020**, *9*, 12416–12427. [CrossRef]
32. Feng, D.C.; Liu, Z.T.; Wang, X.D.; Chen, Y.; Chang, J.Q.; Wei, D.F.; Jiang, Z.M. Machine learning-based compressive strength prediction for concrete: An adaptive boosting approach. *Constr. Build. Mater.* **2020**, *230*, 117000. [CrossRef]
33. Zhou, Z.-H. Ensemble learning. In *Machine Learning*; Springer: Singapore, 2021; pp. 181–210.
34. Gonzalez-Recio, O.; Jimenez-Montero, J.A.; Alenda, R. The gradient boosting algorithm and random boosting for genome-assisted evaluation in large data sets. *J. Dairy Sci.* **2013**, *96*, 614–624. [CrossRef] [PubMed]
35. Ghafouri-Kesbi, F.; Rahimi-Mianji, G.; Honarvar, M.; Nejati-Javaremi, A. Predictive ability of Random Forests, Boosting, Support Vector Machines and Genomic Best Linear Unbiased Prediction in different scenarios of genomic evaluation. *Anim. Prod. Sci.* **2017**, *57*, 229–236. [CrossRef]
36. Bui, X.N.; Nguyen, H.; Le, H.A.; Bui, H.B.; Do, N.H. Prediction of Blast-induced Air Over-pressure in Open-Pit Mine: Assessment of Different Artificial Intelligence Techniques. *Nat. Resour. Res.* **2020**, *29*, 571–591. [CrossRef]
37. Song, H.W.; Ahmad, A.; Farooq, F.; Ostrowski, K.A.; Maslak, M.; Czarnecki, S.; Aslam, F. Predicting the compressive strength of concrete with fly ash admixture using machine learning algorithms. *Constr. Build. Mater.* **2021**, *308*, 125021. [CrossRef]
38. Rathakrishnan, V.; Beddu, S.B.; Ahmed, A.N. Predicting compressive strength of high-performance concrete with high volume ground granulated blast-furnace slag replacement using boosting machine learning algorithms. *Sci. Rep.* **2022**, *12*, 9539. [CrossRef]
39. Feng, D.C.; Wang, W.J.; Mangalathu, S.; Taciroglu, E. Interpretable XGBoost-SHAP Machine-Learning Model for Shear Strength Prediction of Squat RC Walls. *J. Struct. Eng.* **2021**, *147*, 04021173. [CrossRef]
40. Mangalathu, S.; Hwang, S.H.; Jeon, J.S. Failure mode and effects analysis of RC members based on machine-learning-based SHapley Additive exPlanations (SHAP) approach. *Eng. Struct.* **2020**, *219*, 110927. [CrossRef]
41. Friedman, J.H. Greedy function approximation: A gradient boosting machine. *Ann. Stat.* **2001**, *29*, 1189–1232. [CrossRef]
42. Friedman, J.H. Stochastic gradient boosting. *Comput. Stat. Data* **2002**, *38*, 367–378. [CrossRef]
43. Ke, G.; Meng, Q.; Finley, T.; Wang, T.; Chen, W.; Ma, W.; Ye, Q.; Liu, T.-Y. Lightgbm: A highly efficient gradient boosting decision tree. *Adv. Neural Inf. Process. Syst.* **2017**, *30*, 3149–3157.
44. Ju, Y.; Sun, G.; Chen, Q.; Zhang, M.; Zhu, H.; Rehman, M.U. A model combining convolutional neural network and LightGBM algorithm for ultra-short-term wind power forecasting. *IEEE Access* **2019**, *7*, 28309–28318. [CrossRef]
45. Zhou, B.; Xu, J.; Han, F.; Yan, F.Z.; Peng, S.J.; Li, Q.X.; Jiao, F. Pressure of different gases injected into large-scale coal matrix: Analysis of time-space dependence and prediction using light gradient boosting machine. *Fuel* **2020**, *279*, 118448. [CrossRef]
46. Cai, W.; Wei, R.; Xu, L.; Ding, X. A method for modelling greenhouse temperature using gradient boost decision tree. *Inf. Process. Agric.* **2022**, *9*, 343–354. [CrossRef]
47. Freund, Y.; Schapire, R.E. A decision-theoretic generalization of on-line learning and an application to boosting. *J. Comput. Syst. Sci.* **1997**, *55*, 119–139. [CrossRef]
48. Kang, M.C.; Yoo, D.Y.; Gupta, R. Machine learning-based prediction for compressive and flexural strengths of steel fiber-reinforced concrete. *Constr. Build. Mater.* **2021**, *266*, 121117. [CrossRef]
49. Scott, M.; Su-In, L. A unified approach to interpreting model predictions. In Proceedings of the 31st Conference on Neural Information Processing Systems (NIPS 2017), Long Beach, CA, USA, 4–9 December 2017.
50. Ahmad, A.; Ahmad, W.; Aslam, F.; Joyklad, P. Compressive strength prediction of fly ash-based geopolymer concrete via advanced machine learning techniques. *Case Stud. Constr. Mat.* **2022**, *16*, e00840. [CrossRef]
51. Moon, G.D.; Oh, S.; Choi, Y.C. Effects of the physicochemical properties of fly ash on the compressive strength of high-volume fly ash mortar. *Constr. Build. Mater.* **2016**, *124*, 1072–1080. [CrossRef]
52. Yerramala, A.; Desai, B. Influence of fly ash replacement on strength properties of cement mortar. *Int. J. Eng. Sci. Technol.* **2012**, *4*, 3657–3665.
53. Supit, S.W.M.; Shaikh, F.U.A.; Sarker, P.K. Effect of ultrafine fly ash on mechanical properties of high volume fly ash mortar. *Constr. Build. Mater.* **2014**, *51*, 278–286. [CrossRef]
54. Maltais, Y.; Marchand, J. Influence of curing temperature on cement hydration and mechanical strength development of fly ash mortars. *Cem. Concr. Res.* **1997**, *27*, 1009–1020. [CrossRef]

55. Rais, M.S.; Shariq, M.; Masood, A.; Umar, A.; Alam, M.M. An experimental and analytical investigation into age-dependent strength of fly ash mortar at elevated temperature. *Constr. Build. Mater.* **2019**, *222*, 300–311. [CrossRef]
56. Thongsanitgarn, P.; Wongkeo, W.; Chaipanich, A. Hydration and Compressive Strength of Blended Cement Containing Fly Ash and Limestone as Cement Replacement. *J. Mater. Civ. Eng.* **2014**, *26*, 040140. [CrossRef]
57. Chindaprasirt, P.; Kanchanda, P.; Sathonsaowaphak, A.; Cao, H.T. Sulfate resistance of blended cements containing fly ash and rice husk ash. *Constr. Build. Mater.* **2007**, *21*, 1356–1361. [CrossRef]
58. Han, F.H.; Wang, Q.; Feng, J.J. The differences among the roles of ground fly ash in the paste, mortar and concrete. *Constr. Build. Mater.* **2015**, *93*, 172–179.
59. Tangpagasit, J.; Cheerarot, R.; Jaturapitakkul, C.; Kiattikomol, K. Packing effect and pozzolanic reaction of fly ash in mortar. *Cem. Concr. Res.* **2005**, *35*, 1145–1151. [CrossRef]
60. Chindaprasirt, P.; Rukzon, S. Strength, porosity and corrosion resistance of ternary blend Portland cement, rice husk ash and fly ash mortar. *Constr. Build. Mater.* **2008**, *22*, 1601–1606. [CrossRef]
61. Paya, J.; Monzo, J.; Borrachero, M.V.; Peris-Mora, E.; Amahjour, F. Mechanical treatment of fly ashes Part IV. Strength development of ground fly ash-cement mortars cured at different temperatures. *Cem. Concr. Res.* **2000**, *30*, 543–551.
62. Atis, C.D.; Kilic, A.; Sevim, U.K. Strength and shrinkage properties of mortar containing a nonstandard high-calcium fly ash. *Cem. Concr. Res.* **2004**, *34*, 99–102. [CrossRef]
63. Mardani-Aghabaglou, A.; Sezer, G.I.; Ramyar, K. Comparison of fly ash, silica fume and metakaolin from mechanical properties and durability performance of mortar mixtures view point. *Constr. Build. Mater.* **2014**, *70*, 17–25. [CrossRef]
64. Arenas-Piedrahita, J.C.; Montes-Garcia, P.; Mendoza-Rangel, J.M.; Calvo, H.Z.L.; Valdez-Tamez, P.L.; Martinez-Reyes, J. Mechanical and durability properties of mortars prepared with untreated sugarcane bagasse ash and untreated fly ash. *Constr. Build. Mater.* **2016**, *105*, 69–81. [CrossRef]
65. Feng, J.J.; Sun, J.W.; Yan, P.Y. The Influence of Ground Fly Ash on Cement Hydration and Mechanical Property of Mortar. *Adv. Civ. Eng.* **2018**, *2018*, 4023178. [CrossRef]
66. Elkhadiri, I.; Diouri, A.; Boukhari, A.; Aride, J.; Puertas, F. Mechanical behaviour of various mortars made by combined fly ash and limestone in Moroccan Portland cement. *Cem. Concr. Res.* **2002**, *32*, 1597–1603. [CrossRef]
67. Celik, O.; Damei, E.; Piskin, S. Characterization of fly ash and it effects on the compressive strength properties of Portland cement. *Indian J. Eng. Mater. Sci.* **2008**, *15*, 433–440.
68. Wang, H.W.; Zentar, R.; Wang, D.X. Predicting the compaction parameters of solidified dredged fine sediments with statistical approach. *Mar. Georesour. Geotechnol.* **2023**, *41*, 195–210. [CrossRef]
69. Jain, R.; Ganesan, R.A. Reliable sleep staging of unseen subjects with fusion of multiple EEG features and RUSBoost. *Biomed. Signal Process. Control.* **2021**, *70*, 1746–8094. [CrossRef]
70. Baumann, D.; Baumann, K. Reliable estimation of prediction errors for QSAR models under model uncertainty using double cross-validation. *J. Cheminform.* **2014**, *6*, 1758–2946. [CrossRef]

materials

MDPI

Article

Marginal Aggregates: The Role of Clays

Arnon Bentur * and Pavel Larianovsky

National Building Research Institute, Faculty of Civil and Environmental Engineering, Technion, Israel Institute of Technology, Haifa 3200003, Israel; lpavel@technion.ac.il
* Correspondence: bentur@technion.ac.il

Abstract: Clays are components in the fine portion of aggregates, less than 75 microns in size (micro-fines), which are usually washed away when producing coarse or fine (manufactured sand) aggregates in quarries. When marginal sources of aggregates are being used, the content of this washed portion can be quite high, and there is an incentive to keep as much of it in the aggregate, including the clays. The present paper presents a comprehensive treatment of the role of clays in terms of the characterization of their composition and quantification of their effects on the rheological and mechanical properties of cementitious systems, as well as the means to mitigate deleterious influences. It is shown that the strategy for neutralizing the effect of micro-fines containing clays on increased water demand in concrete can be quantified in terms of the combination of their content in concrete and their nature as characterized by the methylene blue value (MBV); this is a more rational approach to considering their influence than their content in specific aggregates as specified in standards. The effect of low and medium MBV aggregates on the water requirement can be neutralized by lignosulfonates when their content in the concrete is below a threshold value of about 150 kg/m^3; polycarboxylates (PC) are required at higher contents; for high MBV aggregates, a combination of PC and clay mitigating admixture (CMA) is required. It is also demonstrated that with proper treatment, such micro-fines can be turned into useful fillers, enhancing the strength of concrete and thus also serving as a means for reducing cement content.

Keywords: marginal aggregates; micro-fines; clays; admixtures; clay mitigation

1. Introduction

The current environmental and ecological constraints lead to the need to extensively use marginal aggregates. As a result, there is interest in looking into the influence of fine materials smaller than 75 microns in size (micro-fines, or fines as termed in the EU standard EN 12620 [1]), which usually have been washed away during processing, with the intention of allowing incorporation of higher levels of them in the concrete, e.g., [2,3].

Several studies dealt separately with a range of issues, focusing mainly only on the standard methylene blue test [4–7], the influence of clays on the rheological behavior of mortars and concretes [8–13], and the mitigation of deleterious effects of clays by the use of admixtures [14–16]. All these issues are interrelated, and it was felt that there is a need for a comprehensive treatment encompassing all these aspects simultaneously in order to advance the fundamental knowledge base, including (i) the mineral composition of the clays, (ii) their characterization by standard test methods, (iii) their influence on rheological properties and their engineering quantification, (iv) the impact on hardened properties, and finally (v) the modifications by use of admixtures. An integrative project with this scope in mind has been carried out at the Technion [17,18].

This comprehensive treatment is required to deal with the gap that exists between the approach taken by the standards and the industry at large, viewing the clays as deleterious to concrete, while at the same time there is considerable advance in the field of admixtures, which may be mobilized towards enabling the use of aggregates with a higher content of clay. Bridging this gap is essential for enhancing the successful application of aggregates, which are currently considered marginal.

The approach taken by standards and specifications is to limit the clay content in aggregates by setting limiting maximum values for the methylene blue test, which provides indications to the content of clays and their nature (in particular swelling clays) in the aggregates. It is backed by studies that address the deleterious effects of clays in concrete, which are usually characterized by their influence on workability, strength, and dimensional stability [5,6,19–22]. The advent of methods to mitigate deleterious influences of clays in aggregates, e.g., [13,14,16,23], has not had an impact on specifications and standards. This is to a large extent the result of the traditional approach, which is focused on aggregate quality rather than concrete performance, where admixture technologies can be applied to mitigate the influence of concrete constituents and enable the use of marginal aggregates. The current paper is intended to provide a rational approach to meet the challenge in this gap, based on a systematic study of the influences of micro-fines on concrete performance in fresh and hardened states.

2. Materials and Methods

2.1. Overview of the Research Plan and Testing Program

The present report is based on a comprehensive study focusing on the evaluation of mortar and concrete systems in which clay containing micro-fines was incorporated with a size range of less than 75 microns.

The concept advanced, different from the current approach, is that focus should be given to the evaluation of the micro-fines since they are the portion of fine aggregate material (e.g., sand, smaller than 4.75 mm) that carries clays that might be deleterious to concrete. The overall methodology to deal with these effects should be based on an evaluation of the micro-fines quality in combination with their content in the concrete, rather than the sand quality as influenced by the clays in it. The final outcome, with respect to concrete standardization, should thus not be the sand quality but rather the micro-fines quality in combination with their content in the concrete. This should be the technological approach to mobilizing the use of marginal fine aggregate (sand) material. This approach is also consistent with the current initiatives to develop and design high performance concretes, such as UHPC (Ultra High Performance Concrete) and SCC (Self Consolidating Concrete), where micro-fines play a critical role. Mobilization of micro-fines, which are considered marginal, could also be useful for high-end applications.

The study consisted of two parts, the first of which was intended to characterize micro-fines derived from various processes in quarries and determine their effects on the behavior of mortars using a range of test methods, in particular rheological means. For that purpose, "engineered sand" samples were prepared, made up of various contents of micro-fines mixed with high quality siliceous sand.

The second part focused on the study of concretes in which micro-fines were incorporated and their fresh and hardened properties were characterized. The properties of the concretes were modified by the use of admixtures intended to mitigate the deleterious influences of the micro-fines. These included lignosulfonate (LS) and polycarboxylate (PC) to control the dispersion of the micro-fines, as well as clay mitigating admixture (CMA) to neutralize the influence of swelling clays. The concretes with optimal admixture type and content to control workability were also evaluated for strength and shrinkage.

The research strategy for both parts of the study was based on the concepts of preparing micro-fines of controlled compositions, which were obtained by using pure clay components and micro-fines recovered from by-products in quarries. In the testing of mortars and concretes, washed aggregates (sand and coarse aggregates) were used, which contained no micro-fines. Micro-fines were added directly to the mortar or concrete separately, and thus the aggregates used are referred to as "engineered". The micro-fine characteristics are outlined below:

- Micro-fines recovered from quarries by-products were obtained by either taking material from clay rich lenses in quarries or fine waste materials in quarries and separating the micro-fines from them by a process that involved water washing and sieving through a 75 microns mesh, followed by drying to remove the water.
- The micro-fines obtained were characterized for their mineralogical composition by quantitative XRD and particle size distribution tests.
- For the characterization of the clay nature using the standard methylene blue test, which is adequate for aggregates containing clays, "engineered sand" was prepared. It consisted of 5% of the micro-fines mixed with 95% of high-quality siliceous sand, with particles smaller than 1 mm having a fineness modulus of 1. Thus, the Methylene Blue Value (MBV) for this engineered sand was used to characterize the clay nature of the micro-fines, which contained, in addition to clay, some other minerals, mainly Calcite, Dolomite and Quartz. The calibration curves of the MBV value against swelling clay, e.g., Montmorillonite, served as a way to provide the "equivalent Montmorillonite" content in the micro-fines.
- The mortars prepared and evaluated consisted of 1:3 cement:sand mixes. The cement was CEM I 52.5. The sand was "engineered" to contain the high-quality siliceous sand described above, in which portions were replaced by different micro-fine contents in the range of 0 to 30%.
- The concrete prepared consisted of washed aggregates containing no micro-fines. The micro-fines were added in controlled amounts, up to 300 kg/m^3 in the concrete mix. To compare with the presence of micro-fines, which are "pure" fillers, high-quality calcium carbonate filler was used.
- The properties of the fresh mortars and concretes were characterized by rheological tests using a rotational ICAR viscometer for concrete and a ConTec rheometer for mortars. The characteristic value used in these tests was the yield value, which was shown to linearly correlate with the standard flow table test for mortars and the slump test for concrete.
- The properties of the mature mixes evaluated were strength and drying shrinkage.

2.2. Characterization of Micro-Fines

The micro-fine mineralogical composition was evaluated by XRD, their clay content and type were characterized by the methylene blue test using the ASTM C 1777 procedure [24], and their particle size distribution was determined by a laser-type instrument in which a dispersion solution was employed.

A systematic evaluation of the effects of these micro-fines containing clays was carried out by investigating the fresh and hardened properties of 1:3 cement:sand mortars in which the sand was "engineered", being composed of high quality siliceous sand with a fineness modulus of 1, and replacing portions of it with fine material with a size smaller than 75 microns. This micro-fine material was either pure clay minerals, Kaolinite and Montmorillonite, or the fine materials obtained from the clay lenses in quarries, which were rich in clay but contained other minerals as well, such as Calcite, Dolomite and Quartz. This procedure provided mixes with varying contents and types of clays.

The composition of the micro-fine material obtained from the quarry clay lenses was determined by XRD, and the quantitative mineral composition was estimated by analysis using the Rietveld method. The physical properties of these fines were characterized by the ASTM D 4318 test [25], which is used for soils. The parameters quantified were the liquid limit, plasticity limit, and plasticity index.

The clay nature of the micro-fines was determined by the methylene blue test using the ASTM C 1777 procedure. This test is based on the quantification of the adsorption of methylene blue by the tested material using the colorimetric method. For that test, which is intended for sand, the "engineered sand" outlined above was prepared, consisting of 95% high quality siliceous sand with 5% of the micro-fines to be characterized.

The properties of the mortar mixes in the fresh state were evaluated by means of a rheological test to determine the yield stress and apparent viscosity values, as well as by the flow table test using the ASTM C 1437 [26] and ASTM C 230 procedures [27] to determine the spread and flow values. These tests were carried out with mixes of different w/c ratios. The compressive strength of these mixes under wet curing for 1, 7, and 28 days was also determined.

A special series of tests using these methodologies was carried out in order to characterize the effect and effectiveness of different admixtures, including conventional water reducing admixtures, lignosulfonates, and PC, as well as CMA and Clarena, a product of WR Grace.

2.3. Micro-Fines Mineralogical Composition and Particle Size Distribution

The mineralogical composition of "pure" components (calcium carbonate, Montmorillonite, and Kaolinite clays) and micro-fines is presented in Table 1.

Table 1. Mineralogical composition and MBV values of micro-fines.

	E3 *	E4 **	E2	E5	E6	E7	E8	E9	E10	E11	E12	E13	E14	E15
Quartz		2.1	39.9	38.2	8.1	6.2	23.2	6.2	21.4	0.8	15.2	37.2	58.8	40.7
Christobalite	51.6													
Calcite			32.6	9.1	67.1	11.0	9.8	60.9	55.8	45.0	22.0	46.1	13.0	7.7
Dolomite			1.3	25.9	12.9	66.6	37.6	7.9	1.8	14.4			4.7	8.2
Ankerite	2.5						4.7	18.2	4.2	39.8	55.3	6.2		5.8
Rutile			2.1	2.3									3.8	
Anhydrite			0.7	0.6									1.5	
Montmorillonite (clay)	45.9		6.3	1.7	0.3		0.3	0.3	1.1		6.5	3.4	6.7	
Kaolinite (clay)		84.6	7.4	10.7	9.4	1.8	19.8	5.4	14.1		20.7	7.0	5.6	18.5
Halloysite (clay)			9.7	11.5	1.4	14.4	4.7	1.2	1.5				6.0	19.1
Illite (clay)		6.7												
Muscovite (clay)		6.7												

* Source for high content Bentonite, technical grade. ** Source for high content Kaolinite, technical grade.

The particle size distribution data are presented in Table 2 and Figure 1. For comparison, the data also includes calcium carbonate, which was used in this study as a reference for "pure" filler in concrete. There is a drastic difference between the nature of the particle size distributions of the micro-fines, which is much wider, relative to the calcium carbonate, which is more of a mono-size distribution. The table also includes the clay content determined by the XRD test and the MBV values determined by the methylene blue test (see Section 2.2), which provide a quantitative indication for the presence of swelling clays, in particular Montmorillonite. This test is very sensitive to the presence of clays, which are usually in small amounts, a few percent or less, and can thus detect clays that are not always readily detected by the quantitative XRD.

Table 2. Statistical parameters of the particle size distribution of the micro-fines and their mineralogical nature with respect to the presence of clays.

Notation	Nature of Sample	D (10%) μm	D (50%) μm	D (90%) μm	<2 μm %wt.	<8 μm %wt.	Clay Content, XRD, %wt.	MBV
E1	Limestone	1.03	7.89	19.41	16.5	50.6	0.0	0.00
E3	Montmorillonite	2.09	10.30	43.58	9.3	43.2	94.3	7.48
E4	Kaolinite	0.91	7.77	27.92	21.2	50.9	94.0	0.22
E8	Micro-fines of sand	1.17	12.83	56.41	16.0	40.6	8.5	1.64
E10	Micro-fines of sand	1.68	12.99	57.71	21.4	40.0	35.8	5.26
E12	Micro-fines of sand	0.66	3.98	50.72	33.9	63.4	66.0	7.00

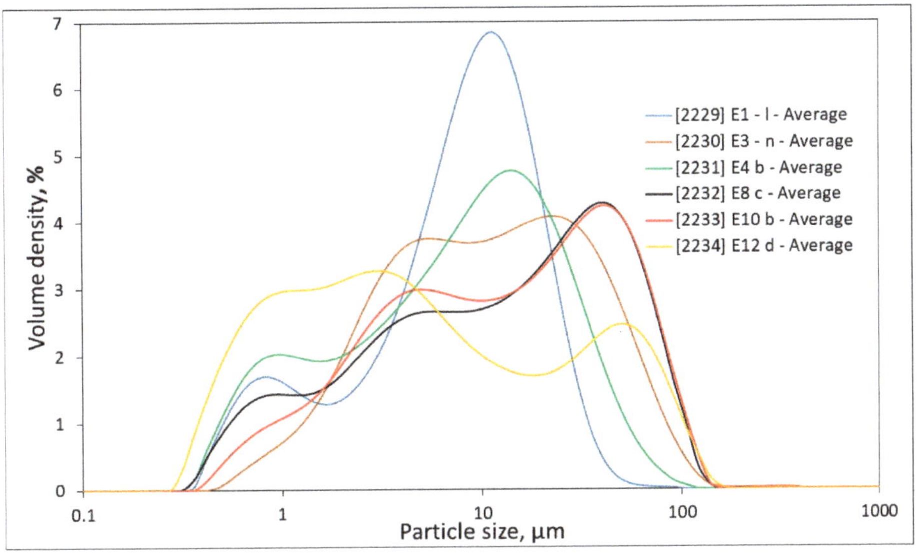

Figure 1. Particle size distributions of the micro-fines (notations as in Table 1).

2.4. The Methylene Blue Test

Since the work here was intended to quantify the effects of clays in terms of their content and type, calibration runs of the methylene blue (MB) test were carried out to determine the Methylene Blue Value (MBV) as defined by ASTM C 1777. Calibration curves of "engineered sand" samples with varying content of pure clay minerals are presented in Figure 2.

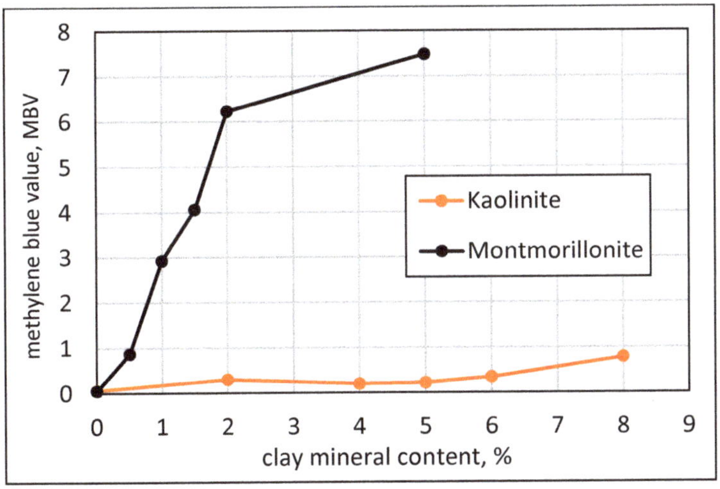

Figure 2. Relations between methylene blue value and the content of clay minerals in "engineered sand" samples with "pure" clay minerals, Kaolinite and Montmorillonite, using the ASTM C 1777 procedure with 20 g sample.

The curve is very shallow for the Kaolinite, demonstrating that the adsorption capacity of the methylene blue is extremely low. For the Montmorillonite, there is a steep linear relationship up to about 2%, reaching a relatively high value of MBV of 6, and thereafter the increase is mild. The high slope up to 2% of the Montmorillonite is indicative of the high adsorption capacity of this clay compound. The mild increase afterwards might be interpreted as an art effect, which is due to the fact that the 0.5% concentration of the methylene blue solution dictated by the standard is already adsorbed with 2% Montmorillonite; an increase in the content of this mineral beyond the 2% value cannot be accompanied by a marked change in the adsorbed solution.

It can be shown by a simple calculation, taking into consideration the 0.5% concentration of the solution, its content of 30 g, and the 20 g quantity of sand mixed with it, that the maximum methylene blue value that can be reached under these conditions is 7.5. This is the value approached in Figure 2 with less than 5% Montmorillonite. The implication is that if higher concentrations of Montmorillonite are expected in the sand, the test procedure should be modified to enable higher levels of adsorption by reducing the quantity of the tested sand. Indeed, the standard recommends using a smaller amount of the tested material (10 g instead of 20 g) when the methylene blue value is approaching 7.5 to account for this effect. The data in Figure 2 suggest that this change should take place even at a lower methylene blue value of about 6. In the following sections, the methylene blue values obtained by the two procedures, the standard with 20 g and the one recommended for high values of methylene blue using 10 g samples, will be reported. They will be addressed as the 20 g and 10 g procedures.

3. Results

3.1. Methylene Blue Values

The results of the characterization of the micro-fine material of the quarry lenses using the MBV values of "engineered sand" containing 5% of the micro-fines and 95% of high-quality siliceous sand are shown in Figure 3. It also includes engineered sands with 5% of the pure clay minerals, Kaolinite and Montmorillonite. The data in Figure 3 are for materials obtained from different quarries in Israel, and it is sorted by the increased level of the MBV.

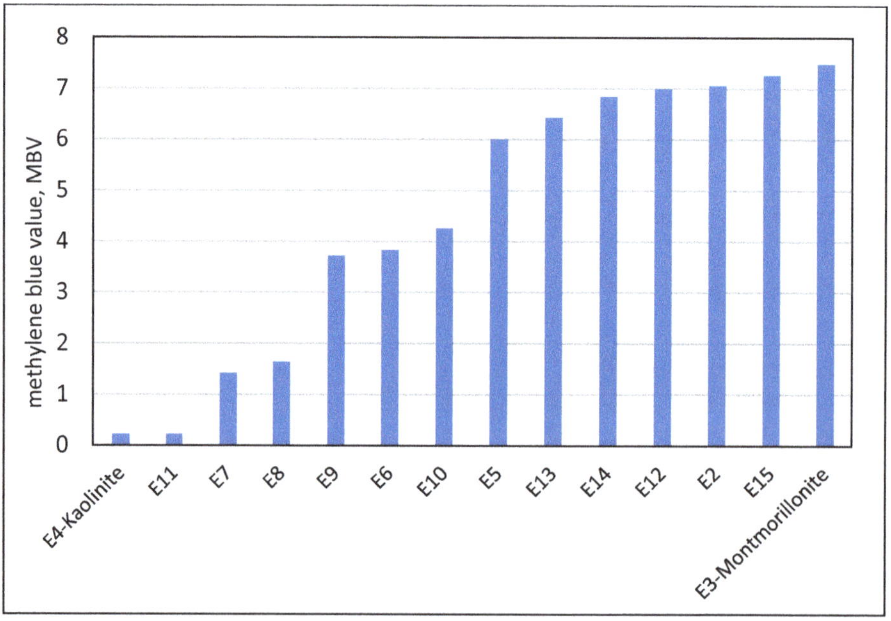

Figure 3. The methylene blue values (20 g procedure) of engineered sands containing 5% of micro-fine (<75 micron) clay type material, either pure minerals or material obtained from clay lenses in quarries.

It can be seen that the lower bound is engineered sand with Kaolinite and the upper bound is with Montmorillonite, almost 7.5. In between, there is a range of values from about 0.2 to slightly more than 7 of the sands engineered from the quarry lenses.

3.2. Mineralogical Composition

The Rietveld quantitative mineralogical analysis based on the XRD patterns confirmed that the pure Kaolinite and Montmorillonite were indeed pure, having contents of about 95% of the relevant mineral. The mineralogical analysis of the clay micro-fine material obtained from the quarry lenses showed a variable composition consisting of Kaolinite, Montmorillonite, Calcite, Dolomite, and Quartz, with the clay minerals being present in significant quantities, between 20 and 60% (Table 1).

The plot of the clay mineral content in the engineered sands as a function of the MBV value, Figure 4, highlights some of the mineralogical variables and their relations with the MBV value.

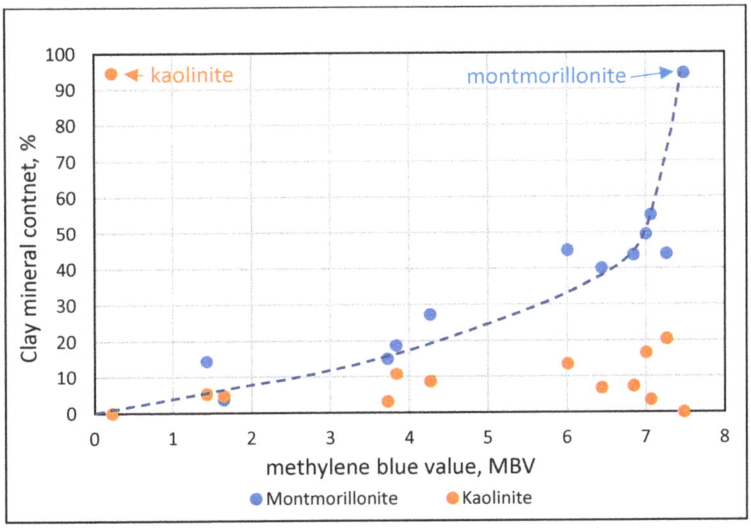

Figure 4. Content of the clay minerals Kaolinite and Montmorillonite as a function of the methylene blue values (20 g test procedure) of the engineered sands containing 5% of micro-fines, either pure clay minerals or <75 micron material obtained from clay lenses in quarries.

The pure clay engineered sands have similar contents of the relevant minerals, Kaolinite and Montmorillonite, 5% of the total sand, but their methylene blue value (MBV) is drastically different, very low for Kaolinite, 0.22, and very high for Montmorillonite, 7.48. All the other engineered sand samples fell in between. Their MBV could be correlated with an increase in their Montmorillonite content but not with changes in the Kaolinite content. The very steep increase in the Montmorillonite–MBV relation, starting at an MBV of about 6, is a reflection of the reduced sensitivity of the methylene blue test at this range when it is carried out according to the ASTM procedure with 20 g of material. This is demonstrated when plotting the MBV values of the 20 g procedure as a function of the Montmorillonite content, Figure 5, showing a plateau for the "20 g procedure" curve when exceeding the contents of about 2% Montmorillonite. The increase in the Montmorillonite content above 2% should have been accompanied by an increase in the uptake of methylene blue, but this does not take place as almost all of the methylene blue in the 0.5% solution is apparently adsorbed at about 2% of Montmorillonite. When plotting the relation between the Montmorillonite content and the MBV obtained by the 10 g procedure, a linear relation is obtained with R^2 = 0.96, demonstrating the significant correlation between the MBV value and the mineralogical composition, as shown in Figure 5 ("10 g procedure" curve). A comparison of the two curves in Figure 5 suggests that when the MBV value in the standard test procedure with 20 g exceeds about 6, the procedure with 10 g is the one that should be employed, as it better reflects the changes in the mineralogical composition.

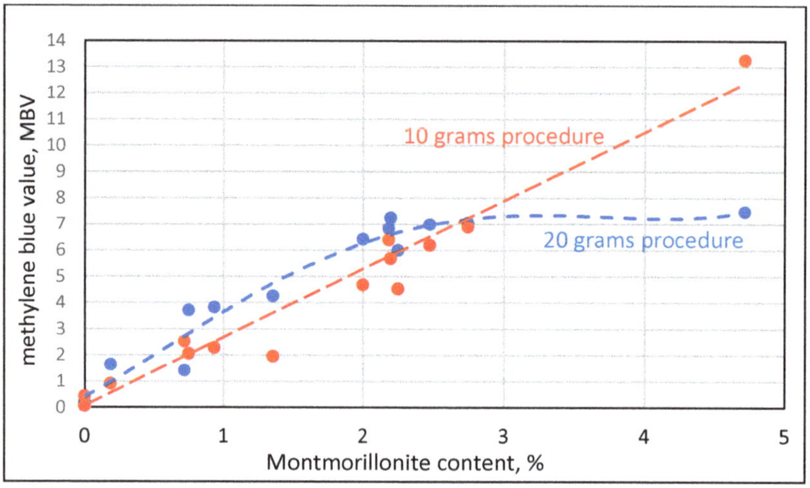

Figure 5. Relation between the methylene blue value obtained by the 20 g and 10 g procedures and the Montmorillonite content in engineered sands containing 5% of micro-fines, either pure clay minerals or material obtained from clay lenses in quarries.

3.3. *Rheological Behavior Characterization*

The rheological behavior was characterized in terms of the yield value, which was found to correlate quite well with the spread obtained in the ASTM flow table test for mortars (Figure 6).

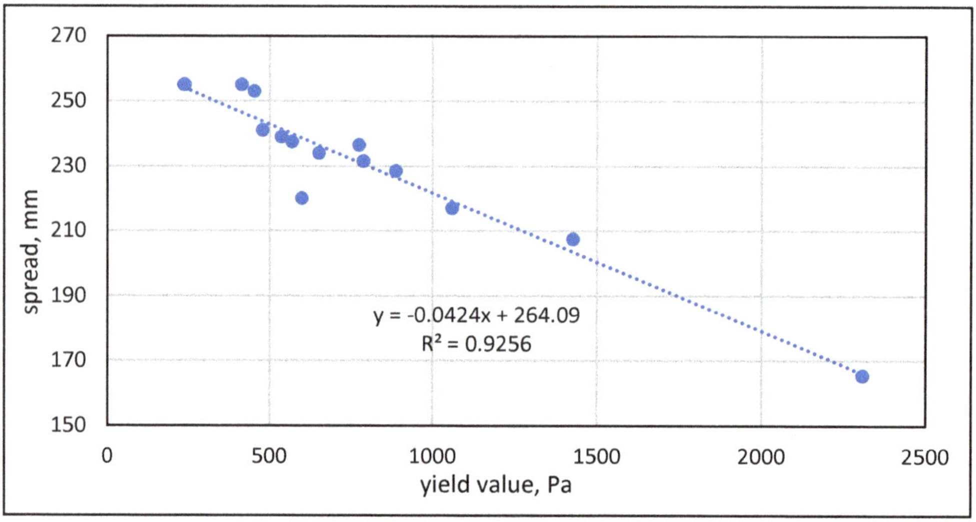

Figure 6. Relations between the yield value obtained in the rheological test and the spread obtained in the ASTM flow table test of mortars with 0.92 w/c ratio.

The effect of the nature of the micro-fines in the engineered sands was quantified for mortars with a 0.92 w/c ratio. This high w/c ratio was chosen for comparison since, at this w/c level, a wide range of micro-fines could be investigated, ranging from ones with a very high yield value (small spread) to ones with a very low yield value (high spread) without the need to use flow enhancing admixtures. The specimens investigated exhibited differences of more than one order of magnitude in the yield values, and these were presented as a function of the MBV of the engineered sands, Figure 7, for the MBV obtained by the 20 g and 10 g procedures.

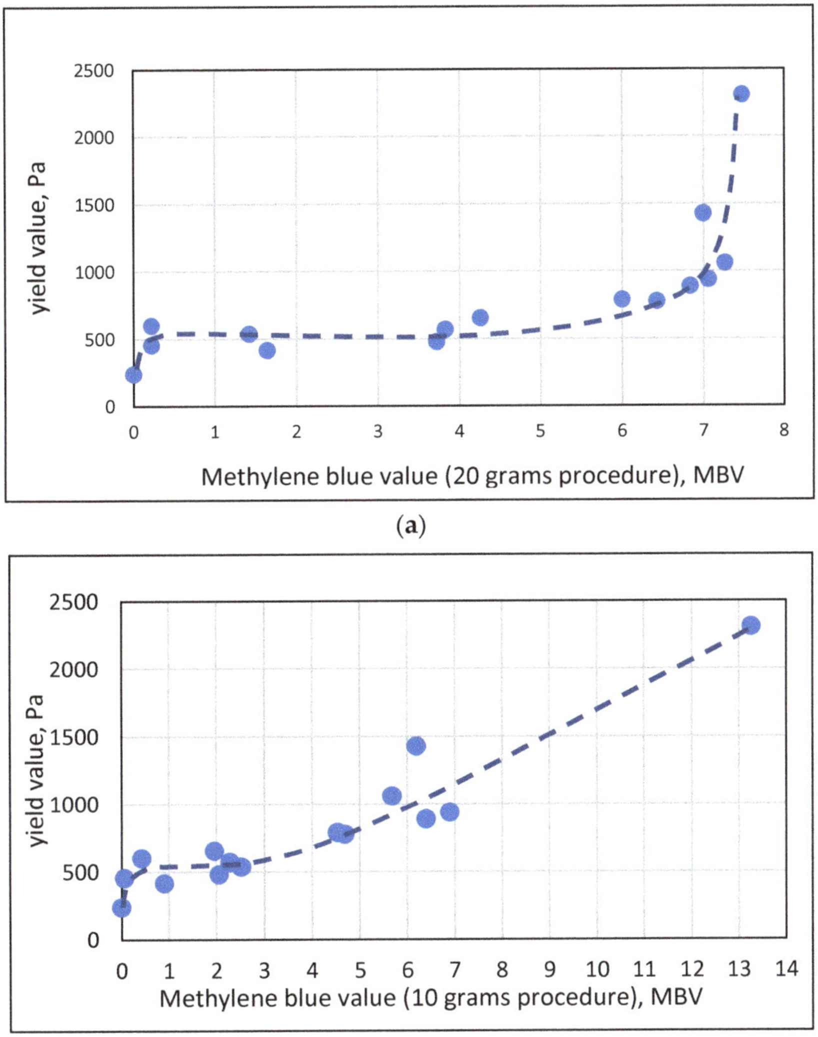

Figure 7. Relation between the yield value and the methylene blue value (**a**) 20 g procedure and (**b**) 10 g procedure for mortars of 0.92 w/c ratio with engineered sands containing 5% of micro-fines, either pure clay minerals or material obtained from clay lenses in quarries.

At the lower range of the MBV, there is a doubling of the yield value once 5% of the sand is replaced with Kaolinite, as the MBV value increases from 0 (for the straight sand) to 0.22 for the Kaolinite. At the high range, the yield value increases by more than an order of magnitude when 5% of the sand is replaced with Montmorillonite. Between these two extremes, there seems to be a plateau in the yield value up to about MBV of 3. Thereafter, there is an increase, monotonic for the 10 g procedure (Figure 7b) and a very mild one for the 20 g procedure, in the range of MBV of 3 to 6. Afterwards, a very steep increase takes place towards an MBV of 7.5. Obviously, the trend in the 10 g procedure is a better reflection of the nature of the effect of the MBV value.

One may try to interpret the changes in the slope of the yield value vs. MBV curves in terms of different mechanisms. If one plots the yield value as a function of the Montmorillonite content, Figure 8, a continuous increasing curve is observed, following an exponential relationship. The nature of this relationship is such that in the range of up to about 1% Montmorillonite, the change in the yield value is relatively small, and beyond that, it increases markedly with the increase in the Montmorillonite content. This trend is consistent with the observations in Figure 7 that, up to an MBV value of about 3, the change in the yield value is relatively small. It may be interpreted in terms of two mechanisms: one dominant up to 1% Montmorillonite, which is perhaps less affected by the adsorption capacity of the clay, and the other above 1% Montmorillonite, where the adsorption influence of the Montmorillonite is becoming much more dominant.

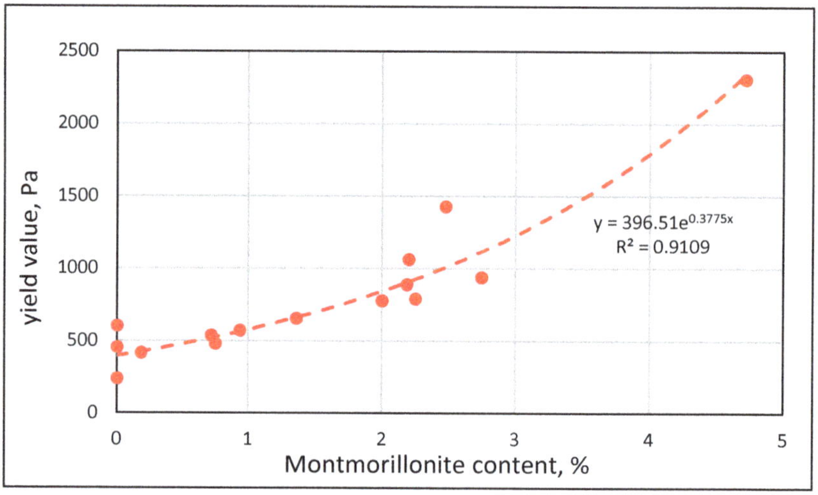

Figure 8. Relation between the yield value and the Montmorillonite content in mortars of 0.92 w/c ratio with engineered sands containing 5% of micro-fines, either pure clay minerals or material obtained from clay lenses in quarries.

3.4. Characterization of Micro-Fines by Plasticity Tests

The results of the plasticity tests, liquid limit, and plasticity index, which are common for soils, are presented in Figures 9 and 10 as a function of the MBV values. The trends resemble those observed in Section 3.3, with regards to the effects taking place at different ranges of MBV. Up to values of about 3 to 4, there is a very mild increase in the plasticity parameters; thereafter, a more marked exponential increase is being observed, which becomes very steep at methylene blue values in the range of 6 to 7.

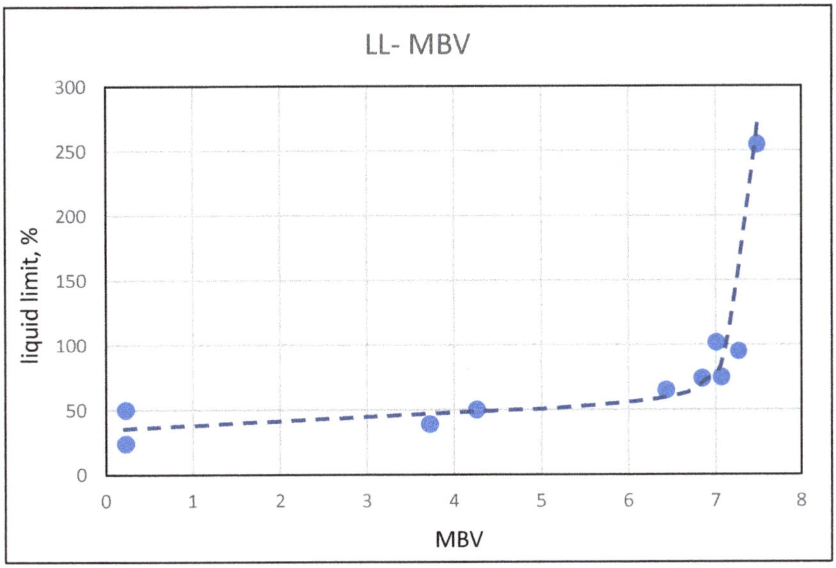

(a)

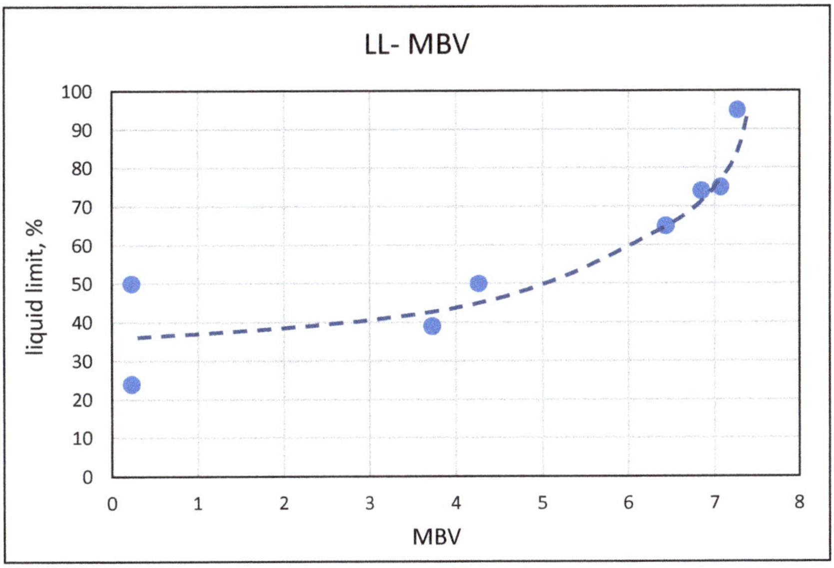

(b)

Figure 9. Relationship between the liquid limit and the methylene blue value (**a**) in the range up to 300 liquid limits and (**b**) in the range up to 100 liquid limit.

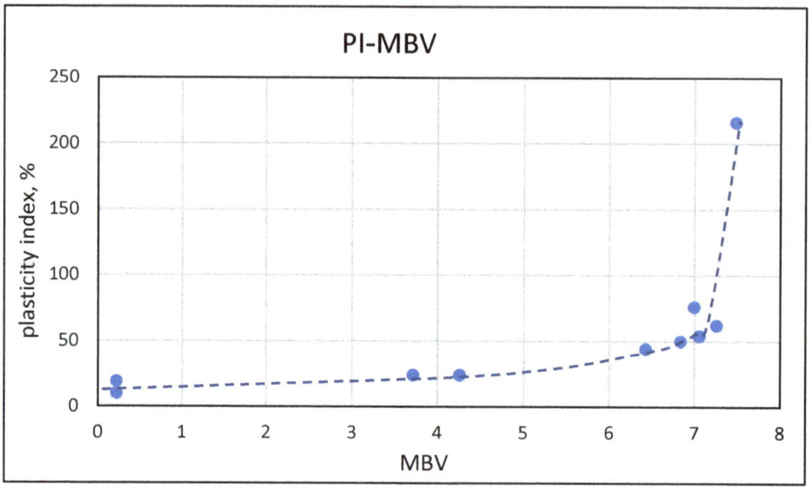

(a)

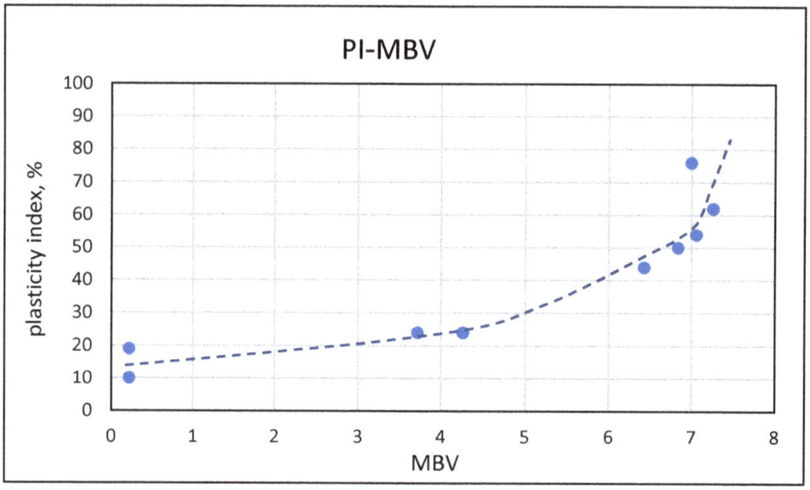

(b)

Figure 10. Relationship between the plasticity index and the methylene blue value (**a**) in the range up to 250 plasticity index and (**b**) in the range up to 100 plasticity index.

4. Discussion

The mortar part of the study was intended to provide a simulation for the concrete. Its purpose was to explore the influence of the clays in the macro-fines on the behavior of the fresh system in order to define several categories of micro-fines quality. The concept is that for each category, a different strategy for admixture control should be applied to mitigate detrimental influences. Three such categories were established. The effectiveness of this classification was verified for concretes, in which micro-fines of the different categories were incorporated at the expense of the sand. The verification dealt with influences on the fresh and hardened properties of the concrete.

4.1. Effects of Admixtures

The obvious means to neutralize the detrimental influence of the clay containing sand on the rheological behavior is the use of admixtures. A series of tests was carried out using different types of admixtures and their combinations, evaluating their influence in w/c ratio mortars of 0.92 and 0.70 w/c. The lower w/c ratio mortar is a better simulation of the mortar phase in normal strength concrete. In particular, the influence of PC admixtures and CMA, which adsorb into the clay surfaces, was studied. Typical results are presented in Figure 11, comparing the influence of the admixtures in mortars with the reference sand and in mortars with engineered sand having an MBV value of 4.26 (sand containing 1.36% Montmorillonite and 0.43% Kaolinite). The comparison in Figure 11 is for 0.8% by weight of PC, with and without 1.2% of CMA.

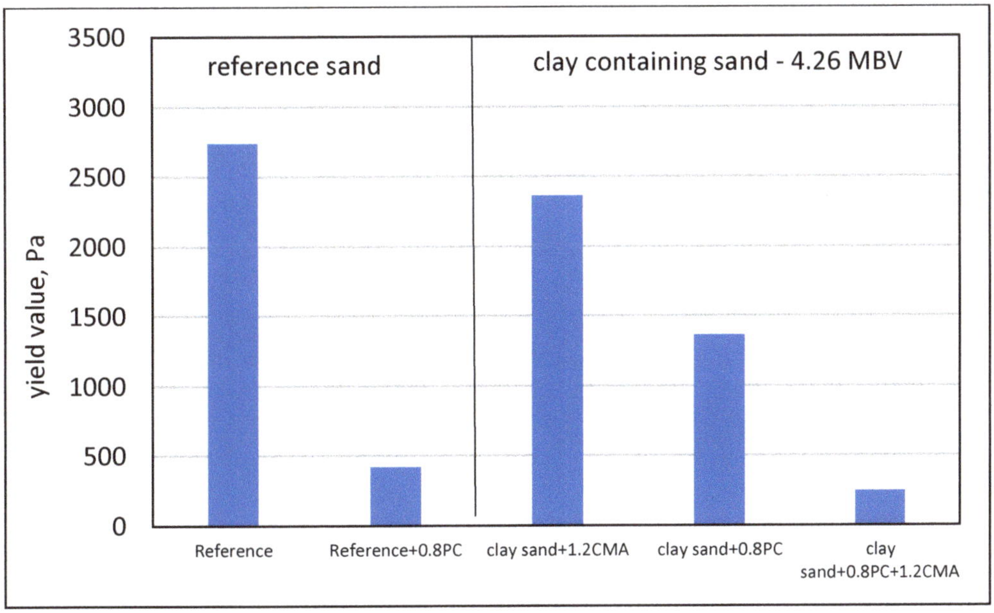

Figure 11. The yield values of 0.70 w/c ratio mortars with reference sand (MBV of 0) and with engineered sands (MBV of 4.26), as influenced by the presence of 0.8% ploycarboxylate admixture (PC), 1.2% clay mitigating admixture (CMA), and combination of these two.

It can be seen that the 0.8% PC is very efficient in reducing the yield value in the reference sand, yet in the clay containing sand, the yield value still remains much higher with the same content of PC. The effect of the CMA seems to be rather small when it is on its own, as the yield value remains rather high in the clay containing sand mortar. However, the combined influence of the PC and CMA in the clay containing sand seems to be very significant, suggesting perhaps a synergistic effect in which the CMA is expected to neutralize the adsorbing effect of the clay, thus allowing the full effectiveness of the PC by presumably preventing its adsorption by the clay that contains Montmorillonite.

4.2. Mortar Rheological Behavior

As outlined in Section 2.1, the approach to be taken is a comprehensive one, considering at the same time the quality of the micro-fines, as estimated by the MBV, and their content in mortars. This should be the basis for developing the admixture strategy to mitigate detrimental influences, rather than looking at the sand's overall behavior. This strategy could enable us to mobilize the positive influences of the micro-fines in high end applications to generate filler effects.

The underlying hypothesis of this study [28] was that the effect of the micro-fines should be evaluated based on a combined effect of their nature (i.e., content and type of clay) and their total content in the concrete rather than in a specific aggregate (usually sand) as currently addressed in standards. To verify this approach, all the data were presented in terms of the relation between the yield value of mortars prepared from engineered sands with varying contents of micro-fines of different natures, up to 30% content of the sand, and their content in the engineered sand, Figure 12. The data in Figure 12 can be characterized by three categories of families of curves: Type I with an MBV smaller than 3, Type II with an MBV between 4 and 6, and Type III with an MBV bigger than 7. It should be emphasized that the MBV of the micro-fines is determined on the basis of the MBV of an engineered sand consisting of 5% of the micro-fines with 95% of high quality siliceous fine sand.

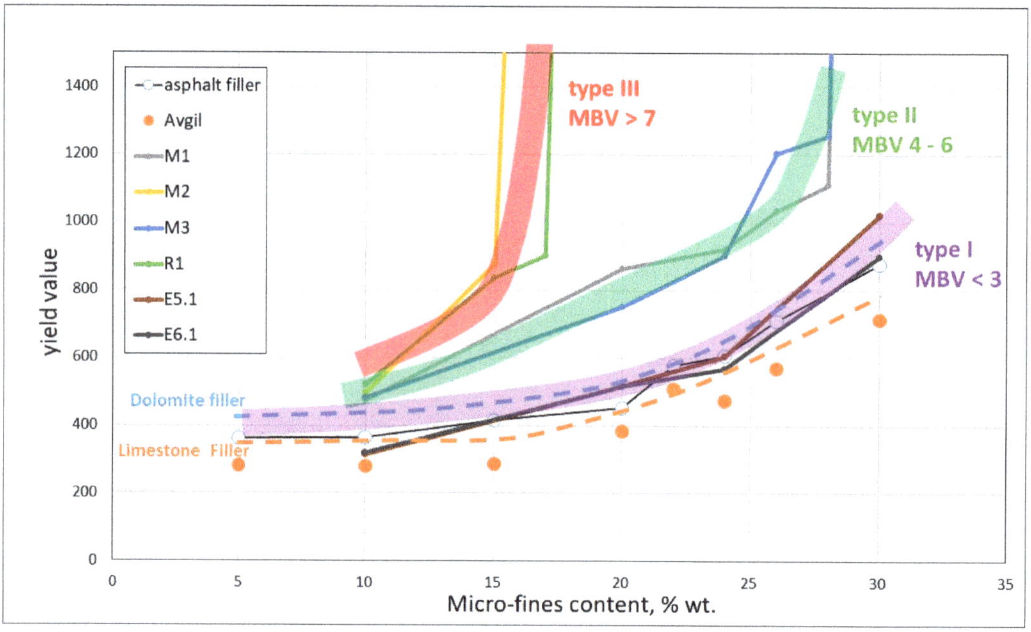

Figure 12. Relations between yield value and micro-fines content in engineered sand mortars for different levels of MBV.

The difference between these types of families of curves is in the initial slope of the curves and the micro-fines content at which the slope tends to shift to much higher values, taking place at about 30% for Type I, 25% for Type II, and 15% for Type III.

The differences in nature between the three types of curves also show up in the quantification of their influences on the water demand of mortars. This is demonstrated in Figure 13 for 0.70 w/c ratio 1:3 cement:sand mortars, with engineered sand containing 20% of micro-fines of different natures (quantified by the MBV value). The differences between the categories are quantified by their influence on the increase in water demand: (i) up to 8% for type I; (ii) a gradual increase with MBV from about 8% to 13% for Type II; and (iii) a marked increase with MBV for Type III.

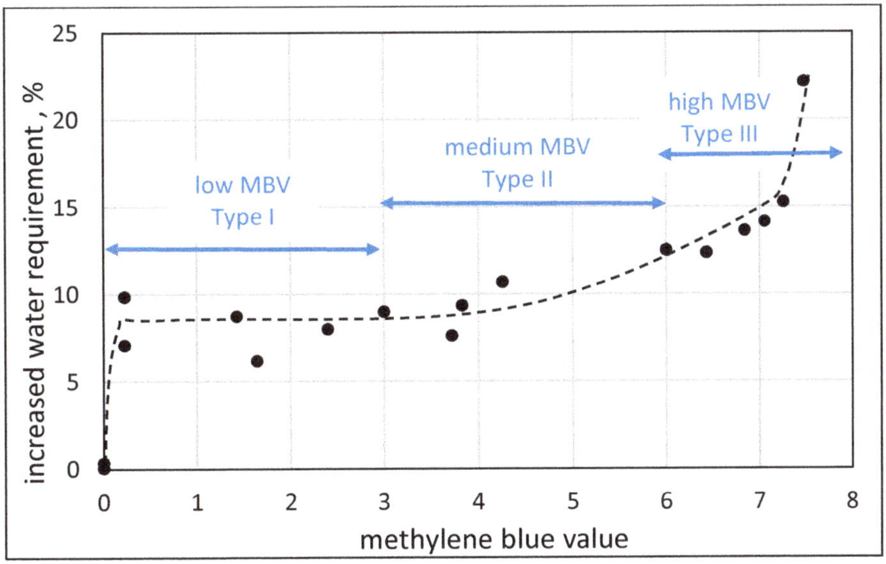

Figure 13. The effect of the methylene blue value of the engineered sand and the increase in water demand in 0.7 w/c ratio mortars made from this sand.

4.3. Concrete Workability Control

The different nature of the three types of families in Figure 12, with respect to their influence on the increase in water demand, suggests that various admixture strategies need to be adjusted to mitigate these influences. These include water reducing admixtures for Types I and II, either lignosulfonates or PC, whereas for Type III, there is also a need to combine CMA.

To study the implication of these three types in concrete, three samples of micro-fines, characterized by differences in their MBV values, were used to be in the range outlined in Figure 13: Type I (MBV in the range smaller than 3), Type II (MBV in the range of 4 to 6), and Type III (MBV bigger than 7). They were derived from by-products in quarries by a procedure of separating the micro-fines through a 75 microns mesh. The three were different in their composition, in particular in the equivalent content of Montmorillonite, as calculated from MBV tests. The mineralogical composition, in addition to the clays, consisted mainly of Dolomite and Calcite, as shown in Table 3, which also includes a Dolomite micro-fine (filler). All these micro-fines served for the preparation of concrete with various quantities of micro-fines and a range of admixtures. The admixture content and type were adjusted to obtain similar workability at a given w/c ratio of 0.75. This concrete composition yields a grade of 30 MPa using CEM I 52.5 cement. Workability was characterized by slumps and yield values.

Table 3. Characterization of the four types of micro-fines.

Micro-Fine Type	MBV 20% Dilution	MBV 30% Dilution	Montmorillonite Equivalent Content, % wt.	Dolomite, % wt.	Calcite, % wt.
"Clean" (R)	0	0	0	98.6	1.4
Low MBV(E37)	3.0	4.6	5.5	88.8	5.8
Medium MBV (E27)	4.4	5.6	8.0	82.5	9.5
High (E)	7.2	10.5	13.7 (11.0 by XRD)	79.5	6.8

The concrete base mix composition is outlined in Table 4. The effect of the micro-fines was evaluated by studying mixes in which 75 to 300 kg/m^3 of micro-fines of different qualities (MBV values) which were incorporated at the expense of the sand. In all the mixes, the contents and combination of admixtures (L—Lignosulfonate, PC—Polycarboxylate, and CMA—Clay Mitigating Admixture) were adjusted to keep the same workability (about 200 mm slump).

Table 4. Base concrete mix composition.

Content, kg/m^3				
Cement	**Coarse Aggregate**	**Graded Sand**	**Sand**	**Water**
330	645	745	337	250

The test results were presented in terms of a diagram showing the effect of the micro-fines content and quality (quantified by their MBV) on the content and type of admixture required to maintain the workability (slump and yield value) of the reference concrete having no micro-fines. The total content and the MBV of the micro-fine aggregate serve as guidance for the admixture strategy required, as shown in Figure 14.

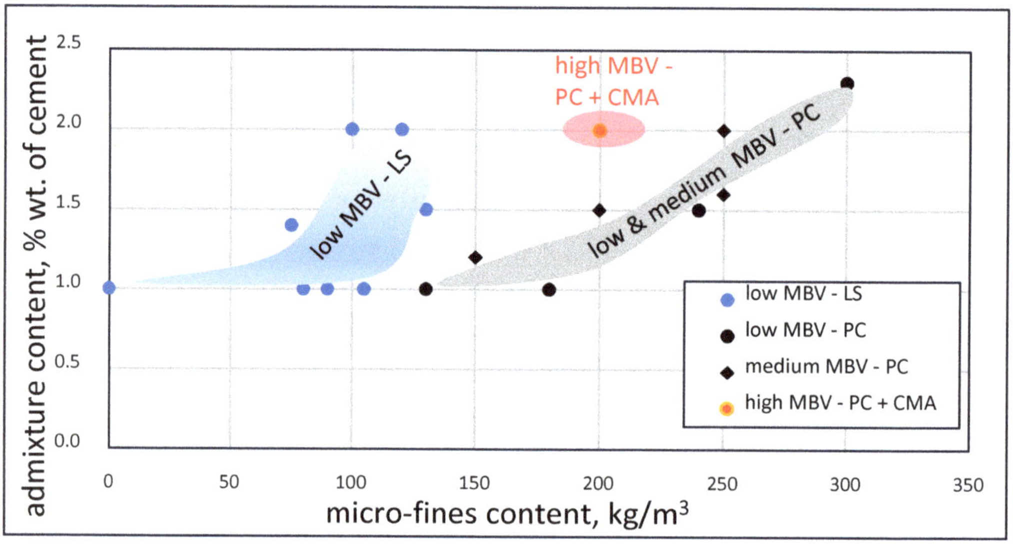

Figure 14. The types of admixtures and admixture combinations required to maintain workability in concretes containing different types (MBV values) and contents of micro-fines.

The diagram in Figure 14 can serve as a demonstration for the strategy suggested in this paper to deal with clays in concrete aggregates by considering the micro-fine content and nature. It calls for considering the effect of the micro-fines in the concrete at large rather than their presence in the various aggregate fractions. It can provide guidelines for the mix design with respect to the type and content of admixtures that need to be incorporated. It shows that for low and medium MBV values of the micro-fines, Lignosulfantes can be sufficient to mitigate deleterious effects on workability, up to about 150 kg/m^3 of micro-fines. Above that value, the content of the Lignosulfonates required would approach 1.5–2.0%, which is the range where they cause retardation, and therefore there is a need to resort to PCs. The PCs could be effective even at levels of 300 kg/m^3 of micro-fines, as long as the micro-fines are of Type I and II (low and medium MBV). If, however, the micro-fines are of type III, there is a need for a combination of PC and CMA if the micro-fines content is 200 kg/m^3.

The results of this study also indicate that the quantity of CMA required is related to the equivalent Montmorillonite content in the clay. It is estimated that the CMA content should be approximately 1/3 of the Montmorillonite content in the micro-fines in the concrete. The equivalent content of the Montmorillonite can be estimated based on the MBV using the calibration curves, such as those presented in Figure 5.

4.4. Concrete Mature Properties: Strength and Shrinkage

The strength tests in this study indicate that if the effect of clays on workability is mitigated by the proper admixture strategy, filler effects could be mobilized, as shown by the increase in strength accompanying the presence of micro-fines, Figure 15. The figure shows that for 200 kg/m^3 of micro-fines in concrete, a 28 days strength increase could be observed, up to about 30%, with higher values obtained with Type II and III micro-fines. This enhanced value for the Type II and III micro-fines suggests that the clays possess added value in addition to the Carbonate and Dolomite particles in the micro-fines. This trend was even more evident at one day.

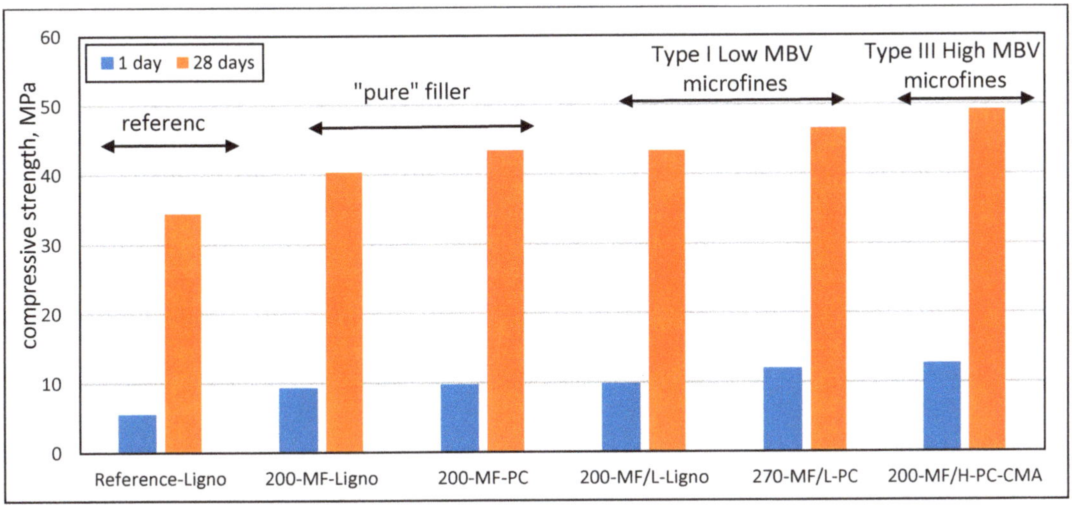

Figure 15. Effect of content and type of micro-fines, in combination with admixtures to maintain constant w/c ratio and workability, on concrete strength at 1 and 28 days; the numerical values indicate the micro-fines contents in kg/m^3 units.

Shrinkage tests demonstrated that at the optimum admixture contents, the shrinkage was practically independent of the micro-fines content, Figure 16. This is a further support for the hypothesis that when the micro-fines are properly dispersed and the detrimental effects of swelling clays in them are neutralized, they act as fillers rather than detrimental "dust" as was assumed in the past.

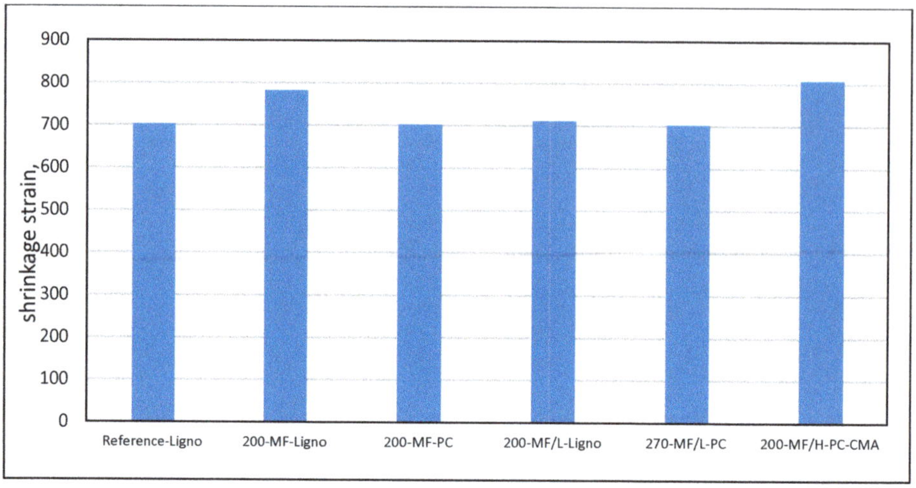

Figure 16. Effect of content and type of micro-fines in combination with admixtures to maintain constant w/c ratio and workability on concrete after 90 days of drying shrinkage; the numerical values indicate the micro-fines and filler contents in kg/m^3 units.

5. Summary and Conclusions

Clays are components in the fine portion of aggregates, less than 75 microns in size, which is usually washed away when producing coarse or fine (manufactured sand) aggregates. When marginal sources of aggregates are being used, this washed portion can be quite high, and there is an incentive to keep as much of it in the aggregate, as is reflected in current standards, in which the allowed content of micro-fines has increased in recent years. In order to avoid deleterious effects of clays that might remain in these micro-fines, the standards have adopted limitations on the clay content that are based on the MBV value. There is thus significance for the appraisal of the nature of quantification obtained by the MBV test and the scope of influences that should be considered in attempting to use aggregates with higher levels of clay. Some of the issues and conclusions highlighted in this study are outlined below:

- The deleterious influence of clays on their rheological properties is largely due to the presence of swelling clays like Montmorillonite, whereas Kaolinite has only a small effect. This difference in behavior can be attributed to the swelling and water uptake characteristic of Montmorillonite. The methylene blue test is indicative of this influence.
- The current study indicates that when the clay content, and in particular the Montmorillonite component in it, is above about 2%, the MBV value obtained by the standard 20 g procedure, which is about 6, is not sufficiently sensitive to quantify the presence and influence of the clay. There is thus a need to resort to the 10 g procedure.
- There is potential for using high clay content aggregates based on a strategy of utilizing admixtures. Yet, for that to be effective, there is a need to develop formulations of blends of admixtures that should be adjusted for the composition and content of the clay.
- These concepts were mobilized for providing guidance on the use of high contents of micro-fines containing clays in concrete, as summarized below:

1. Micro-fines can be categorized into three types, depending on their MBV; low (less than 3), medium (4–6), and high (more than 7), showing differences in increased water demand in each category.
2. The strategy for neutralizing the effect of micro-fines on increased water demand in concrete can be quantified in terms of the combination of their content in concrete and

their MBV; this is a more rational approach to considering their influence than their content in specific aggregates as specified in standards.

3. The effect of low and medium MBV aggregates on the water requirement can be neutralized by lignosulfonates when their content in the concrete is below a threshold value of about 150 kg/m^3; PC is required at higher contents; for high MBV aggregates, a combination of PC and CMA is required.
4. At the optimum admixture content and type, high micro-fines content in aggregates can provide a beneficial effect on strength, suggesting that the filler effect is being mobilized without having a detrimental effect on shrinkage. This beneficial effect can be mobilized to reduce cement contents in concrete.

Author Contributions: Conceptualization, P.L.; Writing—original draft, A.B. All authors have read and agreed to the published version of the manuscript.

Funding: This research was funded by the Israeli Ministry of Construction and Housing, grant no. 4500938257 and the Israeli Ministry of Energy and Infrastructure, grant no. 21717026.

Institutional Review Board Statement: Not applicable.

Data Availability Statement: The original contributions presented in the study are included in the article, further inquiries can be directed to the corresponding author.

Acknowledgments: This paper is an overview of some of the issues identified in a comprehensive study at the Technion on the influence of clays in marginal aggregates on the performance of cementitious systems. The study was supported by the Israeli Ministry of Construction and Housing, The Ministry of Energy and Infrastructure and the cooperation of the Israeli industry, Nesher Cement Company, Hanson, ReadyMix and Even Vesid. The authors appreciate and acknowledge the support and cooperation of these organizations.

Conflicts of Interest: The authors declare no conflict of interests.

References

1. EN 12620. Available online: https://www.google.com.hk/url?sa=t&source=web&rct=j&opi=89978449&url=https://dl.azmanco.com/standards/EN/EN%252012620.pdf&ved=2ahUKEwiipJ_kiIWIAxX_oa8BHdNjFsYQFnoECCUQAQ&usg=AOvVaw2IjeRwxWWFyMUmrNTHZvYT (accessed on 10 July 2023).
2. Katz, A.; Baum, H. Effect of high levels of fines content on concrete properties. *ACI Mater. J.* **2006**, *103*, 474.
3. Quiroga, P.N.; Ahn, N.; Fowler, D.W. Concrete mixtures with high microfines. *ACI Mater. J.* **2006**, *103*, 258–264.
4. Nehdi, M.I. Clay in cement based materials: Critical review of state of the art. *Constr. Build. Mater.* **2014**, *51*, 372–382. [CrossRef]
5. Nikolaides, A.; Manthos, E.; Sarafidou, M. Sand equivalent and methylene blue value of aggregates for highway engineering. *Found. Civ. Environ. Eng.* **2007**, *10*, 111–121.
6. Yool, A.J.G.; Lees, T.P.; Fried, A. Improvements to the methylene blue dye test for harmful clay in aggregates for concrete and mortar. *Cem. Concr. Res.* **1998**, *28*, 1417–1428. [CrossRef]
7. Yukselen, Y.; Kaya, A. Suitability of the methylene blue tests for surface area, cation exchange capacity and swell potential of clayey soils. *Eng. Geol.* **2008**, *102*, 38–45. [CrossRef]
8. Fernandes, V.A.; Purnell, P.; Still, G.T.; Thomas, T.H. The effect of clay content in sands used for cementitious materials in developing countries. *Cem. Concr. Res.* **2007**, *37*, 751–758. [CrossRef]
9. Tregger, N.A.; Pakula, M.E.; Shah, S.P. Influence of clays on the rheology of cement pastes. *Cem. Concr. Res.* **2010**, *40*, 384–391. [CrossRef]
10. Kaci, A.; Chaouche, M.; Andreani, P.A. Influence of bentonite clay on the rheological behavior of fresh mortars. *Cem. Concr. Res.* **2011**, *41*, 373–379. [CrossRef]
11. Chang, T.P.; Shih, J.Y.; Yang, K.M.; Hsiao, T.C. Material properties of Portland cement paste with nano-montmorillonite. *J. Mater. Sci.* **2007**, *42*, 478–487. [CrossRef]
12. Kuo, W.Y.; Huang, J.S.; Lin, C.H. Effects of organo-modified montmorillonite on strength and permeability of cement mortars. *Cem. Concr. Res.* **2006**, *36*, 886–895. [CrossRef]
13. Norvell, J.K.; Stewart, J.G.; Juenger, M.C.G.; Fowler, D.W. Influence of clay-sized particles on concrete performance. *ASCE J. Mater. Mater. Civ. Eng.* **2007**, *19*, 1053–1059. [CrossRef]
14. Lei, L.; Plank, J. A study on the impact of different clay minerals on the dispersing force of conventional and modified vinyl ether based polycarboxylate superplasticizers. *Cem. Concr. Res.* **2014**, *60*, 1–10. [CrossRef]
15. Quintero, L. An overview of surfactant applications in drilling fluids for the petroleum industry. *J. Dispers. Sci. Technol.* **2002**, *23*, 393–404. [CrossRef]

16. Jeknavorian, A.A.; Jardine, L.; Ou, C.C.; Koyata, H.; Folliard, K. Interaction of superplasticizers with clay-bearing aggregates. In Proceedings of the 7th CANMET/ACI International Conference on Superplasticizers and Other Chemical Admixtures in Concrete, Berlin, Germany, 20–23 October 2003; Special Publication (SP-217). Malhotra, V.M., Ed.; American Concrete Institute: Farmington Hills, MI, USA, 2003.
17. Bentur, A.; Larianovsky, P. *Use of High Clay Content Aggregates, Parts I and II*; National Building Research Institute, Technion-Israel Institute of Technology: Haifa, Israel, 2020.
18. Bentur, A.; Larianovsky, P. *Use of Marginal Aggregates for Concrete Production, Parts I and II*; National Building Research Institute, Yechnion-Israel Institute of Technology: Haifa, Israel, 2020.
19. Choundhary, H.; Siddiqui, M.A.; Saha, S. Effect of clay fines on strength and workability of concrete. *J. Civ. Eng. Environ. Technol.* **2019**, *6*, 311–315.
20. Beixing, L.; Mingkai, Z.; Jiliang, W. Effect of the methylene blue value of manufactured sand on performance of concrete. *J. Adv. Concr. Technol.* **2011**, *9*, 127–132. [CrossRef]
21. Pitre, B.P. Application of the Modified Methylene Blue Test to Detect Clay Minerals in Coarse Aggregate Fines. Master's Thesis, Texas A&M University, College Station, TX, USA, 2012.
22. Zhan'ao, L.; Mingkai, Z.; Beixing, L. Relationships between modified methylene blue value of microfines in manufactures sand and concrete properties. *J. Wuhan Univ. Technol.-Mater. Sci. Ed.* **2016**, *31*, 574–581.
23. Sahin, H.G.; Biricik, O.; Mardani-aghabaglou, A. Polycarboxylate-based water reducing admixture–clay compatibility; literature review. *J. Polym. Res.* **2022**, *29*, 33. [CrossRef]
24. ASTM C 1777. Available online: https://www.astm.org/c1777-20.html (accessed on 10 July 2023).
25. ASTM D 4318. Available online: https://www.astm.org/d4318-17e01.html (accessed on 10 July 2023).
26. ASTM C 1437. Available online: https://www.astm.org/c1437-20.html (accessed on 10 July 2023).
27. ASTM C 230. Available online: https://www.astm.org/c0230_c0230m-20.html (accessed on 10 July 2023).
28. Bentur, A.; Larianovsky, P. Turning Marginal Aggregates into Useful Concrete Components. In *Concrete Durability and Service Life Planning. ConcreteLife 2020*; Kovler, K., Zhutovsky, S., Spatari, S., Jensen, O., Eds.; RILEM Bookseries; Springer: Cham, Switzerland, 2020; Volume 26, pp. 3–7.

materials

MDPI

Article

Binary and Ternary Blended Portland Cements Containing Different Types of Rice Husk Ash

Luis Miguel Ordoñez [1], María Victoria Borrachero [2], José Monzó [2], Lourdes Soriano [2] and Jordi Payá [2,*]

1 KhemeChemical S.L., Puerto de Sagunto, 46520 Valencia, Spain; lmordonez@khemechemical.com
2 Institute of Science and Technology of Concrete (ICITECH), Universitat Politècnica de València, 46022 Valencia, Spain; vborrachero@cst.upv.es (M.V.B.); jmmonzo@cst.upv.es (J.M.); lousomar@upvnet.upv.es (L.S.)
* Correspondence: jjpaya@cst.upv.es

Abstract: Rice husk ash (RHA) is agricultural waste with high silica content that has exhibited proven technical feasibility as a pozzolanic material since the 1970s. Notwithstanding, its use in mortars and concrete is limited by the standards currently utilized in some countries where RHA production is high and the aforementioned pozzolanic material is not standardized. This is the case in Spain, one of the main rice producers in Europe. Nowadays, the high pressure placed on the Portland cement production sector to reduce its energy use and CO_2 emissions has given rise to a keen interest in mineral admixtures for cement manufacturing. In this research, we intended to establish the contributions of different RHA types to the final blended Portland cement properties ("H" is used to identify RHA in standardized cements). The experimental results demonstrated that RHA with good pozzolanic properties (large specific surface and high amorphous silica content) had to be limited to 10% cement replacement because of the severe reduction in workability at higher replacement percentages. RHA with lower reactivity, such as crystalline RHA, or fly ash (FA) can be used to prepare binary and ternary blended cements with reactive RHA. It is possible to design the following cements: CEM II/A-H and CEM II/A-(H-V). It would also be possible to design cement (CEM II/B-(H-V) with replacement values of up to 30% and the same 28-day mechanical performance as observed for the Portland cement without mineral addition.

Keywords: rice husk ash; pozzolanic activity; workability; compressive strength; ternary cement

1. Introduction

The construction industry is a large consumer of non-renewable raw materials. In addition, cement manufacturing produces 5–7% of the global industrially emitted CO_2 [1]. The cement production in 2022 amounted to 4160 Mt, and this production is expected to reach 4260 Mt in 2030. The clinker-to-cement ratio (ton per ton) in 2022 was 0.71, and it is likely to lower to a value of 0.57 by 2050 [2]. De Brito and Kurda [3] point out several solutions to reduce the total binder amount: (i) the use of supplementary cementing (pozzolanic or hydraulic) materials; (ii) the use of fillers; (iii) the impact on the water/cement ratio and the use of chemical additives; and (iv) an indirect reduction in the binder amount.

Regarding the use of supplementary cementing materials—specifically pozzolanic materials—one of the most exciting groups is that composed of biomass-burning ashes [4]. Agricultural waste is a sustainable energy resource, and some ash types that are derived from biomass burning have pozzolanic characteristics [5,6]. Some, like rice husk ash (RHA), have been studied for decades [7–9].

Rice husk is an agro-waste material that is produced worldwide in large amounts. According to the literature, the global forecast for 2021/2022 was 714.2 million tons, which is equivalent to around 142.84 million tons (ca. 20% of paddy rice) of rice husk [10].

Most of this waste is produced in undeveloped Asian countries (India, China, Indonesia, Bangladesh and Thailand). In Europe, Spain is one of the most important rice producers, and 381.3 thousand tons of paddy rice grain were produced in 2022 [11]. This means that around 80,000 tons of rice husk is available in Spain for valorization. Rice husk contains [12,13] organic substances (cellulose, lignin, fiber) and 20% of inorganic material, mainly silicon dioxide (SiO_2 > 85%). There are two means of removing organic matter from rice husk: (a) combustion with energy recovery and (b) simple incineration. Endale et al. [10] report that, for 0.2 tons of rice husk, around 0.05 tons (ca. 20% of the yield) of RHA is produced. This means that around 15,000–20,000 tons of ash would be produced in Spain. Thus, the potential RHA production worldwide would be 25–30 million tons.

RHA's pozzolanic activity depends on several parameters, mainly those related to the reached temperature and the residence time at high temperatures. The most important properties of RHA that determine its pozzolanic activity are the amorphous phase content and specific surface area. When the combustion temperature exceeds 800 °C and the residence time at temperatures higher than 800 °C is long, partially/totally crystallized RHA is obtained, and the specific surface area diminishes. Consequently, its pozzolanic activity reduces. RHA is obtained by controlled burning methods with the monitoring of the temperature, where the resulting ash is quickly removed from the furnace. The controlled conditions make it possible to obtain RHA with excellent pozzolanic behavior. Under uncontrolled burning conditions, the temperature could reach more than 1000 °C due to the intense exothermic oxidation of the rice husk's organic matter, and the crystallization process is easily achieved, resulting in the formation of cristobalite and tridymite [14–16]. Regarding the fresh properties, concrete or mortar mixtures containing RHA usually obtain lower slump and fresh and dry density values [17,18]. In the hardened state, mixtures containing RHA have higher compressive strength values and improved durability [19–21].

The design of ternary blended cements is a focus of interest because this combination enhances the sustainability [22], workability [23] and mechanical/durability properties [24].

Depending on the RHA's properties, it is possible to prepare different blended cements based on a Portland cement clinker. Furthermore, distinct strength category types of cement could be designed. The goal of this research is to assess the workability and compressive strength of mortars prepared with cements containing different types of RHA and other mineral admixtures. Based on these characterizations, a proposal for the denomination of potential standardized blended cements is given. Different RHA blending percentages are studied to prepare the optimum blended cements. The synergistic effects on the blended cements based on highly reactive RHA and other mineral admixtures, such as low-reactivity RHA, fly ash (FA) or ground granulated blast furnace slag, are evaluated. Mortars prepared with the RHA-blended cements, with good performance in terms of workability and strength, are proposed.

2. Materials and Methods

The ordinary Portland cement (OPC) used in the experiment was the UNE-EN 197-1 CEM I-52.5-R type [25]. It was supplied by Cemex (Buñol, Spain), while the crystalline (RHA-1 and RHA-2) and semicrystalline (RHA-3) RHA samples were supplied by Agrocítrica (Alzira, Spain) and NK Enterprises (Jharsuguda, Odisha, India), respectively. The amorphous RHA sample (RHA-4) was obtained under controlled combustion conditions using a small incinerator prototype (Figure 1). FA (F type, Andorra Thermoelectric Power Plant, Andorra-Teruel, Spain), non-densified silica fume (NDSF; Ferroatlántica S.A., A Coruña, Spain), ground granulated blast furnace slags (GGBFS-1 and GGBFS-2; Dortmund, Germany) and silica flour QF (quartz S100 type, Sibelco, Arcos de la Frontera, Spain) were also used. The fine aggregate was standard siliceous sand according to the CEN standards (>99% quartz) [26].

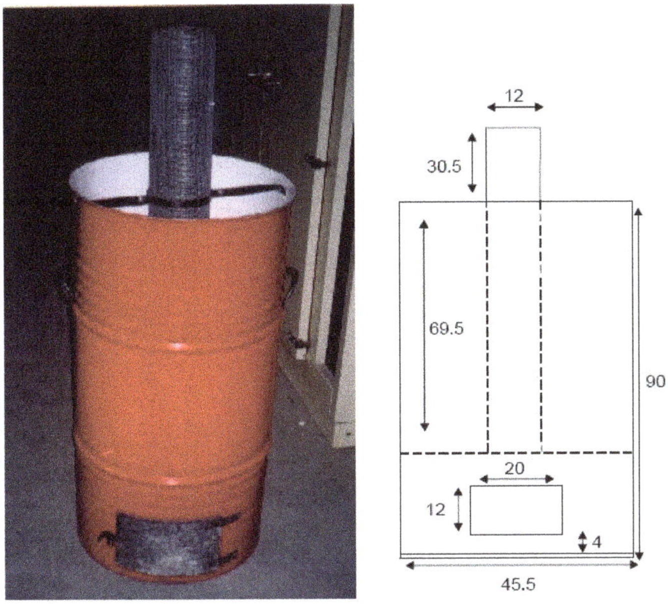

Figure 1. Small incinerator prototype (on the right; all data are expressed as mm).

The RHA samples were ground using a laboratory ball-mill (Gabbrielli, Calenzano, Italy) with 18-mm-diameter alumina balls. The blended cements were prepared by previously mixing ground additions (if necessary) and OPC, following the recommendations of Cook [7].

The total silica content in the mineral admixtures was determined by X-ray fluorescence using the Philips MagiXPRO equipment (Philips Analytical, Almelo, The Netherlands). Lithium tetraborate was employed to prepare the samples and the solid mixture was melted in the Vulcan 4M—Fusion Machine Type VAA 4 equipment (Fluxana GmbH & Co., Bedburg-Hau, Germany). The amorphous silica in the mineral admixtures was determined by the method described by Payá et al. [27]. The BET specific surface area (nitrogen adsorption) was measured by a MicromeriticsTriStar 3000 model (Micromeritics, Norcross, GA, USA).

The compressive strength was determined in prismatic test specimens (40 × 40 × 160 mm) according to [26]. These specimens were cast from a batch of plastic mortar containing cement/water/sand in proportions of 1.0:0.5:3.0 by mass. The mortar was prepared by mechanical mixing. Before molding, the cement's workability was determined according to UNE-EN 413-2:2017 [28]. The specimens in the mold were stored in a moist atmosphere for 24 h at 20 °C. Then, the demolded specimens were stored under lime-saturated water at 20 °C until strength testing. At the required age, the specimens were taken from their wet storage and broken flexurally into two halves, and each half was tested for its strength in compression.

3. Results and Discussion

3.1. Physico-Chemical Properties of RHA Samples

Table 1 summarizes the main physico-chemical properties, which defined the reactivity of the RHA samples and the other materials used therein. Four RHA samples (the chemical compositions of these samples are given in Appendix A—Table A1; the XRD diffractograms are depicted in Appendix B—Figure A1) were selected for the experimental studies: two (RHA-1 and RHA-2) presented high crystallinity because the amorphous silica percentage in relation to the total silica was less than 25%. This characteristic was related to the ash-yielding method: an open-field incineration process without energy recovery

(Figure 2). Under this condition, the temperature reached by the ash was above 900 °C and remained high for several hours. A high combustion temperature and residence time activate amorphous silica crystallization to produce cristobalite and tridymite [7,29]. An RHA sample produced at a controlled combustion temperature (RHA-3) was supplied by NK Enterprises (India). This sample presented 71% amorphous silica in relation to the total silica. Finally, a fourth RHA sample was obtained by controlled combustion in a prototype incinerator (Figure 1). This sample presented 100% amorphous silica in relation to the total silica because the combustion temperatures did not reach 700 °C. RHA's pozzolanic efficiency depends on the amorphous SiO_2 content. Uncontrolled incineration processes provide high temperatures and long residence times for incineration (open-field combustion). The ashes obtained under these conditions are crystalline (formation of tridymite and cristobalite [29]). These samples obtained low loss on ignition (LOI) values because the carbon content was removed as CO_2. RHA-1 and RHA-2 had LOI values below 5% (0.14% and 4.52%, respectively), which demonstrated the effect of the high temperature and long residence time. The LOI values for RHA-3 and RHA-4 were significantly high (8.24% and 17.65%, respectively), which agreed with the low temperature reached during the combustion process. In this case, part of the husk's organic matter was converted into carbon particles [30].

Table 1. Main physico-chemical properties of CEM I 52.5 R (OPC), the RHA samples and the other mineral admixtures tested in this research.

Material	Amorphous Silica (%) #	Specific Surface Area (m²/g) &	Loss on Ignition LOI (%)	Particle Size Distribution (%, by Weight)		
				<3 μm	3–32 μm	>32 μm
CEM I 52.5 R	Nd *	1.6	0.98	10.9	83.8	5.3
RHA-1	11.7	1.6	0.14	29.8	62.4	7.8
RHA-2	12.0	9.3	4.52	23.2	62.6	14.2
RHA-3	71.0	15.2	8.24	13.3	71.5	15.2
RHA-4	100.0	36.5	17.65	13.0	72.8	14.2
FA	Nd *	0.8	0.66	34.0	53.7	12.3
NDSF	100	24.4	2.94	63.0	34.3	2.7
GGBFS-1	Nd *	1.0	<0.1	10.8	75.9	13.3
GGBFS-2	Nd *	1.2	<0.1	14.8	71.1	14.1
QF	Nd *	2.6	<0.1	4.9	83.9	11.2

Amorphous silica obtained by the method described in [27]; amorphous silica in relation to the total silica. * Nd: not determined; &: obtained by BET.

RHA's reactivity also depends on the specific surface area (measured by the BET method), and this parameter is also related to the combustion conditions. Hence, the samples obtained at high temperatures and long residence times had low BET values: 1.6 m^2/g for RHA-1 and 9.3 m^2/g for RHA-2. RHA-1 was collected from the central zone of the ash pile, whereas RHA-2 was collected from the upper zone, which was the reason for the difference in the BET values. The RHA samples obtained under controlled conditions obtained significantly higher BET values: 15.2 m^2/g for RHA-3 and 36.5 m^2/g for RHA-4.

RHA-1 and RHA-4 were compared using SEM. The micrographs (Figure 3a,b) showed differences in the internal structures of the non-ground ash particles. The particle in Figure 3a presents a round-shaped internal structure, which is indicative of silica coalescence [31] due to high temperatures, unlike Figure 3b, which shows an unaltered internal silica skeleton of husk without coalescence. Additional micrographs are presented in Appendix C (Figures A2–A4).

(**a**) (**b**)

Figure 2. Open-field combustion in a very well-ventilated shed for samples RHA-1 and RHA-2: (**a**) at the beginning of burning; (**b**) after burning had ended and part of the ash had been removed. Arrows indicate the zones where RHA-1 and RHA-2 were sampled.

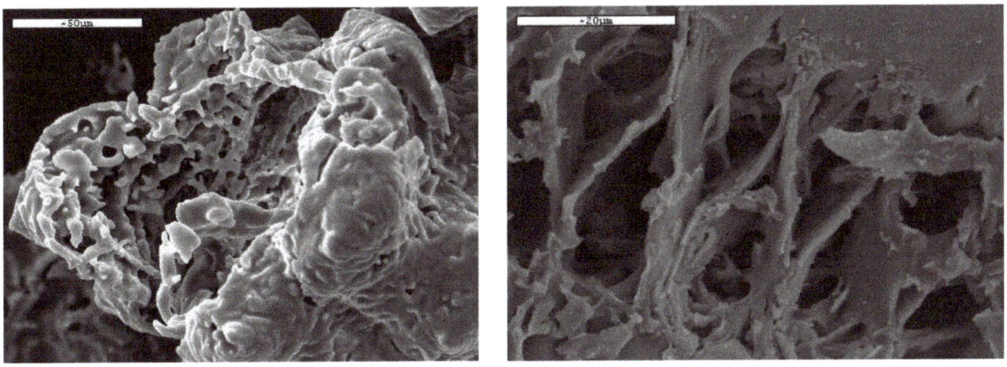

(**a**) (**b**)

Figure 3. SEM micrographs for (**a**) RHA-1, scale bar 50 μm; (**b**) RHA-4, scale bar 20 μm.

As shown in Figure 3, the RHA particles had a high internal porosity, which is not ideal for blending with PC. This is because it leads to the loss of workability of fresh mixes. For this reason, the RHA samples were ground to achieve a similar particle distribution to OPC. Table 1 summarizes the granulometric parameters: the percentage within the 3–32 μm range fell within the 62–73% interval for RHAs and was slightly lower than that for the PC type CEM I 52.5 R (83.8%).

Table 1 also summarizes some physico-chemical parameters for FA, NDSF, GGBFS-1 and GGBFS-2 and QF. There are similarities in the amorphous silica content and the specific surface between RHA-4 and NDSF and also between RHA-1 and QF.

3.2. Influence of RHA on Compressive Strength Development

At this stage, samples RHA-1, RHA-3 and RHA-4 were selected to prepare blended cements by mixing CEM I 52.5R with the RHAs in the RHA/CEM I proportions of 5/95, 10/90, 15/85 and 20/80. The standardized mortars [26] were prepared and tested under compression after 2, 7 and 28 curing days.

Figure 4 shows the compressive strength values obtained for the cured mortars according to the level of replacement of cement with RHA. The mortars containing RHA-4 displayed the best behavior compared to the other RHA samples. Thus, for the 5% and 10% replacement levels, the compressive strengths obtained for all curing ages (2, 7 and

28 days) equaled and/or exceeded the values for the CEM I 52.5R mortar (control), which demonstrated the high pozzolanic reactivity of RHA-4. For the highest replacement levels (15% and 20%), the contribution of RHA-4 for curing ages of 7 and 28 days was very important and they equaled or surpassed the control mortar's strength. At 2 curing days, the compressive strengths for the 15% and 20% replacements were lower than for the control. This was attributed to both the reduction in clinker content and the limited pozzolanic reaction development at an early age due to the low portlandite production from the hydration of calcium silicates in the clinker.

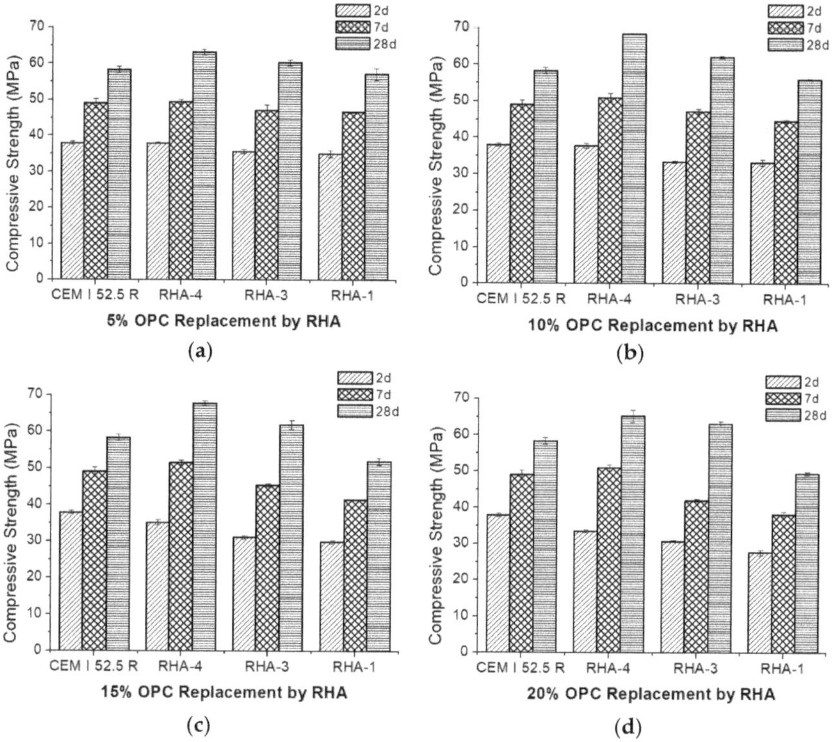

Figure 4. Compressive strength development for the control mortar (CEM I 52.5R) and the RHA mortars with different replacement levels: (**a**) 5%; (**b**) 10%; (**c**) 15%; (**d**) 20%.

The contribution of the strength development of the mortars with RHA-3 and RHA-1 was significantly less, and the control specimen's compressive strength was achieved only for the mortars containing RHA-3 for the 28-day curing period (for all replacement percentages).

The different pozzolanic reactivity of the RHA samples (RHA-4 > RHA-3 > RHA-1) was clearly evidenced by the compressive strength values obtained after 28 curing days and for all replacement percentages. An optimum compressive strength value at 28 curing days when 10% OPC was replaced with RHA-4 was observed. The pozzolanic reaction rate depends on the available portlandite released during OPC hydration. When 20% OPC was replaced with RHA-4, the quantity of released portlandite decreased by approximately 20%. This pozzolan reacts with portlandite at early ages due to its high reactivity. Both effects caused the pozzolanic reaction efficiency to decrease more quickly because the CEM I 52-5R replacement level was higher [15,32].

Ganesan et al. [32] studied the reactivity of an RHA sample with a similar specific surface area (36.47 m^2/g) to RHA-4 and low crystallinity. These authors prepared mortars with the replacement of ordinary Portland cement in the range of 0–35%, and they found that the optimum replacement was 15% for the 28-day compressive strength. Similarly, RHA-4 mortars with 10 and 15% replacement yielded the highest values at 28 days. Distinctively, the behavior found for early ages was different: Ganesan et al. [32] found that, at 1 and 3 days of curing, the optimum was also 15% replacement. However, the compressive strength of RHA-4 mortars at 2 days of curing decreased slightly with the increase in the replacement percentage. This difference at an early curing age is attributed to the RHA fineness: the ash used by Ganesan et al. had a mean particle size of 3.80 μm (six times finer than Portland cement), whereas RHA-4 presented a similar particle distribution to CEM I 52.5 R (Table 1). When the RHA particle size is lower, the pozzolanic reactivity occurs earlier, and the contribution to strength development is more important at early curing ages.

The differences in reactivity among the studied RHAs are related to the potential pozzolanic reactivity derived from the physico-chemical characterization (see Table 1). The granulometric parameters are similar for RHA-1, RHA-3 and RHA-4, and the compressive strength values of the mortars depicted in Figure 4 are attributed to the relative amorphous silica content (100% for RHA-4; 71% for RHA-3; 11.7% for RHA-1) and the specific surface area (36.5 m^2/g for RHA-4; 15.2 m^2/g for RHA-3; 1.6 m^2/g for RHA-1). Thus, higher amorphous silica content and a larger specific surface area yielded better compressive strength in the RHA mortars.

3.3. *Cement Types according to RHA Reactivity and Standards*

The classification of conventional and standardized blended cements is established by the European Standards Association [25]. It considers the following information/descriptions.

- Cement design: CEM type/proportion (types of addition) + strength category (taking into account long and short curing ages).
- Type: I (additions below 5%), II (OPC with additions up to 35%), III (OPC with GGBFS), IV (pozzolanic cements) and V (composed cements).
- Proportion of addition: the classification depends on the type of cement (II, III, IV or V). For CEM II, the CEM II/A type has 6–20% clinker replacement and the CEM II/B type has 21–35% clinker replacement.
- Types of addition: limestone filler (L), FA (V), GGBFS (S), silica fume (D), among others.
- Mechanical strength category: 32.5, 42.5 or 52.5 compressive strength values at 28 days in standardized mortar. Type N and type R, respectively, indicate low- and high-rate strength development after 2 or 7 curing days in standardized mortar (Table 2).

Table 2. Compressive strength categories of standardized cements according to UNE-EN 197-1 [25].

<table>
<tr><th rowspan="2">Category</th><th colspan="4">Mechanical Compressive Strength MPa (N/mm^2)</th></tr>
<tr><th>2 Days</th><th>7 Days</th><th colspan="2">28 Days</th></tr>
<tr><td>32.5 N</td><td>--</td><td>≥16.0</td><td rowspan="2">≥32.5</td><td rowspan="2">≤52.5</td></tr>
<tr><td>32.5 R</td><td rowspan="2">≥10.0</td><td>--</td></tr>
<tr><td>42.5 N</td><td>--</td><td rowspan="2">≥42.5</td><td rowspan="2">≤62.5</td></tr>
<tr><td>42.5 R</td><td rowspan="2">≥20.0</td><td>--</td></tr>
<tr><td>52.5 N</td><td>--</td><td rowspan="2">≥52.5</td><td rowspan="2">--</td></tr>
<tr><td>52.5 R</td><td>≥30.0</td><td>--</td></tr>
</table>

For instance, CEM I-52.5-R was composed of 95–100% clinker and provided more than 30 and 52.5 MPa compressive strength at 2 and 28 days, respectively, in the standard mortars. Another example could be CEM II/B-(L-V) 32.5 N, composed of 21–35% FA plus limestone filler, which provided more than 16 and 32.5 MPa compressive strength at 7 and 28 days, respectively.

However, RHA use is not permitted as mineral addition during cement formulation. In the future, it is likely that RHA will be one of the candidates to be considered for mineral addition in cement because the availability of slag, and especially of FA, is being questioned for sustainability and climate change reasons [33]. If we use the symbol "H" to denote the cements with RHA, the following cements could be prepared after taking into account the RHA samples analyzed in the previous section (Table 3). For the RHA-blended cements, clinker replacement above 20% could not be considered because the resulting cement would perform poorly in terms of workability. This effect has also been observed by Kamau et al. [34]. Thus, only the CEM II/A-type cements can be considered regarding the assignment of nomenclature to the new blended cements with RHA and, consequently, the CEM II/B and CEM IV or V types cannot be contemplated.

Table 3. Designation of the standard blended cements with different RHA reactivity.

RHA	Replacement Percentage (%, by Weight)			
	5%	10%	15%	20%
RHA-4	CEM II/A-H 52.5 R	CEM II/A-H 52.5 R	CEM II/A-H 52.5 R (*)	CEM II/A-H 52.5 R (*)
RHA-3	CEM II/A-H 52.5 R	CEM II/A-H 52.5 R	CEM II/A-H 52.5 N	CEM II/A-H 52.5 N (*)
RHA-1	CEM II/A-H 52.5 R	CEM II/A-H 52.5 N	CEM II/A-H 42.5 R	CEM II/A-H 42.5 R

(*) Mortars with considerably reduced workability (>15%) in relation to the control mortar (see Section 3.5 for more details).

In general, when high-reactivity RHA is used (e.g., RHA-4), CEM II/A-H 52.5 R can be prepared. However, for RHA-3, the strength category lowers to 52.5N when high replacement levels (15–20%) are selected. Finally, for the RHA with the lowest reactivity (RHA-1), CEM II/A-H 52.5 N was prepared with less than 10% RHA, and the strength category was lowered to 42.5 R for the highest replacement levels (15% and 20%).

3.4. Comparison of Contribution of RHA and Other Mineral Admixtures to Compressive Strength Development

Reactivity studies of different mineral admixtures in terms of the mortar's compressive strength performance were carried out to perform a comparison in terms of the RHA contribution. The cements were CEM I 52.5 R (control mortar) and the blended cements in which 10% of CEM I was replaced with the corresponding mineral admixture. Four RHA samples (RHA-1, RHA-2, RHA-3, RHA-4) were tested. Additional cements were prepared using GGBFS-1 (4000 cm^2/g Blaine), GGBFS-2 (5000 m^2/g Blaine), FA, NDSF and QF. Figure 5 depicts the compressive strength values of the blended cements (10% replacement) for 2, 7 and 28 curing days. The compressive strength gain [35] values are also shown. The strength gain (S_G, %) was calculated as follows:

$$S_G\ (\%) = \frac{S_i - [(P * S_c)/100]}{[(P * S_c)/100]} * 100 \tag{1}$$

where S_c: compressive strength of the control mortar (100% CEM I 52.5 R); S_i: compressive strength of the blended cement; P: percentage of replaced CEM I in the mortars containing the mineral admixture.

The 28-day compressive strength values obtained in the blended cements with RHA-1, RHA-2 and FA were similar at all curing ages. However, the compressive strength obtained using quartz flour (QF) was lower. This behavior suggests the minor contribution of the pozzolanic and/or physical effects to the strength gain (5–8% after 28 curing days) when using either 10% RHA-1 or 10%RHA-2.

The compressive strength and mechanical gain values at the earliest curing age (2 days) are noteworthy as they were higher when RHA-4 was used in the blended cements than when employing NDSF. The main reason for this was the larger specific surface area when utilizing RHA-4. Xu et al. [36] reported a higher strength value for a mortar with RHA at 3 days (44.23 MPa) than for a silica fume one (39.43 MPa). The opposite effect was observed at 28 curing days, when the compressive gain was 10% higher when NDSF was used. In this case, the higher LOI value for RHA-4 (17.65%) and the larger number of particles smaller than 3 microns in NDSF (aggregate–paste interface refinement) were the reasons for this behavior at longer curing times. Xu et al. [36] also reported higher strength for a silica fume mortar (75.59 MPa) than for an RHA one (71.02 MPa) at 28 curing days.

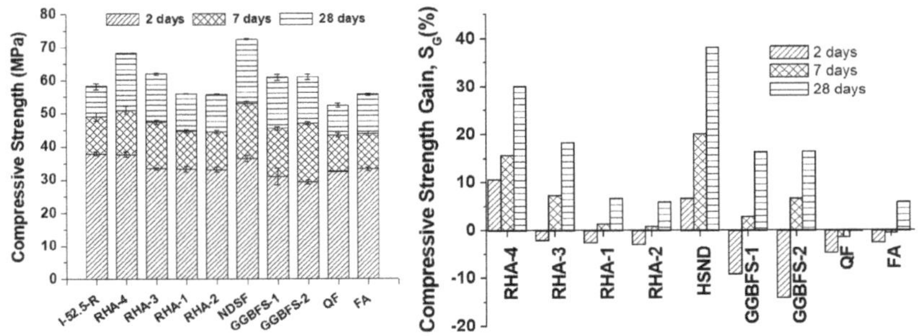

(**a**) (**b**)

Figure 5. Strength development of the blended cements (10% replacement level): (**a**) compressive strength; (**b**) compressive strength gain (S_G).

Applying GGBFS delayed the early clinker hydration stages because a negative compressive strength gain of 10% and 15% for GGBFS-1 and GGBFS-2, respectively, was noted (Figure 5b). However, between 2 and 7 curing days, the compressive strength gain was higher due to the hydraulic reactivity of these admixtures. Finally, at 28 curing days, the compressive strength for the control mortar was achieved using 10% GGBFS replacement, which yielded 15% S_G. The observed mechanical behavior when using 10% GGBFS-2 was similar to that observed for RHA-3 from 7 to 28 curing days. The strength development of the cement containing QF showed that its contribution was negligible: the S_G value for 28 curing days was close to zero.

Thus, in terms of compressive strength development, for the period of 28 curing days, the RHA-4 sample was similar to NDSF, RHA-3 was similar to GGBFS, and RHA-1 and RHA-2 were similar to FA.

3.5. Workability and Mechanical Properties of Standardized Blended Cement Containing RHA

In practical terms, additional information may be supplied for the blending of cements, especially aspects related to the water demand of the new blended cements. It is well known that some mineral additions play a water-reducing role when mixed with OPC, e.g., FAs [37,38]. However, when mineral admixtures with a large specific surface area (e.g., silica fume) are blended, the water demand is likely to increase [39,40], to such an extent that it is not feasible to prepare mixtures with appropriate workability at high replacement percentages. In this context, workability studies have been carried out using a flow table [28] for some RHA-blended cements.

Replacing the cement with different RHA mineral admixtures may modify the mortar's workability due to the distinct specific surface area, particle morphology and texture, water absorption, reactivity and fineness. Figure 6a presents the workability (%W) percentage related to the control OPC mortar, and %W is calculated as follows:

$$\%W = \frac{W_i - W_c}{W_c} * 100 \quad (2)$$

where W_i is the workability in mm for the mortar prepared with the blended cement; W_c is the workability for the control mortar (in this case, using CEM I 52.5 R).

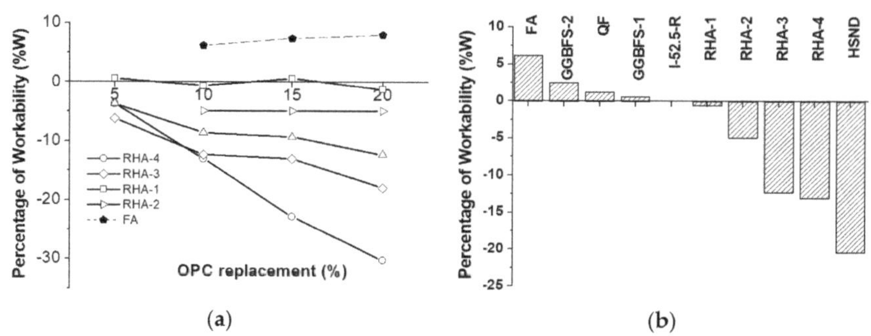

Figure 6. Workability of the mortars using blended cements containing RHA: (**a**) effect of CEM I 52.5 R replacement with RHA and FA; (**b**) effect of the mineral admixture type for 10% cement replacement.

The specific surface area effect was clearly observed when RHA-1 (1.6 m^2/g) was used rather than RHA-4 (36.5 m^2/g). The similar specific surface area of RHA-1 to that of cement CEM I 52.5R (1.6 m^2/g) resulted in the same workability values within the 0–20% replacement range (Figure 6a). RHA-4 (36.5 m^2/g) had a larger specific surface area than RHA-3 (15.2 m^2/g), RHA-2 (9.3 m^2/g) and RHA-1 (1.6 m^2/g). This parameter significantly influenced the mortar workability values for the cement replacement percentages that equaled or exceeded 5%. These replacement percentages (5–20%) allow the production of non-workable mixtures due to the high absorption/reaction of water [41]. The main reasons for this may be (i) the large specific surface area of the mineral additions, (ii) the pozzolanic reaction where CSH formation on the surface of the RHA grains consumes water; and (iii) the acceleration of cement hydration by nucleation in the grains of the mineral additions. Moreover, the presence of unburned components in the RHA could also influence the workability and the specific surface area [42]. For instance, RHA-2 (9.3 m^2/g, 4.52% LOI) provided worse consistency than RHA-1 (1.6 m^2/g, 0.14% LOI).

The workability behavior for the 15–20% blended cements using RHA-4 and RHA-3 means that it is not advisable to prepare this blended cement type with high proportions of these ashes. The upper replacement level could be 10% according to the workability behavior. Thus, some cements, indicated by (*) in Table 3, present problems from a workability point of view.

Figure 4b shows the comparison of the different mineral admixtures in terms of workability for the 10% replacement mortars. The workability of the mortars with 10% RHA-1, RHA-2, RHA-3 and RHA-4 was lower than for those achieved with GGBFS-1 (1.0 m^2/g) and GGBFS-2 (1.2 m^2/g), QF (2.6 m^2/g) and FA (0.8 m^2/g). The main reasons for these results were the smaller specific surface area of the admixtures other than RHA and, specifically for FA, the spherical morphology of the FA particles. However, the worst workability was observed for NDSF (24.4 m^2/g) compared to RHA-4 because of its greater fineness (63% in the volume of NDSF particles < 3 μm).

To prepare blended cements with percentages of cement replacement with RHA higher than 10%, mixtures of the different mineral additions were designed. Mixing RHA-4 and RHA-1 (they presented very different behaviors in workability terms, as previously demonstrated) allowed a higher replacement percentage to be applied, and the workability improved compared to the sole use of reactive RHA-4. Figure 7 shows the workability of the mortars with different proportions of RHA-4 and RHA-1. Low workability (113 mm) was obtained using the 20% RHA-4 replacement. To enhance the workability, the two mixtures of 5% RHA-4 plus 5–15% RHA-1 and 10%RHA-4 plus 5–10% RHA-1 were tested. The workability for the 20% replacement mortars increased, and the following workability values were obtained: 142 mm (for $5\%_{RHA\text{-}4} + 15\%_{RHA\text{-}1}$) and 131 mm (for $10\%_{RHA\text{-}4} + 10\%_{RHA\text{-}1}$). These values were significantly higher than those for $20\%_{RHA\text{-}4}$. For the tested mixtures, the workability of the mortars obtained by mixing RHA-4 and RHA-1 showed a linear decrease with the amount of replaced cement, and no synergic effect took place. This workability study demonstrates that preparing mixtures containing both a large and small specific surface area could be promising for the valorization of different RHA types.

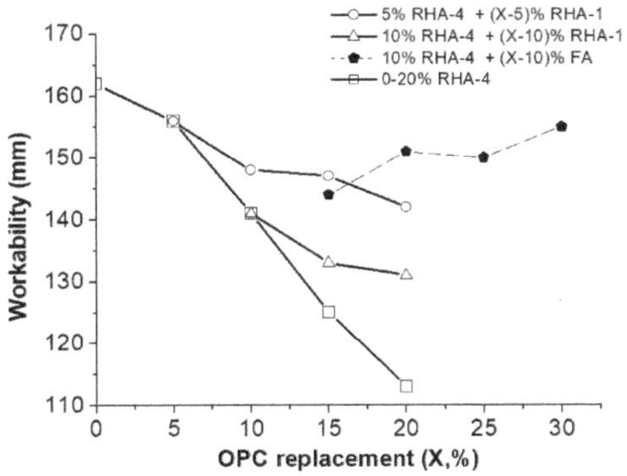

Figure 7. Workability of mortars using ternary blended cements: mixtures RHA4+RHA1 and RHA-4+FA. Comparison to the workability of the mortars containing 5–20% RHA-4.

The mixture of different types of pozzolans can produce synergic effects as a result of the individual properties of each one [43]. This fact was noted when highly reactive RHA-4 and FA were used together. For instance, the worsening of the workability when employing RHA-4 with a large specific surface area could be notably improved by FA blending. Figure 6a shows (dashed line) the workability of the mortars with a 10–20% level of cement replacement with FA: a slight increase in %W occurred with the rising replacement percentage. Figure 7 depicts (dashed line) the workability improvement upon increasing the quantity of FA added to RHA-4. The 20% replacement system with $10\%_{RHA\text{-}4}$ + $10\%_{FA}$ yielded 151 mm workability and the 30% replacement one with $10\%_{RHA\text{-}4}$ + $20\%_{FA}$ yielded 155 mm. These values are equivalent to those obtained for the $5\%_{RHA\text{-}4}$ system. The main reasons for the improved workability with FA addition are the smaller specific surface area and the high sphericity of the FA particles compared to CEM I 52.5 R. The obtained results demonstrate an interesting strategy to compensate for the reduction in workability by using RHA with a large specific surface area and adding a mineral admixture with a ball-bearing effect [44].

From previous workability studies, it can be concluded that employing certain mixtures of mineral additions can be appropriate for the preparation of mortars with good workability. However, compressive strength tests must be carried out. Some mixed mineral admixtures, i.e., RHA-4+RHA-1 and RHA-4+FA, were also studied from a mechanical point of view. The compressive strength values for the mortars cured at 2, 7 and 28 days were measured. Figure 8 offers the compressive strength values for the mortars with 5–20% replacement levels. Two sets of mixtures of RHA-4+RHA-1 were studied: one containing 5% RHA-4 and the other one containing 10% RHA-4. The $10\%_{RHA\text{-}4}$ + $(X - 10\%)_{FA}$ mixtures (where X = 20, 25 and 30%) were also tested. In Figure 8, the lines corresponding to the control mortar (CEM I 52.5 R) represent the theoretical linear decrease according to the cement content in the cement-replaced mortars.

For the 2- and 7-day curing times, most samples obtained similar or lower compressive strength values than those found for the control mortar. However, all of the cement-replaced samples showed equal or higher strength values compared to the corresponding one when taking into account the amount of CEM I 52.5 R in these mixtures. This fact suggests that blending admixtures significantly contributes to the strength at these early ages.

Except for mixture $5\%_{RHA\text{-}4}$ + $15\%_{RHA\text{-}1}$, the compressive strength values at 28 curing days were similar to or higher than those for the 100% OPC control specimen. This means that the strength development contribution of the mixture of pozzolanic additions was very important at the 28-day curing time. Thus, it is possible to prepare ternary blended cements with 30% less OPC using the selected mixtures of types RHA-4+RHA-1 and RHA-4+FA.

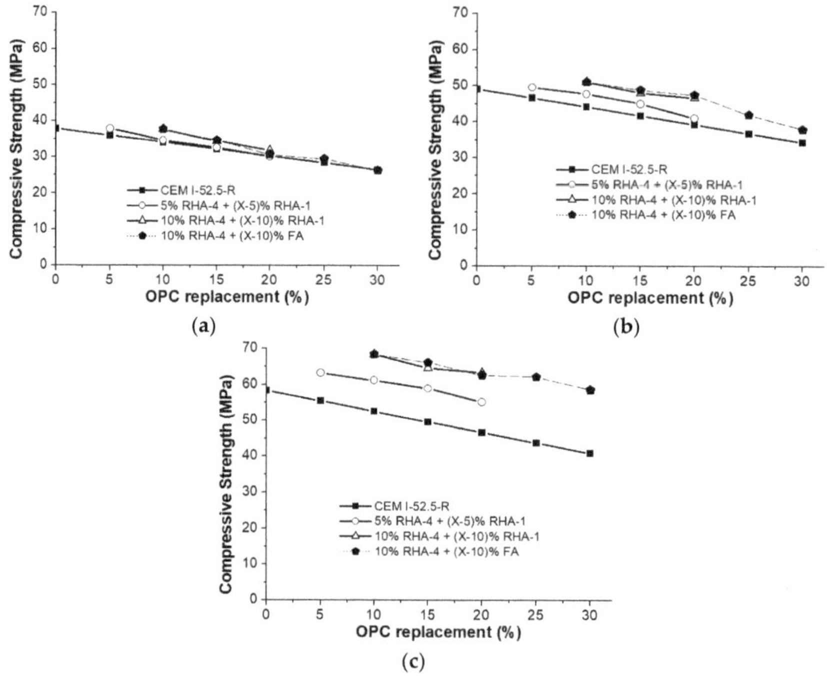

Figure 8. Compressive strength development for the mortars containing ternary blended cements with RHA-4 for the curing times of (**a**) 2 days, (**b**) 7 days and (**c**) 28 days. The values for the CEM I 52.5 R mortars were calculated by multiplying the control mortar by the cement content, i.e., (100 − %OPC) replacement.

The compressive strength values for the ternary blended cements with either siliceous FA or RHA-1 were similar at 28 curing days. However, when using FA, the consistency was better (151 mm, $10\%_{RHA\text{-}4} + 10\%_{FA}$) compared to when RHA-1 was employed (131 mm, $10\%_{RHA\text{-}4} + 10\%_{RHA\text{-}1}$). This fact suggests the positive effect of FA addition on RHA-containing ternary cements.

According to these results, some standard ternary blended cements could be prepared with high-reactivity RHA and with either siliceous FA or low-reactivity RHA; these are listed in Table 4. Some cements are proposed according to the compressive strength values, as well as a minimum workability value for the mortar (the proposal had 140 mm, which corresponded to reduced workability by ca. 15% vs. the CEM I 52.5 R mortar). All of the ternary cements that can be produced belong to the 52.5 strength category. Thus, the CEM II/A-H 52.5 cements can be prepared by blending 5%/5%, 5%/10% or 5%/15% of high-/low-reactivity RHA. This demonstrates that low-reactivity RHA valorization is feasible. Moreover, cements of type CEM II/A-(H-V) can be designed with high-reactivity RHA with 10% content. Additionally, some blended cements with replacement percentages above 20% can be prepared by blending RHA and FA, and, consequently, the CEM-II B (H-V) ternary cements can be prepared.

Table 4. Designation of cements containing high-reactivity RHA and low-reactivity RHA or FA.

Nomenclature	Ternary Blended Cement
CEM II/A-H 52.5 R	5% RHA-4 + 5% RHA-1
	5% RHA-4 + 10% RHA-1
CEM II/A-H 52.5 N	5% RHA-4 + 15% RHA-1
CEM II/A-(H-V) 52.5 R	10% RHA-4 + 5% FA
CEM II/A-(H-V) 52.5 N	10% RHA-4 + 10% FA
CEM II/B-(H-V) 52.5 N	10% RHA-4 + 15% FA
	10% RHA-4 + 20% FA

4. Conclusions

Replacing cement with a mineral admixture modifies the workability of the blended cement: it decreases as more OPC (CEM I 52.5 R) is replaced with RHA-4 with a large specific surface area and high amorphous silica content.

RHA-4 is feasible for the production of standard blended cements of the CEM II/A type if the optimum quantity of 10% replacement is applied due to the workability limitations. CEM II/A-H would be produced (where "H" indicates RHA as a mineral component) with the same strength performance as CEM I.

RHA-4 mixtures with other inorganic additions (FA or low-reactivity RHA) are adequate for the preparation of ternary blended cements, such as CEM II/A and CEM II/B, with improved compressive strength at early (2 and 7 days) and 28-day curing ages. The significant contribution to the compressive mechanical strength due to the pozzolanic activity of high-reactivity RHA-4 favors mixtures with materials with lower pozzolanic reactivity, such as crystalline RHA (RHA-1) or FA. Cement replacement values of up to 30% are proposed and some standardized cements of types CEM II/A (H-V) and CEM II/B (H-V) can be designed.

In summary, in view of these results, which serve as an example, the Portland cement industry has the potential to implement cements with new mineral additions that valorize waste (for example, waste from agricultural activities), in compliance with the required standards of quality.

Author Contributions: Conceptualization, L.M.O. and J.P.; Methodology, J.M., L.M.O. and L.S.; Formal analysis, L.M.O., M.V.B. and L.S.; Investigation, L.M.O. and J.M.; Resources, M.V.B. and J.P.; Data curation, L.M.O. and J.M.; Writing—original draft, L.M.O. and J.P.; Writing—review and editing, L.M.O., J.M., M.V.B. and J.P.; Visualization, L.S., M.V.B. and J.P.; Supervision, J.P.; Project administration, J.M. and J.P.; Funding acquisition, M.V.B. and J.P. All authors have read and agreed to the published version of the manuscript.

Funding: This research was funded by the Spanish Government (Ministerio de Ciencia y Tecnología), grant number AMB1998-0337. We appreciate Leandro Ordoñez's collaboration in the design and manufacture of the incinerator prototype. Finally, we thank the Microscopy Service of the Universitat Politècnica de València.

Institutional Review Board Statement: Not applicable.

Informed Consent Statement: Not applicable.

Data Availability Statement: The original contributions presented in the study are included in the article, further inquiries can be directed to the corresponding author.

Conflicts of Interest: Author Luis Miguel Ordoñez was employed by the company KhemeChemical S.L. The remaining authors declare that the research was conducted in the absence of any commercial or financial relationships that could be construed as a potential conflict of interest.

Appendix A

Table A1. XRF data for RHA samples (weight %).

	RHA-1		RHA-2		RHA-3		RHA-4	
	%	Std	%	Std	%	Std	%	Std
SiO_2	94.35	0.65	89.12	0.45	86.12	0.04	78.00	0.10
Al_2O_3	0.52	0.00	0.27	0.00	0.20	0.00	0.16	0.00
Fe_2O_3	0.08	0.00	0.09	0.00	0.39	0.00	0.31	0.01
CaO	1.02	0.00	1.20	0.01	1.26	0.00	0.98	0.00
MgO	0.47	0.00	0.54	0.00	0.41	0.00	0.45	0.00
SO_3	0.10	0.02	0.19	0.06	0.21	0.01	0.12	0.02
Na_2O	0.14	0.01	0.35	0.01	0.18	0.02	0.24	0.02
K_2O	2.14	0.05	2.95	0.06	2.52	0.02	1.92	0.07
TiO_2	0.09	0.00	0.09	0.00	0.08	0.00	0.07	0.00
Mn_2O_3	0.05	0.00	0.06	0.00	0.09	0.00	0.08	0.00
LOI	0.42	0.02	4.69	0.02	8.51	0.05	17.65	0.02

Appendix B

XRD diffractograms for RHA samples.

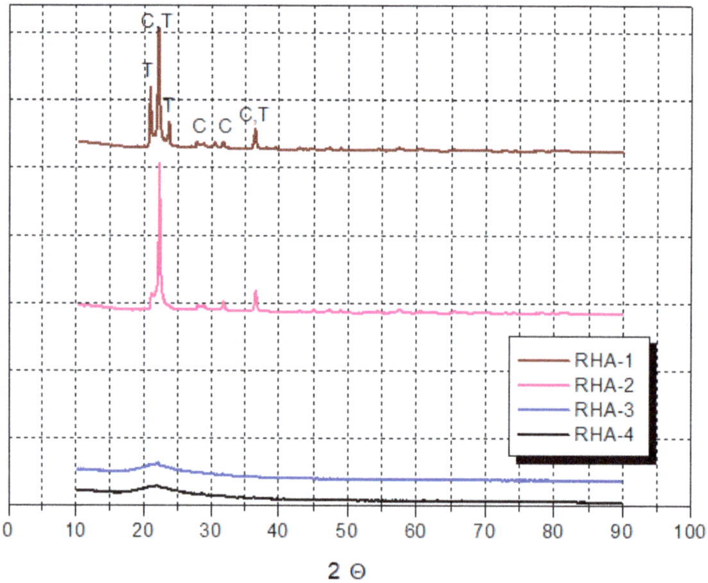

Figure A1. X-ray diffraction patterns of RHA samples (Key: C: Cristobalite; T: Tridymite).

Appendix C

SEM micrographs for non-ground RHA samples.

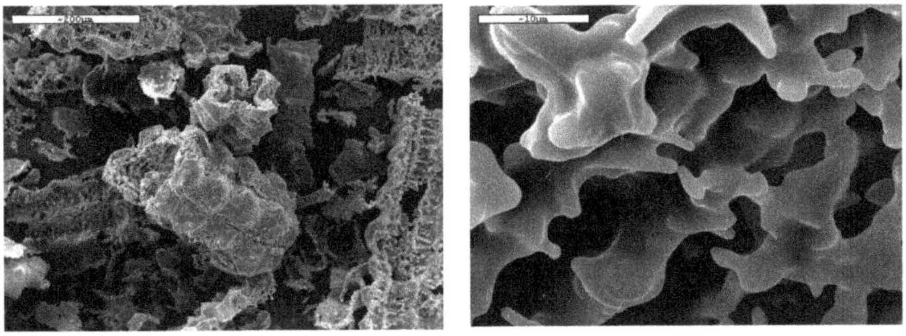

(**a**) (**b**)

Figure A2. SEM micrographs for RHA-1: (**a**) ×200; (**b**) ×3500.

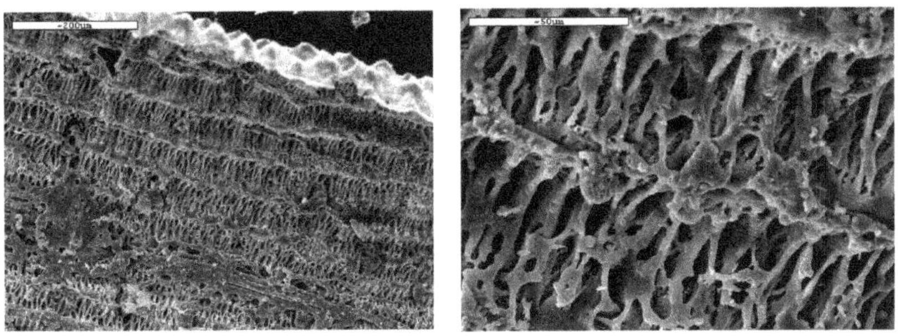

(**a**) (**b**)

Figure A3. SEM micrographs for RHA-2: (**a**) ×200; (**b**) ×2000.

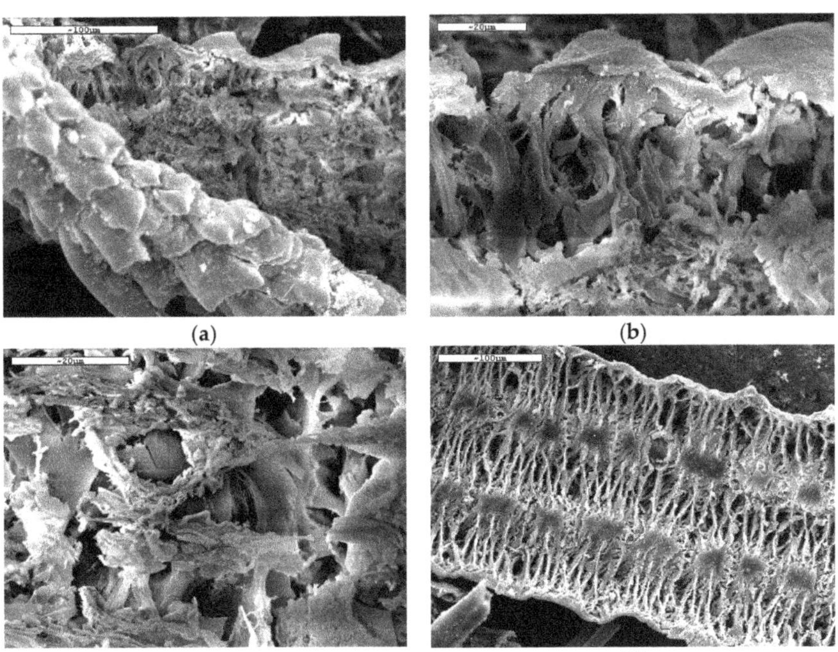

(**a**) (**b**)

(**c**) (**d**)

Figure A4. SEM micrographs for RHA-4: (**a**) ×500; (**b**) ×1500; (**c**) ×2000; (**d**) ×350.

References

1. Mohamad, N.; Muthusamy, K.; Embong, R.; Kusbiantoro, A.; Hashim, M.H. Environmental impact of cement production and solutions: A review. *Mater. Today Proc.* **2022**, *48*, 741–746. [CrossRef]
2. Net Zero Roadmap: A Global Pathway to Keep the 1.5 °C Goal in Reach. 2023. Available online: https://www.iea.org/reports/cement-3#overview (accessed on 28 March 2024).
3. De Brito, J.; Kurda, R. The past and future of sustainable concrete: A critical review and new strategies on cement-based materials. *J. Clean. Prod.* **2021**, *281*, 123558. [CrossRef]
4. Olatoyan, O.J.; Abiodun Kareem, M.; Adebanjo, A.U.; Olawale, S.O.A.; Alao, K.T. Potential use of biomass ash as a sustainable alternative for fly ash in concrete production: A review. *Hybrid. Adv.* **2023**, *4*, 100076. [CrossRef]
5. Marzouk, H.A.; Arab, M.A.; Fattouh, M.S.; Hamouda, A.S. Effect of agricultural phragmites, rice straw, rice husk, and sugarcane bagasse ashes on the properties and microstructure of high-strength self-compacted self-curing concrete. *Buildings* **2023**, *13*, 2394. [CrossRef]
6. Islam, M.H.; Law, D.W.; Gunasekara, C.; Sobuz, M.H.R.; Rahman, M.N.; Habib, M.A.; Sabbir, A.K. Assessing the influence of banana leaf ash as pozzolanic material for the production of green concrete: A mechanical and microstructural evaluation. *Materials* **2024**, *17*, 720. [CrossRef] [PubMed]
7. Cook, D.J. Rice husk ash. In *Cement Replacement Materials*; Swamy, R.N., Ed.; Surrey University Press: Guildford, UK, 1986; pp. 171–196.
8. Capelo, A.R.; Mármol, G.; Rossignolo, J.A. Optimization of the rice husk ash production process for the manufacture of magnesium silicate hydrate cements. *J. Clean. Prod.* **2023**, *425*, 138891. [CrossRef]
9. Salas Montoya, A.; Chung, C.-W.; Kim, J.H. High performance concretes with highly reactive rice husk ash and silica fume. *Materials* **2023**, *16*, 3903. [CrossRef] [PubMed]
10. Endale, S.A.; Taffese, W.Z.; Vo, D.H.; Yehualaw, M.D. Rice husk ash in concrete. *Sustainibility* **2023**, *15*, 137. [CrossRef]
11. Statista. Available online: https://es.statista.com/estadisticas/501804/produccion-de-arroz-cascara-grano-en-comunidades-autonomas-de-espana/ (accessed on 28 March 2024).
12. Mansaray, K.G.; Ghaly, A.E. Thermogravimetric analysis of rice husks in air atmosphere. *Energy Source* **1998**, *20*, 653–663. [CrossRef]
13. Tsai, W.T.; Lee, M.K.; Chang, Y.M. Fast pyrolysis of rice husk: Product yields and compositions. *Biores. Tech.* **2007**, *98*, 22–28. [CrossRef]
14. Camargo-Pérez, N.R.; Abellán-García, J.; Fuentes, L. Use of rice husk ash as a supplementary cementitious material in concrete mix for road pavements. *J. Mater. Res. Tech.* **2023**, *25*, 6167–6182. [CrossRef]
15. Antiohos, S.K.; Papadakis, V.G.; Tsimas, S. Rice husk ash (RHA) effectiveness in cement and concrete as a function of reactive silica and fineness. *Cem. Concr. Res.* **2014**, *61–62*, 20–27. [CrossRef]
16. Vieira, A.P.; Toledo Filho, R.D.; Tavares, L.M.; Cordeiro, G.C. Effect of particle size, porous structure and content of rice husk ash on the hydration process and compressive strength evolution of concrete. *Cons. Build. Mat.* **2020**, *236*, 117553. [CrossRef]
17. Haque, M.; Ray, S.; Mita, A.F.; Bhattacharjee, S.; Shams, M.D.J. Prediction and optimization of the fresh and hardened properties of concrete containing rice husk ash and glass fiber using response surface methodology. *Case Stud. Constr. Mater.* **2021**, *14*, e00505. [CrossRef]
18. Zhang, P.; Wei, S.; Cui, G.; Zhu, Y.; Wang, J. Properties of fresh and hardened self-compacting concrete incorporating rice husk ash: A review. *Rev. Adv. Mater. Sci.* **2022**, *61*, 563–575. [CrossRef]
19. Wang, J.; Xiao, J.; Zhang, Z.; Han, K.; Hu, X.; Jiang, F. Action mechanism of rice husk ash and the effect on main performances of cement-based materials: A review. *Cons. Build. Mat.* **2021**, *288*, 123068. [CrossRef]
20. Lacerda, C.; Valle, P.; Da Silva, A.C.; De Souza, C.; Jacques, F.S. Mechanical and durability properties of concrete produced with construction and demolition waste and rice husk ash. *Cons. Build. Mat.* **2023**, *406*, 133471. [CrossRef]
21. Ambedkar, B.; Alex, J.; Dhanalakshmi, J. Enhancement of mechanical properties and durability of the cement concrete by RHA as cement replacement: Experiments and modeling. *Cons. Build. Mat.* **2017**, *148*, 167–175. [CrossRef]
22. Cordoba, G.; Barquero, M.; Bonavetti, V.; Irassar, E.F. Sustainability of concretes with binary and ternary blended cements considering performance parameters. *Cement* **2023**, *13*, 100077. [CrossRef]
23. Muhammad, A.; Thienel, K.-C. Properties of Self-Compacting Concrete Produced with Optimized Volumes of Calcined Clay and Rice Husk Ash—Emphasis on Rheology, Flowability Retention and Durability. *Materials* **2023**, *16*, 5513. [CrossRef]
24. Han, Y.; Oh, S.; Wang, X.-Y.; Lin, R.-S. Hydration–Strength–Workability–Durability of Binary, Ternary, and Quaternary Composite Pastes. *Materials* **2022**, *15*, 204. [CrossRef]
25. *UNE-EN 197-1*; Cement-Part 1: Composition, Specifications and Conformity Criteria for Common Cements. European Committee for Standardisation, AENOR: Madrid, Spain, 2011.
26. *UNE-EN 196-1*; Methods of Testing Cement–Part 1: Determination of Strength. European Committee for Standardisation, AENOR: Madrid, Spain, 2018.
27. Payá, J.; Monzó, J.; Borrachero, M.V.; Mellado, A.; Ordóñez, L.M. Determination of Amorphous Silica in Rice Husk Ash by a Rapid Analytical Method. *Cem. Concr. Res.* **2001**, *31*, 227–231. [CrossRef]
28. *UNE-EN 413-2*; Masonry Cement-Part 2: Test Methods. European Committee for Standardisation, AENOR: Madrid, Spain, 2017.

29. Payá, J.; Monzó, J.; Borrachero, M.V.; Peris-Mora, E.; Ordóñez, L.M. Studies on crystalline rice husk ashes and the activation of their pozzolanic properties. In *Waste Materials in Construction Wascon 2000 (Waste Management Series)*; Woolley, G.R., Goumans, J.J.J.M., Wainwright, P.J., Eds.; Elsevier: Amsterdam, The Netherlands, 2000; Volume 1, pp. 493–503. [CrossRef]
30. Krishnarao, R.V.; Subrahmanyam, J.; Jagadish Kumar, T. Studies on the formation of black particles in rice husk silica ash. *J. Eur. Ceram. Soc.* **2001**, *21*, 99–104. [CrossRef]
31. Wu, J.; Yu, D.; Zeng, X.; Yu, X.; Han, J.; Wen, C.; Yu, G. Ash formation and fouling during combustion of rice husk and its blends with a high alkali Xinjiang coal. *Energy Fuels* **2018**, *32*, 416–424. [CrossRef]
32. Ganesan, K.; Rajagopal, K.; Thangavel, K. Rice husk ash blended cement: Assessment of optimal level of replacement for strength and permeability properties of concrete. *Cons. Build. Mat.* **2008**, *22*, 1675–1683. [CrossRef]
33. Edwards, G.A.S. Coal and climate change. *WIREs Clim. Change* **2019**, *10*, e607. [CrossRef]
34. Kamau, J.; Ahmed, A.; Hyndman, F.; Hirst, P.; Kangwa, J. Influence of rice husk ash density on the workability and strength of structural concrete. *Eur. J. Eng. Res. Sci.* **2017**, *2*, 36–43. [CrossRef]
35. Payá, J.; Monzó, J.; Borrachero, M.V.; Peris-Mora, E. Comparisons among magnetic and non-magnetic fly ash fractions: Strength development of cement-fly ash mortars. *Waste Manag.* **1996**, *16*, 119–124. [CrossRef]
36. Xu, W.; Lo, T.Y.; Wang, W.; Ouyang, D.; Wang, P.; Xing, F. Pozzolanic reactivity of silica fume and ground rice husk ash as reactive silica in a cementitious system: A comparative study. *Materials* **2016**, *9*, 146. [CrossRef]
37. Peris Mora, E.; Payá, J.; Monzó, J. Influence of different sized fractions of a fly ash on workability of mortars. *Cem. Concr. Res.* **1993**, *23*, 917–924. [CrossRef]
38. Megat Johari, M.A.; Brooks, J.J.; Kabir, S.; Rivard, P. Influence of supplementary cementitious materials on engineering properties of high strength concrete. *Cons. Build. Mat.* **2011**, *25*, 2639–2648. [CrossRef]
39. Baghabra Al-Amoudi, O.S.; Maslehuddin, M.; Ibrahim, M.; Shameem, M.; Al-Mehthel, M.H. Performance of blended cement concretes prepared with constant workability. *Cem. Concr. Comp.* **2011**, *33*, 90–102. [CrossRef]
40. Hamada, H.M.; Abed, F.; Yati Binti Katman, H.; Humada, A.M.; Al Jawahery, M.S.; Majdi, A.; Yousif, S.T.; Thomas, B.S. Effect of silica fume on the properties of sustainable cement concrete. *J. Mat. Res. Tech.* **2023**, *24*, 8887–8908. [CrossRef]
41. Ahsan, M.B.; Hossain, Z. Supplemental use of rice husk ash (RHA) as a cementitious material in concrete industry. *Cons. Build. Mat.* **2018**, *178*, 1–9. [CrossRef]
42. Cordeiro, G.C.; Toledo Filho, R.D.; de Moraes Rego Fairbairn, E. Use of ultrafine rice husk ash with highcarbon content as pozzolan in high performance concrete. *Mater. Struct.* **2009**, *42*, 983–992. [CrossRef]
43. Arshad, M.T.; Ahmad, S.; Khitab, A.; Hanif, A. Synergistic use of fly ash and silica fume to produce high-strength self-compacting cementitious composites. *Crystals* **2021**, *11*, 915. [CrossRef]
44. Du, S.; Zhao, Q.; Shi, X. High-volume fly ash-based cementitious composites as sustainable materials: An overview of recent advances. *Adv. Civil. Eng.* **2021**, *2021*, 4976169. [CrossRef]

materials

MDPI

Article

On the Flow of a Cement Suspension: The Effects of Nano-Silica and Fly Ash Particles

Chengcheng Tao [1] and Mehrdad Massoudi [2,*]

[1] School of Construction Management Technology, Purdue University, West Lafayette, IN 47907, USA; tao133@purdue.edu

[2] National Energy Technology Laboratory (NETL), U. S. Department of Energy, Pittsburgh, PA 15236, USA

* Correspondence: mehrdad.massoudi@netl.doe.gov

Abstract: Additives such as nano-silica and fly ash are widely used in cement and concrete materials to improve the rheology of fresh cement and concrete and the performance of hardened materials and increase the sustainability of the cement and concrete industry by reducing the usage of Portland cement. Therefore, it is important to study the effect of these additives on the rheological behavior of fresh cement. In this paper, we study the pulsating Poiseuille flow of fresh cement in a horizontal pipe by considering two different additives and when they are combined (nano-silica, fly ash, combined nano-silica, and fly ash). To model the fresh cement suspension, we used a modified form of the power-law model to demonstrate the dependency of the cement viscosity on the shear rate and volume fraction of cement and the additive particles. The convection–diffusion equation was used to solve for the volume fraction. After solving the equations in the dimensionless forms, we conducted a parametric study to analyze the effects of nano-silica, fly ash, and combined nano-silica and fly ash additives on the velocity and volume fraction profiles of the cement suspension. According to the parametric study presented here, larger nano-silica content results in lower centerline velocity of the cement suspension and larger non-uniformity of the volume fraction. Compared to nano-silica, fly ash exhibits an opposite effect on the velocity. Larger fly ash content results in higher centerline velocity, while the effect of the fly ash on the volume fraction is not obvious. For cement suspension containing combined nano-silica and fly ash additives, nano-silica plays a dominant role in the flow behavior of the suspension. The findings of the study can help the design and operation of the pulsating flow of fresh cement mortars and concrete in the 3D printing industry.

Keywords: cement; rheology; nano-silica; fly ash; non-Newtonian fluids

1. Introduction

There is a strong demand to improve the rheological properties of cement and concrete materials including ultra-high performance concrete, high-strength concrete, self-consolidated concrete, and concrete for 3D printing in the industry. The cement and concrete design with improved rheology is important to enhance constructability and workability, facilitate processing and mixing, and enhance in situ material performance [1]. The addition of nano-additives, such as nano-silica, graphene oxide, nano-TiO_2, and cellulose nanofibers (CNFs) can significantly modify and improve the rheological properties of cement-based materials; these additives can also affect the yield stress, the plastic viscosity, the rate of structural buildup at rest, the degree of hydration, mechanical properties, and durability [2–7]. Various factors including the nanomaterial type, constituent, degree of dispersion, mixing energy, high-range water reducer, sonication, and water-to-cement ratio (w/c) can affect the rheology of fresh cement and concrete. The combined effects of these various factors would be quite complex. Among different nanoparticles used in cement and concrete materials, nano-silica or nano-SiO_2 is the most commonly used nano-additive for performance improvement, mainly due to the pozzolanic reactivity and the pore-filling effect [8]. According to Lavergne, et al. [9], nano-silica particles accelerated the hydration of the cement paste and increased the shear strength and early strength significantly because of the specific surface of the nano-silica. Kim, et al. [10] indicated that the addition of nano-silica and silica fume to the high-volume fly ash cement (HVFC) accelerated the initial hydration reaction of HVPC, resulting in a shorter setting time, and higher compressive strength during the initial and middle stages of hydration. The high viscosity of alkali silicate-activated cements has been a big issue that prevents the wide application of this type of material [11]. Fly ash additives have been widely used in cement, concrete, and asphalt industries to improve rheology and workability by decreasing yield stress and viscosity [12–18]. Fly ash microspheres (FAMs) with highly spherical particles work as ball bearings in fresh cement suspensions by reducing the internal friction between the additives, mitigating the floc agglomeration, and reducing locked water [11]. Wang, et al. [13] concluded that the addition of fly ash increased the fluidity of graphene oxide-cement paste and improved the rheological properties of the cement paste with fly ash and graphene oxide additives. Fly ash plays an economical and effective role in achieving the intended properties of graphene oxide-cement paste because of its ball effect, size gradation, and less water requirement. Golewski [19] illustrated the combined effect of different additives, including siliceous fly ash, silica fume, and nano-silica, which accelerated the hardening of concrete after 28 days of curing and reduced the microcracks in the interfacial transition zone between the aggregates and the cement paste. García-Taengua, et al. [20] investigated the effect of nano-silica and fly ash on the rheology of fresh cement mortars. The study showed that the viscosity reached its maximum at 2% nano-silica content, and larger fly ash content increased the viscosity when used together with the nano-silica additives. In addition, researchers have used various additives such as viscosity modifying agents, superplasticizers, and organic chemical admixtures [21–24]. In this paper, we are focusing on nano-additives and fly ashes since they have been widely used in cementitious materials and exhibit effectiveness in improving the rheological properties of fresh cement and concrete.

Computational modeling has been considered a useful tool to simulate the behavior of cement and concrete materials [25,26]. Rheological models can provide some necessary information for the fluid behavior of fresh cement slurries [27]. In general, fresh cement slurries are considered to behave as non-Newtonian fluids [28]. In our previous studies [29–31], we reviewed the rheological models for fresh cement by considering various effects on viscosity and yield stress, and conducted parametric studies for steady and pulsating Poiseuille flow of fresh cement modeled as non-Newtonian fluids. The study of the pulsating flow of fresh cement materials provides useful information for the 3D-printed cement and concrete industry by considering the vibration/oscillatory effect. Researchers have conducted experimental investigations and summarized the effect of different additives on the viscosity and yield stress of fresh cementitious suspensions. However, most of the existing models of cement suspensions with additives are obtained from experimental data in one-dimensional form. There is a lack of a comprehensive rheological model with three-dimensional (3D) tensor notation where the non-Newtonian nature of the cement suspensions with additives is considered. In addition, there is no constitutive relation that considers the effect and the significance of shear rate, volume fraction, and additives such as nano-silica, fly ashes, and combined additives on the flow of fresh cement suspensions.

In this paper, we look at the pulsating flow of a cement suspension in a pipe with two additives and when they are combined: (1) nano-silica, (2) fly ash, and (3) nano-silica and fly ash. The fresh cement is modeled as a non-Newtonian fluid. We use a modified form of the power-law model to study the dependency of the cement suspension viscosity on the shear rate and volume fraction of the cement and additive particles. The volume fraction of particles is solved through a convection–diffusion equation. In Section 2, the governing equations of motion are presented, assuming no thermal or chemical effects (isothermal conditions). In Section 3, we briefly discuss the constitutive equations for the viscous stress tensor and the diffusive flux vector. In Section 4, we solve for the dimensionless form of the governing equations and provide numerical solutions in the dimensionless form. A limited parametric study is performed for a limited range of different dimensionless numbers. In Section 5, we present a few concluding remarks.

2. Governing Equations

A fresh cement suspension, in general, behaves like a multiphase fluid composed of different materials such as cement particles, water, additives, etc. To model it mathematically, there are many approaches; for example, the multicomponent (Mixture Theory) approach can be used, where there is a two-way coupling between the two phases (constituents). In this approach, the governing equations are written for each component. Consequently, constitutive equations are needed for the stress tensors, along with the interaction forces (see [32,33]). Alternatively, to simplify things, the cement suspension can be assumed to behave like a non-homogeneous fluid suspension with microstructure because of the presence of the cement particles and other added particles. In this approach, the material parameters, such as viscosity, may depend on the particle concentration, also known as the volume fraction of the particles. This approach has been advocated by Tao et al. [34] and we will use the same approach here. For an isothermal case, the governing equations are the balance equations for mass, linear momentum, angular momentum, and the convection–diffusion for the particle flux [35].

2.1. Conservation of Mass

$$\frac{\partial \rho}{\partial t} + div(\rho \boldsymbol{v}) = 0 \tag{1}$$

where $\partial/\partial t$ is the partial derivative with respect to time, div is the divergence operator, ρ is the density of the fresh cement suspension, which is related to the density of water and the cement particles, and $\boldsymbol{v}$ is the velocity vector. Notice that now we have a new parameter, namely the volume fraction ϕ of the (cement) particles. This function has the property that $0 \leq \phi(\mathbf{x}, t) \leq \phi_{max} < 1$; that is, its value is either one or zero. Since $\phi(\mathbf{x}, t)$ is a smooth function, we, in a sense, homogenize the solid particle distribution; thus, the shape and size of the particles do not have any direct impact on the constitutive formulation. As a result, the density of the suspension is given by $\rho = \rho_{f0}(1 - \phi) + \rho_{s0}\phi$, where ρ_{f0} and ρ_{s0} are the pure density of water and the cement particles before mixing with water, respectively. The fresh cement suspension is assumed to behave as an incompressible fluid, that is

$$div\ \boldsymbol{v} = 0 \tag{2}$$

2.2. Conservation of Linear Momentum

$$\rho \frac{d\boldsymbol{v}}{dt} = div\boldsymbol{T} + \rho \boldsymbol{b} \tag{3}$$

where d/dt is the total time derivative, given by $\frac{d(.)}{dt} = \frac{\partial(.)}{\partial t} + [grad(.)]\boldsymbol{v}$, $grad$ is the gradient operator, $\boldsymbol{b}$ represents the body force, and $\boldsymbol{T}$ is the stress tensor.

2.3. Conservation of Angular Momentum

The angular momentum equation indicates that the stress tensor is symmetric if there are no couple stresses, namely

$$\boldsymbol{T} = \boldsymbol{T}^T \tag{4}$$

2.4. Convection–Diffusion Equation

For the motion of the particles, a convection–diffusion equation is used; the particles move with the velocity of the suspension (see [36,37]).

$$\frac{\partial \phi}{\partial t} + div(\phi \boldsymbol{v}) = -div\boldsymbol{N} \tag{5}$$

where $\partial\phi/\partial t$ is the rate of accumulation of particles, $div(\phi\boldsymbol{v})$ represents particle motion due to the bulk flow, and $div\boldsymbol{N}$ is the diffusive flux.

In addition to the governing equations above, we also need constitutive relations for $\boldsymbol{T}$ and $\boldsymbol{N}$. In the next section, we discuss the constitutive relations of the stress tensor $\boldsymbol{T}$ and the flux vector $\boldsymbol{N}$.

3. Constitutive Relations

3.1. Stress Tensor

A cement suspension generally behaves as a complex fluid that has a yield stress, and oftentimes exhibits thixotropy, viscoelasticity, etc. For cement suspensions, the formulation of the stress tensor significantly affects the properties and the workability of fresh and hardened materials. It is usually assumed that the stress tensor has two parts: $\boldsymbol{T} = \boldsymbol{T_y} + \boldsymbol{T_v}$, with $\boldsymbol{T_y}$ being the yield stress and $\boldsymbol{T_v}$ the viscous stress. Tao et al. [29] gave a comprehensive review of the constitutive relations for fresh cement suspensions. Tao et al. [34], based on the model proposed by Massoudi and Phuoc [38], suggested a general constitutive relation [39,40]. In this paper, a simplified version of this model for the viscous stress $\boldsymbol{T_v}$ is used,

$$\boldsymbol{T_v} = -p\boldsymbol{I} + \mu_{\text{eff}}(\phi, \boldsymbol{A_1})\boldsymbol{A_1} \tag{6}$$

where p is the pressure (the mean value of the stress tensor), $\boldsymbol{I}$ is the identity tensor, and μ_{eff} represents the effective shear viscosity (a function of the volume fraction ϕ and the shear rate). The tensor $\boldsymbol{A_1}$ is given as

$$\boldsymbol{A_1} = grad\boldsymbol{v} + (grad\boldsymbol{v})^T \tag{7}$$

Bridges and Rajagopal [41] used an effective viscosity relationship:

$$\mu_{\text{eff}}(\phi, \boldsymbol{A_1}) = \mu^*(\phi)\left[1 + \alpha tr\boldsymbol{A_1}^2\right]^m \tag{8}$$

where $\mu^*(\phi)$ is assumed to be given by the Batchelor and Green [42]:

$$\mu^*(\phi) = \mu_0(1 + 2.5\phi + 7.6\phi^2) \tag{9}$$

where μ_0 is the viscosity of the base fluid [43]. Equation (8) follows the traditional power-law model, where the viscosity depends on the shear rate. When $m < 0$, the suspension behaves as a shear-thinning fluid, and when $m > 0$, it behaves as a shear-thickening one; and α is related to the consistency index. There are other correlations for $\mu^*(\phi)$ (see Krieger and Dougherty's model [44]).

By substituting Equations (8) and (9) into (6), the viscous stress tensor is given:

$$\boldsymbol{T_v} = -p\boldsymbol{I} + \mu_0\left(1 + 2.5\phi + 7.6\phi^2\right)\left[1 + \alpha tr\boldsymbol{A_1}^2\right]^m \boldsymbol{A_1} \tag{10}$$

Equation (10) will be used as the starting point of our modeling. We will modify this equation by incorporating the effects of the nano-silica and the fly ash particles. In this paper, we ignore the thixotropic effect and the yield stress [28,45].

3.1.1. Nano-SiO_2 Additive

With the emergence of nanotechnology as a new and promising technology, nanomaterials have been introduced to improve the strength of cementitious materials. Nano-SiO_2 (NS) particle is one of the most commonly used nanomaterials in cement. In this paper, we use a viscosity relationship that considers the effect of a hydrophilic NS [46]. The mean diameter of the NS is 12 nm, with a specific surface area of 200 m^2/g, density of 2.2 g/cm^3, pH of 4.5, with a purity greater than 99.8%. The cement suspension used in the model has the following composition: CaO–65.0%, SiO_2–21.2%, Al_2O_3–5.16%, Fe_2O_3–3.39%, SO_3–2.43%, MgO–1.32%, K_2O–0.63%, TiO_2–0.78%, and Na_2O–0.07%.

Equation (10) is now modified to include the effect of NS particles, obtaining the equation suggested by Ouyang et al. [46]. This viscous stress tensor is shown in Equation (11),

$$T_v = -pI + \mu_0\left(\frac{1}{1-2.5\left(1+\frac{r_{ns}}{R_{ns}}\right)^3 \phi_{NS/W}}\right)(1+2.5\phi+7.6\phi^2)\left[1+\alpha trA_1{}^2\right]^m A_1 \quad (11)$$

where, r_{ns} is the adsorbed water layer thickness for the nanoparticle, R_{ns} is the diameter of nanoparticles, $\phi_{NS/W}$ is the ratio of nano-silica particle volume to the total volume of nano-silica suspension.

3.1.2. Fly Ash Additive

Fly ash (FA) is another commonly used additive in cement; it is primarily used to enhance rheology and the workability of cement. Based on the fly ash study by Sybertz and Reick [18], the viscous stress tensor as given in Equation (10) is modified to include the effect of fly ash additive, as shown in Equation (12). The fly ash we consider in this paper has the following properties: density of Fly Ash I–2320 kg/m^3, density of Fly Ash II–2380 kg/m^3, surface area from Blaine analysis of Fly Ash I is 309 m^2/kg, and surface area from Blaine analysis of Fly Ash II is 332 m^2/kg. The cement suspension model now has the following form:

$$T_v = -pI + \mu_0\left(Q_1 e^{Q_2\phi}\left(1-\phi_f\right) + Q_3 e^{Q_4\phi}\phi_f\right)\left(1+2.5\phi+7.6\phi^2\right)\left[1+\alpha trA_1{}^2\right]^m A_1 \quad (12)$$

where, ϕ_f is the fly ash content and Q_n (n = 1, 2, 3, 4) are the parameters from experimental fitting, where $Q_1 = 2.31 \times 10^{-3}$ (Fly Ash I) and 4.75×10^{-3} (Fly Ash II), $Q_2 = 20.7$ (Fly Ash I) and 19.1 (Fly Ash II), $Q_3 = 7.96 \times 10^{-5}$ (Fly Ash I) and 1.29×10^{-5} (Fly Ash II), and $Q_4 = 17$ (Fly Ash I) and 18.9 (Fly Ash II) [18].

3.1.3. Combined Nano-SiO_2 and Fly Ash Additives

Considering that both the nano-SiO_2 particles and fly ash particles are added to the cement suspension, the viscous stress tensor for the NSFA-added case is shown in Equation (13),

$$\begin{aligned} T_v = -pI + \mu_0\left(\frac{1}{1-2.5\left(1+\frac{r_{ns}}{R_{ns}}\right)^3 \phi_{NS/W}}\right)&(1+2.5\phi+7.6\phi^2)((Q_1 e^{Q_2\phi}(1-\phi_f) \\ &+Q_3 e^{Q_4\phi}\phi_f)\left[1+\alpha trA_1{}^2\right]^m A_1 \end{aligned} \quad (13)$$

where the meaning of the parameters is as described previously. We use these three stress tensor equations in our analysis.

3.2. Particle Flux

In our earlier paper, we used a correlation based on the work of Philips et al. [47]. Here, we assume that N is given by [41]:

$$N = -D\nabla\phi \quad (14)$$

where D is the diffusion coefficient or the diffusivity, and ∇ is the gradient operator. We also assume that D has two parts: a shear-rate-dependent part and a concentration-dependent part:

$$D = D(\dot{\gamma}, \phi) = D_1(\dot{\gamma})D_2(\phi) \quad (15)$$

For $D_1(\dot{\gamma})$, we have [41]:

$$D_1(\dot{\gamma}) = \eta\left\|A_1{}^2\right\| \quad (16)$$

where η is a constant, and $\|\cdot\|$ is the trace-norm, where $\|A_1{}^2\| = \left(\frac{\partial v}{\partial r}\right)^2$ for the one-dimensional flow of a suspension in a pipe. For $D_2(\phi)$, based on [48], we use the relation suggested by [42,49], where for a dense suspension:

$$D_2(\phi) = D_0[(1-\phi)^{2.55}(1+4\phi+4\phi^2-4\phi^3+\phi^4)] \tag{17}$$

where D_0 is a constant. Substituting Equations (16) and (17) in Equation (14), we have the following:

$$\mathbf{N} = -D_0\eta\left\|A_1{}^2\right\|[(1-\phi)^{2.55}(1+4\phi+4\phi^2-4\phi^3+\phi^4)]\nabla\phi \tag{18}$$

We use Equation (18) in this paper.

4. Results and Discussion

Here, we study the pulsating flow of a cement suspension in a horizontal pipe. Figure 1 shows the schematic of the horizontal pipe flow of cement suspension with nano-silica and fly ash additives.

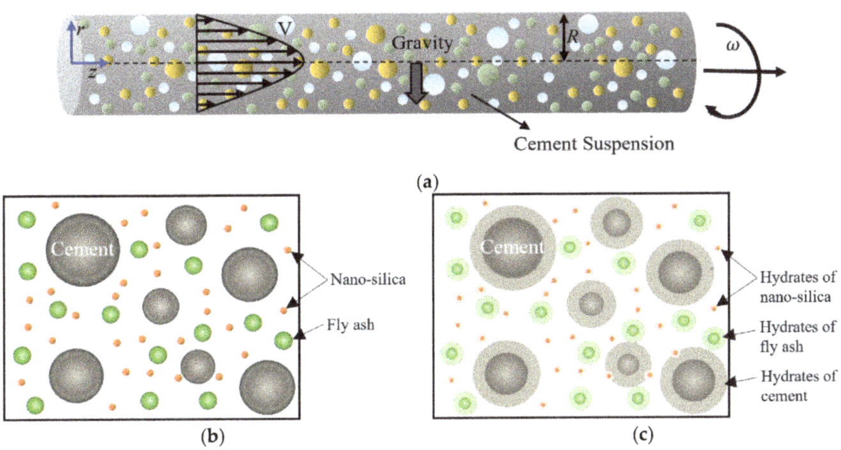

Figure 1. (**a**) Schematic of the horizontal pipe flow of cement suspension; (**b**) cement suspension with nano-silica and fly ash additives; (**c**) hydration of cement, nano-silica, and fly ash.

4.1. Problem Statement

We assume that the velocity and the volume fraction fields can be expressed as follows:

$$\begin{cases} \phi = \phi(r,t) \\ \mathbf{v} = v(r,t)\mathbf{e_z} \end{cases} \tag{19}$$

Indicating that an unsteady flow is in the z-direction, where the velocity and the volume fraction depend on time and the radius. With the assumptions above, Equation (1) is satisfied.

By substituting the constitutive Equations (11)–(13) into the governing Equation (3), the linear momentum equations are reduced to (using the cylindrical coordinate system)

$$\frac{\partial p}{\partial r} + \rho g = 0 \tag{20}$$

$$-\frac{1}{r}\frac{\partial p}{\partial \theta} = 0 \tag{21}$$

$$\frac{1}{r}\frac{\partial}{\partial r}\left\{r\mu^{**}(\phi)\left[1+2\alpha\left(\frac{\partial v}{\partial r}\right)^2\right]^m\frac{\partial v}{\partial r}\right\}-\frac{\partial p}{\partial z}=\rho\frac{\partial v}{\partial t} \tag{22}$$

When NS particles are added to the cement suspension, $\mu^{**}(\phi)$ becomes

$$\mu^{**}(\phi)=\mu_0\left(\frac{1}{1-2.5\left(1+\frac{r_{ns}}{R_{ns}}\right)^3\phi_{NS/W}}\right)(1+2.5\phi+7.6\phi^2) \tag{23}$$

When FA particles are added to the cement suspension, $\mu^{**}(\phi)$ becomes

$$\mu^{**}(\phi)=\mu_0(Q_1e^{Q_2\phi}(1-\phi_f)+Q_3e^{Q_4\phi}\phi_f)(1+2.5\phi+7.6\phi^2) \tag{24}$$

And, when both NS and FA particles are added to the cement suspension, $\mu^{**}(\phi)$ becomes

$$\mu^{**}(\phi)=\mu_0\left(\frac{1}{1-2.5\left(1+\frac{r_{ns}}{R_{ns}}\right)^3\phi_{NS/W}}\right)(Q_1e^{Q_2\phi}(1-\phi_f)+Q_3e^{Q_4\phi}\phi_f)(1+2.5\phi+7.6\phi^2) \tag{25}$$

Equations (20) and (21) show the variation of the pressure in the r- and θ-directions, respectively. To look at the effects of the pressure, based on previous studies, it is assumed that the axial pressure gradient is expressed as follows [41]:

$$\frac{\partial p}{\partial z}=-[A_0+B_0\sin\omega t] \tag{26}$$

where A_0 and B_0 are constants and ω is the pulsating frequency. By substituting Equation (26) into (22), we have

$$\frac{1}{r}\frac{\partial}{\partial r}\left\{r\mu^{**}(\phi)\left[1+2\alpha\left(\frac{\partial v}{\partial r}\right)^2\right]^m\frac{\partial v}{\partial r}\right\}+A_0+B_0\sin\omega t=\left[\rho_{f0}(1-\phi)+\rho_{s0}\phi\right]\frac{\partial v}{\partial t} \tag{27}$$

By substituting Equation (15) into the convection–diffusion Equation (5), we obtain

$$\frac{1}{r}\frac{\partial}{\partial r}\left\{r\cdot\eta\cdot\left(\frac{\partial v}{\partial r}\right)^2\cdot D_0\left[(1-\phi)^{2.55}(1+4\phi+4\phi^2-4\phi^3+\phi^4)\right]\frac{\partial\phi}{\partial r}\right\}=\frac{\partial\phi}{\partial t} \tag{28}$$

The above-coupled equations need to be solved numerically. Before solving Equations (27) and (28), we will obtain their dimensionless forms by non-dimensionalizing the equations through:

$$\overline{r}=\frac{r}{R}\quad\overline{v}=\frac{v}{V}\quad\overline{t}=\frac{Vt}{R}\quad\overline{\omega}=\frac{R\omega}{V} \tag{29}$$

where R is a reference length (the radius) and V is a reference velocity. The dimensionless forms of Equations (27) and (28) become

$$\begin{aligned}\left[(1-\phi)+\frac{\rho_{s0}}{\rho_{f0}}\phi\right]&\frac{\partial\overline{v}}{\partial\overline{t}}\\&=\frac{1}{Re}\frac{1}{\overline{r}}\frac{\partial}{\partial\overline{r}}\left\{\overline{r}X_{add}(1+2.5\phi+7.6\phi^2)\left[1+\delta\left(\frac{\partial\overline{v}}{\partial\overline{r}}\right)^2\right]^m\frac{\partial\overline{v}}{\partial\overline{r}}\right\}\\&+P(1+\zeta\sin\overline{\omega}\overline{t})\end{aligned} \tag{30}$$

$$\frac{\partial\phi}{\partial\overline{t}}=\frac{\kappa}{\overline{r}}\frac{\partial}{\partial\overline{r}}\left\{\overline{r}\cdot\left[(1-\phi)^{2.55}(1+4\phi+4\phi^2-4\phi^3+\phi^4)\right]\left(\frac{\partial\overline{v}}{\partial\overline{r}}\right)^2\frac{\partial\phi}{\partial\overline{r}}\right\} \tag{31}$$

where

$$Re = \frac{\rho_{f0} V R}{\mu_0}; \delta = 2\alpha \frac{V^2}{R^2}; P = \frac{A_0 R}{\rho_{f0} V^2}; G_0 = \frac{gR}{V^2}; \kappa = \frac{\eta D_0}{RV}; \zeta = \frac{B_0}{A_0}; \chi = \frac{\rho_{s0}}{\rho_{f0}} \tag{32}$$

where Re is the Reynolds number, the ratio of the forces due to inertia to the viscous effects, δ is related to the shear rate, P to the pulsation pressure, G_0 is related to the gravity term, κ to the cement suspension diffusivity and the volume fraction, ζ is related to the terms in the pulsating pressure, and χ represents the ratio of the two densities.

In Equation (30), when NS particles are added to the cement suspension, $X_{add} = \left(\frac{1}{1-2.5\left(1+\frac{r_{ns}}{R_{ns}}\right)^3 \phi_{NS/W}} \right)$. And, when FA particles are added to the cement suspension, $X_{add} = \left(Q_1 e^{Q_2\phi}(1-\phi_f) + Q_3 e^{Q_4\phi}\phi_f\right)$. When NS and FA particles are added to the cement suspension, $X_{add} = \left(\frac{1}{1-2.5\left(1+\frac{r_{ns}}{R_{ns}}\right)^3 \phi_{NS/W}} \right)\left(Q_1 e^{Q_2\phi}(1-\phi_f) + Q_3 e^{Q_4\phi}\phi_f\right)$.

From the above equations, it is obvious that two boundary conditions are needed for the velocity and two for the volume fraction. Initial conditions also need to be specified for v and ϕ.

4.2. Boundary and Initial Conditions

We impose the no-slip boundary condition and we use the symmetry condition as indicated in Equation (33).

$$\overline{v}(1,\overline{t}) = 0 \quad \frac{\partial \phi}{\partial \overline{r}}(1,\overline{t}) = 0 \tag{33}$$

For the volume fraction ϕ, an average value is assumed, ϕ_{avg} [34], where

$$\int_0^1 \phi d\overline{r} = \phi_{avg} \tag{34}$$

For the initial conditions, based on [41], we apply a uniform value for the velocity at $t = 0$ and impose a parabolic profile for the volume fraction [41].

$$\overline{v}(\overline{r},0) = 1 \quad \phi(\overline{r},0) = \overline{r}^2(3-2\overline{r}) \tag{35}$$

According to Equation (35), this initial concentration near the centerline is smaller than the concentration of the particles at the wall. We carried out some tests for a uniform volume fraction profile at $t = 0$ and we did not notice any noticeable deviation from the suggested equation given in Equation (35).

Also, the flow rate or the flow output is defined as,

$$\overline{Q}(\overline{t}) = 2\pi \int_0^1 \overline{v}(\overline{r},\overline{t})\overline{r} d\overline{r} \tag{36}$$

4.3. Numerical Results and Discussion

MATLAB (R2022a) *pdepe* is a function that is used to solve initial-boundary value problems for time-dependent partial differential equations (PDEs) that describe various physical phenomena. This solver is particularly useful for problems with one spatial dimension and time evolution. The above dimensionless partial differential equations are solved using the MATLAB *pdepe* solver for one spatial variable $\overline{r}$ and time $\overline{t}$. The step size is the default value used in the solver. In this section, we will present the results for the velocity and volume fraction profiles for a limited range of dimensionless numbers. In this paper, we focus on the effects of $\phi_{NS/W}$ (the ratio of nano-silica particle volume to the total volume of nano-silica) and ϕ_f (the fly ash content); by varying these parameters, we look at the effects on the velocity and the volume fraction fields. In this section, a parametric study is performed to investigate the effect of different volume fractions of the additives at different time cycles on the behavior of fresh cement suspension. Table 1 shows the values of the volume fraction of the additives and the dimensionless parameters.

Table 1. Values of the dimensionless parameters.

Parameters	Values
$\phi_{NS/W}$	0.001, 0.005, 0.01, 0.02, 0.04
ϕ_f	0.01, 0.05, 0.1, 0.2, 1
ϕ_{avg}	0.4
k	6.5
χ	3.15
ζ	1
m	0.5
Re	100
P	10
G_0	0
κ	0.001
Q_1	4.04×10^{-5}
Q_2	19
Q_3	0.54×10^{-5}
Q_4	20.9

4.3.1. NS-Added Particles to the Cement Suspension

In this section, we demonstrate the effect of nano-silica content on the velocity and the cement volume fraction. As an example, we chose the following dimensionless values for our numerical simulations: $\phi_{avy} = 0.4$, k = 6.5, $\chi = 3.15$, $\zeta = 1$, m = 0.5, $Re = 100$, $P = 10$, $G_0 = 0$, $\kappa = 0.001$. Figure 2a–d show the effects of different $\phi_{NS/W}$ on the velocity at four different time cycles (cycle #2, #10, #50, and #220). According to Figure 2a–d, larger $\phi_{NS/W}$ values produce a lower centerline velocity. The velocity profile reaches a steady state and does not change much after 50 time cycles. Figure 3a–d show the effects of different $\phi_{NS/W}$ on the cement volume fraction at four different time cycles (cycle #2, #10, #50, and #220). The results show that higher values of $\phi_{NS/W}$ results in larger non-uniformity of the volume fraction, especially at the early stages. The volume fraction profiles reach a steady state after 220 time cycles. Figure 4a,b show the effects of $\phi_{NS/W}$ on the velocity versus time cycle and the flow rate versus time cycle. The fresh cement suspension with higher nano-silica content reaches the steady state condition faster than those with lower nano-silica content. The results align well with the findings in the literature [9,50–53] for the effect of nano-silica additives on the early-stage performance of fresh cement suspension. According to Lavergne, et al. [9] and Kim, et al. [10], nano-silica particles accelerated the initial hydration reaction of the cement suspension and increased the shear strength and the early strength significantly because of the specific surface of the nanoparticles, leading to a shorter time of set and a higher compressive early strength.

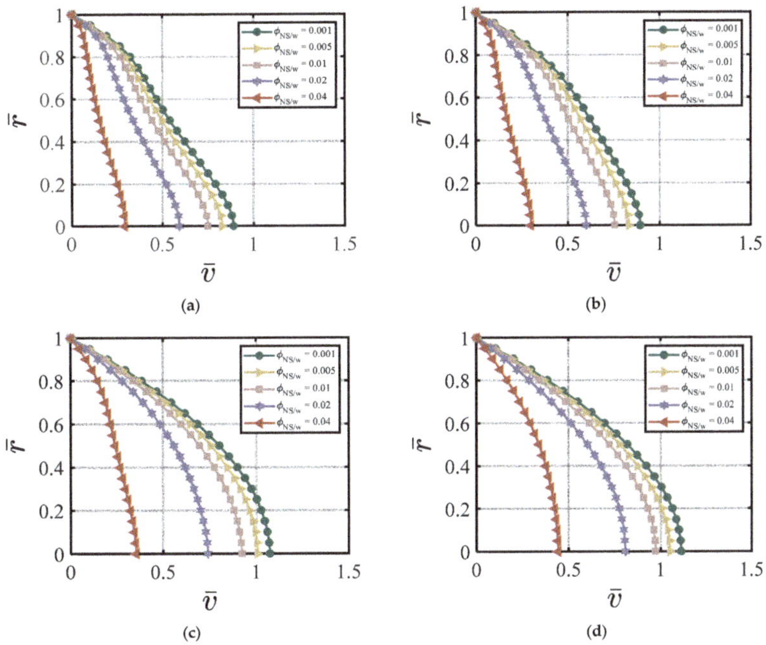

Figure 2. Effect of $\phi_{NS/W}$ on the velocity at (**a**) cycle #2; (**b**) cycle #10; (**c**) cycle #50; (**d**) cycle #220 with $\phi_{avg} = 0.4$, k = 6.5, χ = 3.15, ζ = 1, m = 0.5, *Re* = 100, *P* = 10, G_0 = 0, κ = 0.001.

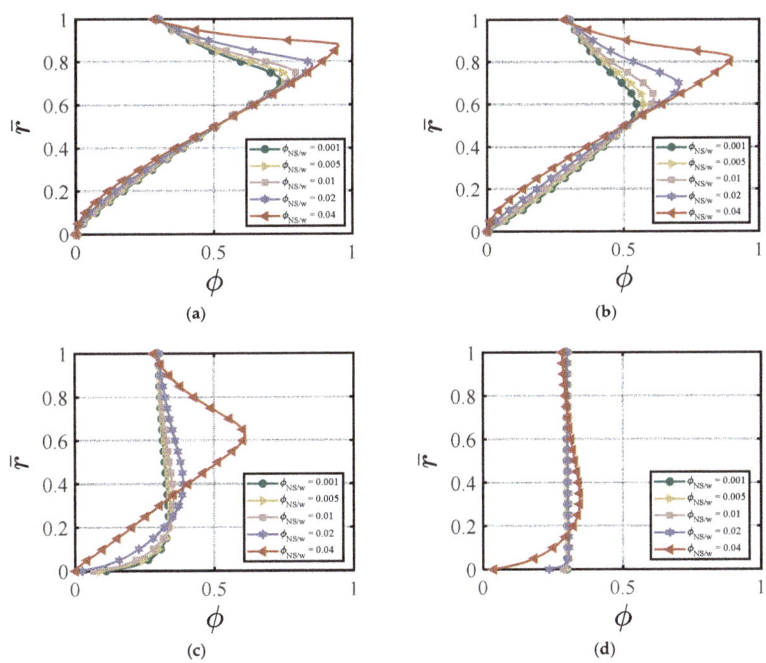

Figure 3. Effect of $\phi_{NS/W}$ on the volume fraction at (**a**) cycle #2; (**b**) cycle #10; (**c**) cycle #50; (**d**) cycle #220 with $\phi_{avg} = 0.4$, k = 6.5, χ = 3.15, ζ = 1, m = 0.5, *Re* = 100, *P* = 10, G_0 = 0, κ = 0.001.

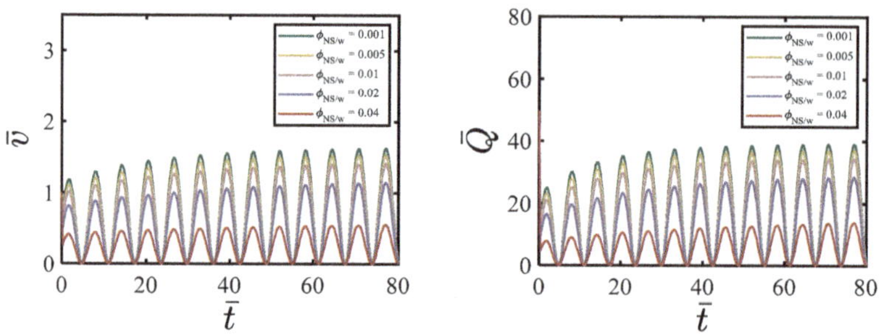

(a) (b)

Figure 4. Effect of $\phi_{NS/W}$ on (**a**) $\overline{v}$ vs. t; (**b**) $\overline{Q}$ vs. t with $\phi_{avg} = 0.4$, k = 6.5, χ = 3.15, ζ = 1, m = 0.5, Re = 100, P = 10, G_0 = 0, κ = 0.001.

4.3.2. FA-Added Particles to the Cement Suspension

Here, we use the following values for our numerical simulations: $\phi_{avy} = 0.4$, $Q_1 = 4.04 \times 10^{-5}$, $Q_2 = 19$, $Q_3 = 0.54 \times 10^{-5}$, $Q_4 = 20.9$, χ = 3.15, ζ = 1, m = 0.5, Re = 100, P = 10, G_0 = 0, κ = 0.001. Figure 5a–d show the effects of different values of ϕ_f on the velocity and the volume fraction at four different time cycles #5, #10, #50, and #200. According to Figure 5a–d, a higher ϕ_f value causes a higher centerline velocity, which exhibits an opposite trend compared to the nano-silica additives in Figure 2a–d. The velocity profile reaches a steady state after 20 time cycles. Figure 6a–d show the effects of different values of ϕ_f on the volume fraction at four different time cycle #5, #10, #50, and #200. The effect of the fly ash particles on the volume fraction is not that obvious. And, the volume fraction field reaches its steady state value very fast (within 5 time cycles). Figure 7a,b show the effects of ϕ_f on the velocity versus time cycle and the flow rate versus time cycle. In contrast to the cement suspension with nano-silica additives, the cement suspension with lower fly ash content reaches the steady state condition faster than those with higher fly ash content. The behavior agrees with the findings by Wang, et al. [50], who indicated that the effect of nano-silica on cement hydration and performance happened in the early stages while high content of fly ash contributed to the strength growth in the later stages. The cement suspension with higher nano-silica content reaches the steady state condition faster than those with lower nano-silica content. In addition, the findings align well with the study by García-Taengua, et al. [20], who indicated that fly ash affected the viscosity only when used together with nano-silica additives, and the fly ash did not affect the dormant period of cement hydration.

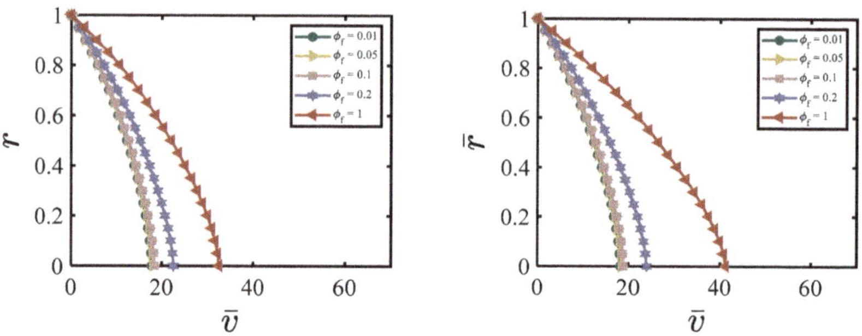

Figure 5. *Cont.*

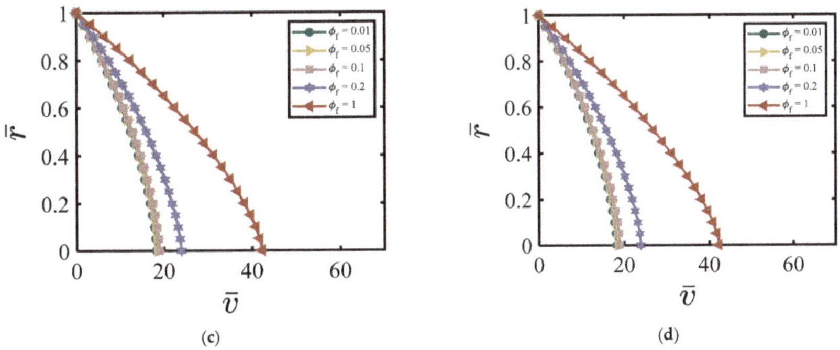

Figure 5. Effect of ϕ_f on the velocity at (**a**) cycle #5; (**b**) cycle #10; (**c**) cycle #50; (**d**) cycle #200 with $\phi_{avg} = 0.4$, $Q_1 = 4.04 \times 10^{-5}$, $Q_2 = 19$, $Q_3 = 0.54 \times 10^{-5}$, $Q_4 = 20.9$, $\chi = 3.15$, $\zeta = 1$, m = 0.5, $Re = 100$, $P = 10$, $G_0 = 0$, $\kappa = 0.001$.

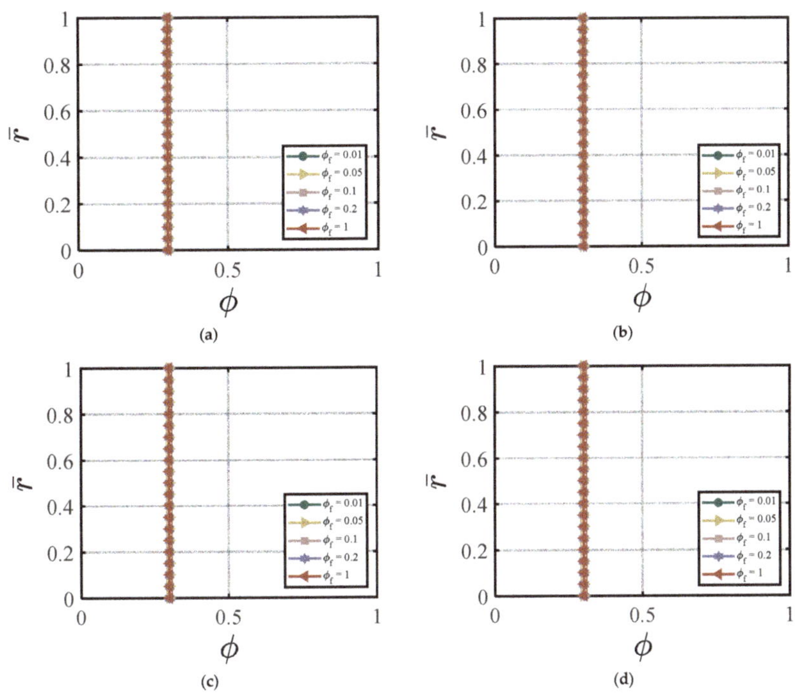

Figure 6. Effect of ϕ_f on the volume fraction at (**a**) cycle #5; (**b**) cycle #10; (**c**) cycle #50; (**d**) cycle #200 with $\phi_{avg} = 0.4$, $Q_1 = 4.04 \times 10^{-5}$, $Q_2 = 19$, $Q_3 = 0.54 \times 10^{-5}$, $Q_4 = 20.9$, $\chi = 3.15$, $\zeta = 1$, m = 0.5, $Re = 100$, $P = 10$, $G_0 = 0$, $\kappa = 0.001$.

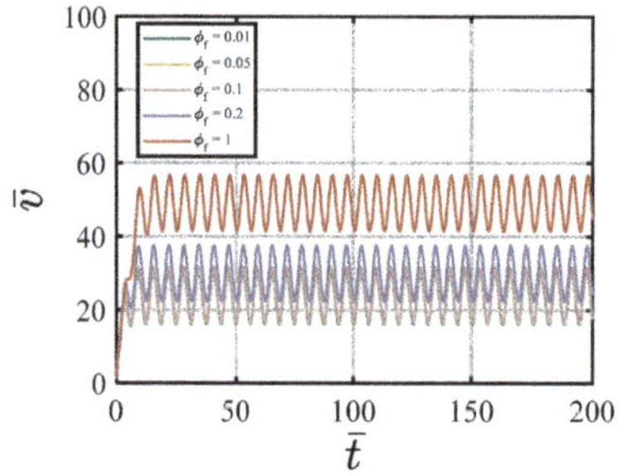

(a)

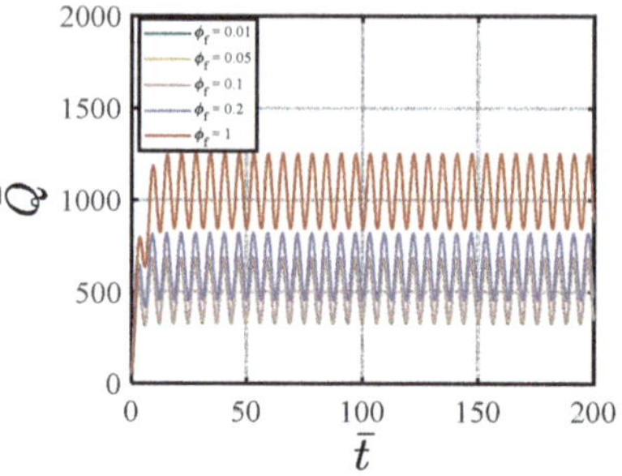

(b)

Figure 7. Effect of ϕ_f on (**a**) $\bar{v}$ vs. t; (**b**) $\overline{Q}$ vs. t with $\phi_{avg} = 0.4$, $Q_1 = 4.04 \times 10^{-5}$, $Q_2 = 19$, $Q_3 = 0.54 \times 10^{-5}$, $Q_4 = 20.9$, $\chi = 3.15$, $\zeta = 1$, m = 0.5, $Re = 100$, $P = 10$, $G_0 = 0$, $\kappa = 0.001$.

4.3.3. Combined NS and FA-Added Particles to the Cement Suspension

Here, we use the following values for our numerical simulations: $\phi_{avy} = 0.4$, $Q_1 = 4.04 \times 10^{-5}$, $Q_2 = 19$, $Q_3 = 0.54 \times 10^{-5}$, $Q_4 = 20.9$, $\chi = 3.15$, $\zeta = 1$, m = 0.5, $Re = 100$, $P = 10$, $G_0 = 0$, $\kappa = 0.001$. Figure 8a–d show the effects of $\phi_{NS/W}$ on the velocity of the cement suspension with combined nano-silica and fly ash additives at four different time cycles # 2, #10, #50, and #200. From Figure 8a–d, the effect of different values of $\phi_{NS/W}$ on velocity is similar to the NS-added cement suspension case. Larger $\phi_{NS/W}$ values cause smaller centerline velocity and with the increasing time cycle value, the velocity field does not change much. Figure 9a–d show the effects of $\phi_{NS/W}$ on the volume fraction of cement suspension with combined additives at four different time cycles #2, #10, #50, and #200. The volume fraction becomes more uniform in the early time cycle (#10). As shown in Figure 9a–d, after the time cycle 10, the volume fraction profile becomes uniform. Figure 10a,b show the effects of $\phi_{NS/W}$ of the cement suspension with combined additives on the velocity and flow rate versus time. According to Figures 8a–d and 10a,b, the dominance of nano-silica is also exhibited with a similar trend in Section 4.3.1. For the case of combined nano-silica and fly ash particles, we find that the nano-silica particles play a dominant effect, where the velocity and the volume fraction fields show similar trends to the case of only nano-silica-added cement suspension. The finding is validated by Wang et al. [50], that nano-silica particles enhance hydration in early stages and provide favorable hydration conditions for fly ash in later stages. García-Taengua, et al. [20] also showed that fly ash did not affect the dormant period of hydration, while the addition of nano-silica shortened the cement hydrate phase. In addition, the synergistic effect of the two additives increases non-evaporative water content and improves pozzolanic activity. Similar findings are found by Kim, et al. [10], indicating the dominant effect of nano-silica added to high-volume fly ash cement composites, resulting in shorter setting time, higher initial and middle stages of hydration rate, and higher compressive strength.

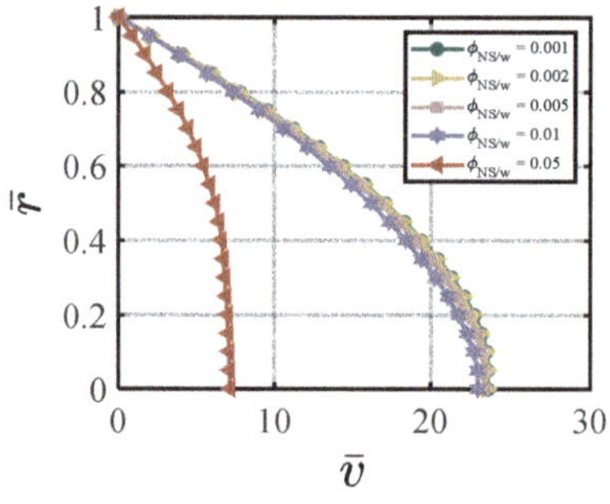

(a)

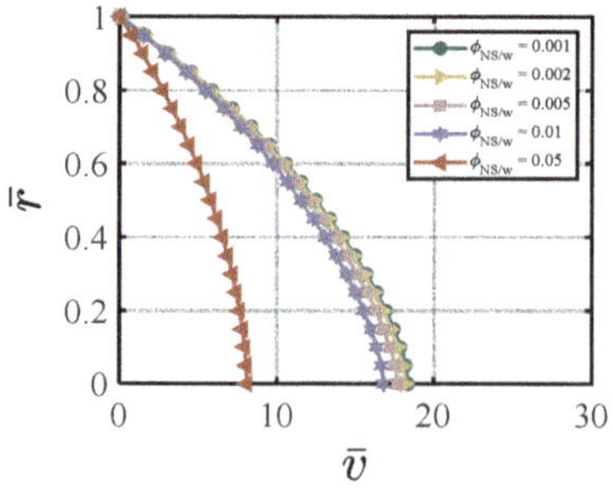

(b)

Figure 8. *Cont.*

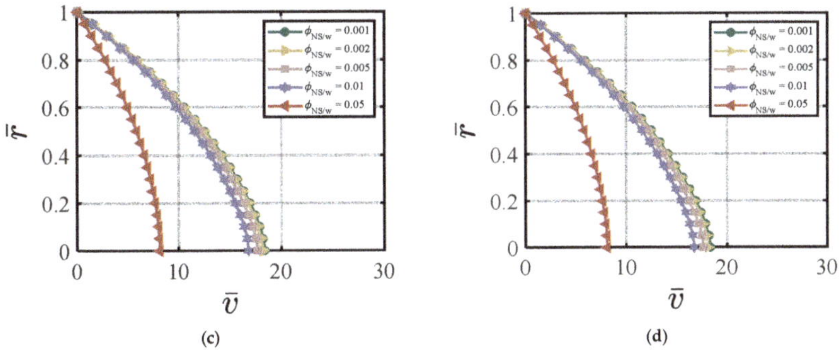

Figure 8. Effect of $\phi_{NS/W}$ of the combined NS FA-added cement suspension on velocity at (**a**) cycle #2; (**b**) cycle #10; (**c**) cycle #50; (**d**) cycle #200 with $\phi_{avy} = 0.4$, k = 6.5, $Q_1 = 4.04 \times 10^{-5}$, $Q_2 = 19$, $Q_3 = 0.54 \times 10^{-5}$, $Q_4 = 20.9$, $\chi = 3.15$, $\zeta = 1$, m = 0.5, $Re = 100$, $P = 10$, $G_0 = 0$, $\kappa = 0.001$.

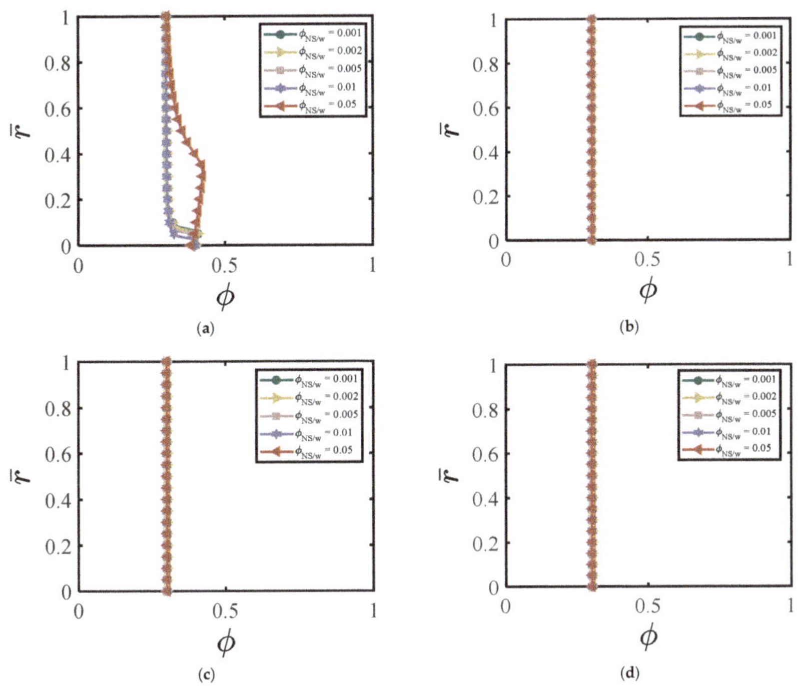

Figure 9. Effect of $\phi_{NS/W}$ of the combined NS FA-added cement suspension on volume fraction at (**a**) cycle #2; (**b**) cycle #10; (**c**) cycle #50; (**d**) cycle #200 with $\phi_{avy} = 0.4$, k = 6.5, $Q_1 = 4.04 \times 10^{-5}$, $Q_2 = 19$, $Q_3 = 0.54 \times 10^{-5}$, $Q_4 = 20.9$, $\chi = 3.15$, $\zeta = 1$, m = 0.5, $Re = 100$, $P = 10$, $G_0 = 0$, $\kappa = 0.001$.

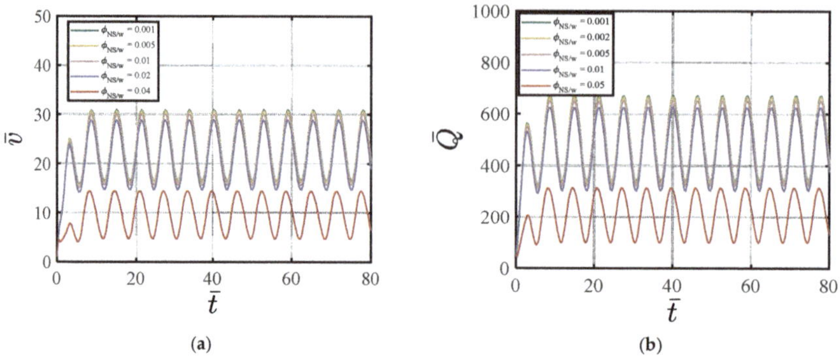

Figure 10. Effect of $\phi_{NS/W}$ of the combined NS FA-added cement suspension on (**a**) $\overline{v}$ vs. t; (**b**) $\overline{Q}$ vs. t with $\phi_{avy} = 0.4$, k = 6.5, $Q_1 = 4.04 \times 10^{-5}$, $Q_2 = 19$, $Q_3 = 0.54 \times 10^{-5}$, $Q_4 = 20.9$, χ = 3.15, ζ = 1, m = 0.5, Re = 100, P = 10, G_0 = 0, κ = 0.001.

5. Conclusions

In this paper, we look at the flow of a fresh cement suspension in a pipe where two different additives and when they are combined are added to the cement suspension: (1) nano-silica, (2) fly ash, and (3) nano-silica and fly ash. The main findings of the paper are as follows:

1. The cement suspension is modeled as a non-Newtonian fluid, where a modified form of the power-law model is used to study the dependency of the cement suspension viscosity on the shear rate and the volume fraction. The governing equations are made dimensionless, and a limited parametric study is performed for the different dimensionless numbers and parameters.
2. For the parametric study, we focused on the effects of $\phi_{NS/W}$ (the ratio of NS particle volume to the total volume of nano-silica), ϕ_f (the fly ash content), and the combined two additives; by varying these parameters, we looked at their effects on the velocity and the volume fraction fields.
3. Based on the limited numerical simulations, the results indicate that the velocity and the volume fraction profiles are affected by the addition of different additives. For the pulsating flow of a cement suspension with various additives, we can see that larger values of nano-silica particles can cause a smaller centerline velocity and a higher non-uniformity of the volume fraction field. In contrast to nano-silica particles, larger values of fly ash cause a higher centerline velocity. The effect of fly ash content on the volume fraction is not that obvious.
4. It can also be seen that the fresh cement suspensions with smaller values of nano-silica reach the steady state conditions faster, while the cement suspension with added fly ash particles reaches the steady state very fast.
5. For the case of combined nano-silica and fly ash particles, the nano-silica plays a dominant effect, where the velocity and the volume fraction fields show similar trends to the case of nano-silica-added cement suspension.

These findings can help the design and operation of the pulsating flow of fresh cement mortars and concrete in the 3D printing industry. For future studies, we will do a more systematic parametric study by changing the dimensionless numbers. We will also consider the effect of other factors such as the type of nanoparticles, dispersion of the additives, water-to-cement ratio, particle size and shape, temperature, and the direction of flow on the behavior of fresh cement suspension. In addition, we plan to conduct rheological testing in the laboratory by measuring the viscosity and the yield stress of fresh cement suspensions with various additives under different conditions to check the applicability of the modeling results.

Author Contributions: C.T. performed most of the derivations and all the numerical simulations, and M.M. developed the framework of the paper. Both authors contributed to the writing of the paper. All authors have read and agreed to the published version of the manuscript.

Funding: Funding for this work (performed by C.T.) was provided by the ACI Foundation (ACIF) Program under its Concrete Research Council (CRC) through Grant No. P0068, the National Science Foundation (NSF) I-Corps through Grant No. TI-2243641, and the faculty start-up funding from the School of Construction Management Technology at Purdue University. The findings and opinions expressed in this article are those of the authors only and do not necessarily reflect the views of the sponsors.

Institutional Review Board Statement: Not applicable.

Informed Consent Statement: Not applicable.

Data Availability Statement: Not applicable.

Conflicts of Interest: The authors declare no conflicts of interest.

Nomenclature

Symbol	**Explanation**
ρ	Density
g	Acceleration due to gravity
r	Radius of the horizontal pipe
V	Reference velocity
a	Particle radius
r_{ns}	Adsorbed water layer thickness for the nanoparticle
R_{ns}	Radius of nanoparticles
t	Time
D	Diffusion coefficient
D_0	Cement suspension diffusivity parameter
D_1	Diffusion coefficient
D_2	Volume fraction-dependent diffusion coefficient
η	Constant in volume fraction-dependent diffusion coefficient
$\dot{\gamma}$	Local shear rate
ϕ	Volume fraction of particles
$\phi_{NS/W}$	Volume fraction of nano-silica particles
ϕ_f	Volume fraction of fly ash
Q_n	Experimental fitting parameters for different types of fly ash
φ	Capillary pore in cement suspension
μ	Viscosity
μ_0	Coefficient of viscosity
μ_{eff}	Effective viscosity
$\mu^*(\phi)$	Concentration-dependent viscosity
$\mu^{**}(\phi)$	Additive concentration-dependent viscosity
β	Experimental (fitting) parameter for viscosity
ϕ_m	Maximum particle concentration
ϕ_{avg}	Average volume fraction
p	Pressure
α, m	Material parameters
Re	Reynolds number
$\delta, G_0, \kappa, \zeta, \chi$	Dimensionless numbers
$\boldsymbol{v}$	Velocity vector
$\boldsymbol{b}$	Body force vector
$\boldsymbol{N}$	Particle transport flux
$\boldsymbol{T}$	Cauchy stress tensor
$\boldsymbol{L}$	Gradient of the velocity vector
$\boldsymbol{A_n}$	n-th order Rivlin–Ericksen tensor
∇	Gradient symbol
div	Divergence operator

References

1. Khayat, K.H.; Meng, W.; Vallurupalli, K.; Teng, L. Rheological properties of ultra-high-performance concrete—An overview. *Cem. Concr. Res.* **2019**, *124*, 105828. [CrossRef]
2. Meng, W.; Khayat, K.H. Mechanical properties of ultra-high-performance concrete enhanced with graphite nanoplatelets and carbon nanofibers. *Compos. Part B Eng.* **2016**, *107*, 113–122. [CrossRef]
3. Meng, W.; Khayat, K.H. Effect of graphite nanoplatelets and carbon nanofibers on rheology, hydration, shrinkage, mechanical properties, and microstructure of UHPC. *Cem. Concr. Res.* **2018**, *105*, 64–71. [CrossRef]
4. Wu, Z.; Khayat, K.H.; Shi, C.; Tutikian, B.F.; Chen, Q. Mechanisms underlying the strength enhancement of UHPC modified with nano-SiO_2 and nano-$CaCO_3$. *Cem. Concr. Compos.* **2021**, *119*, 103992. [CrossRef]
5. Ghafari, E.; Costa, H.; Júlio, E. Critical review on eco-efficient ultra high performance concrete enhanced with nano-materials. *Constr. Build. Mater.* **2015**, *101*, 201–208. [CrossRef]
6. Praveenkumar, T.R.; Vijayalakshmi, M.M.; Meddah, M.S. Strengths and durability performances of blended cement concrete with TiO2 nanoparticles and rice husk ash. *Constr. Build. Mater.* **2019**, *217*, 343–351. [CrossRef]
7. Long, G.; Li, Y.; Ma, C.; Xie, Y.; Shi, Y. Hydration kinetics of cement incorporating different nanoparticles at elevated temperatures. *Thermochim. Acta* **2018**, *664*, 108–117. [CrossRef]
8. Aggarwal, P.; Singh, R.P.; Aggarwal, Y. Use of nano-silica in cement based materials—A review. *Cogent Eng.* **2015**, *2*, 1078018. [CrossRef]
9. Lavergne, F.; Belhadi, R.; Carriat, J.; Fraj, A.B. Effect of nano-silica particles on the hydration, the rheology and the strength development of a blended cement paste. *Cem. Concr. Compos.* **2019**, *95*, 42–55. [CrossRef]
10. Kim, B.-J.; Lee, G.-W.; Choi, Y.-C. Hydration and mechanical properties of high-volume fly ash concrete with nano-silica and silica fume. *Materials* **2022**, *15*, 6599. [CrossRef]
11. Yang, T.; Zhu, H.; Zhang, Z.; Gao, X.; Zhang, C.; Wu, Q. Effect of fly ash microsphere on the rheology and microstructure of alkali-activated fly ash/slag pastes. *Cem. Concr. Res.* **2018**, *109*, 198–207. [CrossRef]
12. Jiang, D.; Li, X.; Lv, Y.; Zhou, M.; He, C.; Jiang, W.; Liu, Z.; Li, C. Utilization of limestone powder and fly ash in blended cement: Rheology, strength and hydration characteristics. *Constr. Build. Mater.* **2020**, *232*, 117228. [CrossRef]
13. Wang, Q.; Cui, X.; Wang, J.; Li, S.; Lv, C.; Dong, Y. Effect of fly ash on rheological properties of graphene oxide cement paste. *Constr. Build. Mater.* **2017**, *138*, 35–44. [CrossRef]
14. Szostak, B.; Golewski, G.L. Rheology of cement pastes with siliceous fly ash and the CSH nano-admixture. *Materials* **2021**, *14*, 3640. [CrossRef] [PubMed]
15. Sobolev, K.; Vivian, I.F.; Saha, R.; Wasiuddin, N.M.; Saltibus, N.E. The effect of fly ash on the rheological properties of bituminous materials. *Fuel* **2014**, *116*, 471–477. [CrossRef]
16. Laskar, A.I.; Talukdar, S. Rheological behavior of high performance concrete with mineral admixtures and their blending. *Constr. Build. Mater.* **2008**, *22*, 2345–2354. [CrossRef]
17. Provis, J.L.; Duxson, P.; van Deventer, J.S.J. The role of particle technology in developing sustainable construction materials. *Adv. Powder Technol.* **2010**, *21*, 2–7. [CrossRef]
18. Sybertz, F.; Reick, P. Effect of fly as on the rheological properties of cement paste. In *Rheology of Fresh Cement and Concrete*; Banfill, P.F.G., Ed.; Proceedings of an International Conference, Liverpool, 1991; CRC Press: London, UK; pp. 13–22.
19. Golewski, G.L. Concrete composites based on quaternary blended cements with a reduced width of initial microcracks. *Appl. Sci.* **2023**, *13*, 7338. [CrossRef]
20. García-Taengua, E.; Sonebi, M.; Hossain, K.M.A.; Lachemi, M.; Khatib, J. Effects of the addition of nanosilica on the rheology, hydration and development of the compressive strength of cement mortars. *Compos. Part B Eng.* **2015**, *81*, 120–129. [CrossRef]
21. Bessaies-Bey, H.; Khayat, K.H.; Palacios, M.; Schmidt, W.; Roussel, N. Viscosity modifying agents: Key components of advanced cement-based materials with adapted rheology. *Cem. Concr. Res.* **2022**, *152*, 106646. [CrossRef]
22. Papo, A.; Piani, L. Effect of various superplasticizers on the rheological properties of Portland cement pastes. *Cem. Concr. Res.* **2004**, *34*, 2097–2101. [CrossRef]
23. Zhang, Q.; Chen, J.; Zhu, J.; Yang, Y.; Zhou, D.; Wang, T.; Shu, X.; Qiao, M. Advances in Organic Rheology-Modifiers (Chemical Admixtures) and Their Effects on the Rheological Properties of Cement-Based Materials. *Materials* **2022**, *15*, 8730. [CrossRef] [PubMed]
24. Souza, M.T.; Ferreira, I.M.; de Moraes, E.G.; Senff, L.; Arcaro, S.; Pessôa, J.R.C.; Ribeiro, M.J.; de Oliveira, A.P.N. Role of chemical admixtures on 3D printed Portland cement: Assessing rheology and buildability. *Constr. Build. Mater.* **2022**, *314*, 125666. [CrossRef]
25. Tao, C.; Watts, B.; Ferraro, C.C.; Masters, F.J. A Multivariate Computational Framework to Characterize and Rate Virtual Portland Cements. *Comput.-Aided Civ. Infrastruct. Eng.* **2019**, *34*, 266–278. [CrossRef]
26. Watts, B.; Tao, C.; Ferraro, C.; Masters, F. Proficiency analysis of VCCTL results for heat of hydration and mortar cube strength. *Constr. Build. Mater.* **2018**, *161*, 606–617. [CrossRef]
27. Skripkiunas, G.; Karpova, E.; Barauskas, I.; Bendoraitiene, J.; Yakovlev, G. Rheological properties of cement pastes with multiwalled carbon nanotubes. *Adv. Mater. Sci. Eng.* **2018**, *2018*, 8963542. [CrossRef]
28. Wallevik, J.E. Thixotropic investigation on cement paste: Experimental and numerical approach. *J. Non-Newton. Fluid Mech.* **2005**, *132*, 86–99. [CrossRef]

29. Tao, C.; Kutchko, B.G.; Rosenbaum, E.; Massoudi, M. A Review of Rheological Modeling of Cement Slurry in Oil Well Applications. *Energies* **2020**, *13*, 570. [CrossRef]
30. Tao, C.; Rosenbaum, E.; Kutchko, B.; Massoudi, M. Pulsating Poiseuille flow of a cement slurry. *Int. J. Non-Linear Mech.* **2021**, *133*, 103717. [CrossRef]
31. Tao, C.; Rosenbaum, E.; Kutchko, B.G.; Massoudi, M. A Brief Review of Gas Migration in Oilwell Cement Slurries. *Energies* **2021**, *14*, 2369. [CrossRef]
32. Rajagopal, K.R.; Tao, L. *Mechanics of Mixtures*; World Scientific: Singapore, 1995; Volume 35, ISBN 9810215851.
33. Massoudi, M. A Mixture Theory formulation for hydraulic or pneumatic transport of solid particles. *Int. J. Eng. Sci.* **2010**, *48*, 1440–1461. [CrossRef]
34. Tao, C.; Kutchko, G.B.; Rosenbaum, E.; Wu, W.-T.; Massoudi, M. Steady Flow of a Cement Slurry. *Energies* **2019**, *12*, 2604. [CrossRef]
35. Slattery, J.C. *Advanced Transport Phenomena*; Cambridge University Press: Cambridge, UK, 1999; ISBN 1316583902.
36. Probstein, R.F. *Physicochemical Hydrodynamics: An Introduction*; John Wiley & Sons: Hoboken, NJ, USA, 2005; ISBN 0471725129.
37. Li, Y.; Wu, W.-T.; Liu, X.; Massoudi, M. The effects of particle concentration and various fluxes on the flow of a fluid-solid suspension. *Appl. Math. Comput.* **2019**, *358*, 151–160. [CrossRef]
38. Massoudi, M.; Tran, P.X. The Couette–Poiseuille flow of a suspension modeled as a modified third-grade fluid. *Arch. Appl. Mech.* **2016**, *86*, 921–932. [CrossRef]
39. Truesdell, C.; Noll, W. *The Non-linear Field Theories of Mechanics*; Springer-Verlag: Berlin/Heidelberg, Germany, 1992.
40. Rivlin, R.; Ericksen, J. Stress-Deformation Relations for Isotropic Materials. *J. Ration. Mech. Anal.* **1955**, *4*, 323–425. [CrossRef]
41. Bridges, C.; Rajagopal, K.R. Pulsatile flow of a chemically-reacting nonlinear fluid. *Comput. Math. Appl.* **2006**, *52*, 1131–1144. [CrossRef]
42. Batchelor, G.K.; Green, J.T. The determination of the bulk stress in a suspension of spherical particles to order c2. *J. Fluid Mech.* **1972**, *56*, 401–427. [CrossRef]
43. Asaga, K.; Roy, D.M. Rheological properties of cement mixes: IV. Effects of superplasticizers on viscosity and yield stress. *Cem. Concr. Res.* **1980**, *10*, 287–295. [CrossRef]
44. Krieger, I.M.; Dougherty, T.J. A mechanism for non-Newtonian flow in suspensions of rigid spheres. *Trans. Soc. Rheol.* **1959**, *3*, 137–152. [CrossRef]
45. Barnes, H.A. Thixotropy—A review. *J. Non-Newton. Fluid Mech.* **1997**, *70*, 1–33. [CrossRef]
46. Ouyang, J.; Han, B.; Chen, G.; Zhao, L.; Ou, J. A viscosity prediction model for cement paste with nano-SiO_2 particles. *Constr. Build. Mater.* **2018**, *185*, 293–301. [CrossRef]
47. Phillips, R.J.; Armstrong, R.C.; Brown, R.A.; Graham, A.L.; Abbott, J.R. A constitutive equation for concentrated suspensions that accounts for shear-induced particle migration. *Phys. Fluids A Fluid Dyn.* **1992**, *4*, 30–40. [CrossRef]
48. Carnahan, N.F.; Starling, K.E. Equation of state for nonattracting rigid spheres. *J. Chem. Phys.* **1969**, *51*, 635–636. [CrossRef]
49. Russel, W.B.; Russel, W.B.; Saville, D.A.; Schowalter, W.R. *Colloidal Dispersions*; Cambridge University Press: Cambridge, UK, 1991; ISBN 0521426006.
50. Wang, J.; Liu, M.; Wang, Y.; Zhou, Z.; Xu, D.; Du, P.; Cheng, X. Synergistic effects of nano-silica and fly ash on properties of cement-based composites. *Constr. Build. Mater.* **2020**, *262*, 120737. [CrossRef]
51. Li, H.; Zhang, L.; Ding, S.; Shu, X.; Wang, X.; Han, B. Effects and mechanisms of incorporated nanoparticles on the rheological performance of cement pastes. *J. Build. Eng.* **2023**, *73*, 106694. [CrossRef]
52. Zhang, C.; Kong, X.; Yin, J.; Fu, X. Rheology of fresh cement pastes containing polymer nanoparticles. *Cem. Concr. Res.* **2021**, *144*, 106419. [CrossRef]
53. Garcia-Rios, M.; Luquot, L.; Soler, J.M.; Cama, J. Influence of the flow rate on dissolution and precipitation features during percolation of CO2-rich sulfate solutions through fractured limestone samples. *Chem. Geol.* **2015**, *414*, 95–108. [CrossRef]

materials

MDPI

Article

Preparation of Steel-Slag-Based Hydrotalcite and Its Adsorption Properties on Cl^- and SO_4^{2-}

Zebo Dong [1], Bei Huang [1,2,*], Tao Zhang [1], Na Liu [1] and Zhongyang Mao [1,2]

[1] College of Materials Science and Engineering, Nanjing Tech University, Nanjing 211800, China; 202161103094@njtech.edu.cn (Z.D.); 201961203162@njtech.edu.cn (T.Z.); 202161103117@njtech.edu.cn (N.L.); mzy@njtech.edu.cn (Z.M.)

[2] State Key Laboratory of Materials-Oriented Chemical Engineering, Nanjing 211800, China

* Correspondence: huangbei@njtech.edu.cn; Tel.: +86-138-5160-2803

Abstract: Large amounts of chloride ions (Cl^-) and sulfate ions (SO_4^{2-}) are present in salt-washing wastewater, making it unsuitable for direct release. Adsorption can be used to eliminate Cl^- and SO_4^{2-} from salt-washing wastewater, and hydrotalcite is an excellent adsorbent with high adsorption properties for these ions because of a layered bimetallic hydroxide structure. The selective extraction of various metals, such as calcium, magnesium, aluminum, and iron, from steel slag via acid leaching facilitates the utilization of steel slag in the preparation of hydrotalcite. In this study, the leaching mechanism of metal in steel slag was investigated using steel slag as a raw material and acetic acid as the reaction medium. The study obtained the optimal leaching mechanism for preparing hydrotalcite. Hydrotalcite was synthesized from the steel slag leaching solution by hydrothermal synthesis, and its structure was characterized. The adsorption performance of Cl^- and SO_4^{2-} in salt-washing wastewater was investigated by solution adsorption experiments. The removal rates of Cl^- and SO_4^{2-} in salt-washing wastewater reached 12.8% and 38.0%, respectively. After multiple adsorption cycles, the removal rates increased to 98.0% for Cl^- and 96.4% for SO_4^{2-}.

Keywords: steel slag; hydrotalcite; hydrothermal synthesis; adsorption performance

1. Introduction

Soil salinization in the western region of China not only impedes the sustainable development of the agricultural economy but also presents significant resource issues. The process of "irrigation and salt washing" has been demonstrated to be a highly effective method for enhancing saline soils [1,2]. However, the salt-washing wastewater created from this technique has substantial amounts of Cl^- and SO_4^{2-}, which would be an environmental burden if discharged directly [3]. Adsorption is a common method used for removing Cl^- and SO_4^{2-} from wastewater generated during salt washing. Hydrotalcite, an adsorbent material that has interlayer anion exchangeability [4], has been widely used to remove harmful impurities such as phosphate [5,6], lead [7], and arsenic [8] from water. It is also effective in adsorbing Cl^- and SO_4^{2-} in salt-washing wastewater [9]. There are generally two types of raw materials used in producing hydrotalcite: pure chemicals and solid waste.

Steel slag, a solid waste produced by the iron and steel industry, is currently a focal point of research [10]. Steel slag contains iron, CaO, and MgO, which can be utilized as sintering materials [11]. Additionally, steel slag has been demonstrated to be environmentally friendly and possesses cementing properties when used as a subgrade material [12,13]. Steel slag can also be applied to prepare microcrystalline glass. Zhao Guizhou et al. prepared microcrystalline glass with varying alkalinity using steel slag, via a one-step sintering method [14]. They also discovered that combining steel slag with coal gangue resulted in microcrystalline glass with a heavy metal ion curing efficiency exceeding 90% [15]. In addition, steel slag has potential for use in the preparation of silicate cement and sulphoaluminate cement [16–19]. Yang Kaimin et al. showed that using steel slag as a substitute for iron raw materials did not change the main mineral composition of the clinker and resulted in a more comprehensive development of alite [20]. Deqiang Zhao et al. successfully prepared multiphase cement clinker from granular steel slag using multiphase sintering technology [21,22]. Since calcium silicate in steel slag has cementitious activity, it can be used as a cement admixture in production [23–25]. Nurul Hidayah Roslan and colleagues conducted research on replacing 20% of the cement in concrete with steel slag, which resulted in higher compressive strength at a later stage of curing compared to the control group [26]. Extracting metal elements from steel slag through acid leaching provides an opportunity for comprehensive utilization. This process leaches metal oxides, such as calcium, magnesium, aluminum, and iron, from the steel slag. These metal oxides can then be utilized for preparing hydrotalcite. Steel-slag-derived hydrotalcite significantly enhances the flame-retardant properties of materials [27]; Guanhao Liu et al. employed steel slag after nitric acid acidolysis to prepare roasted hydrotalcite by the co-precipitation method, which yields high catalytic activity and good stability [28]. It is worth noting that most acidolysis methods for steel slag utilize strong acids such as nitric acid and sulfuric acid [27–29]. However, properly treating these strong acid waste products is essential for preventing adverse environmental impacts.

Currently, there are only a few studies that concentrate on producing hydrotalcite from steel slag. This paper employs acetic acid to acidify the steel slag to investigate the leaching mechanism of metal elements in it under various process conditions. The hydrothermal process is used to prepare hydrotalcite from the leach solution of the steel slag. The study examines the adsorption effectiveness of hydrotalcite prepared for Cl^- and SO_4^{2-} found in salt-washing wastewater. The outcomes will help reduce costs linked with hydrotalcite production and enhance the resource utilization of valuable metals within steel slag, thus promoting sustainable and efficient use.

2. Materials and Methods

2.1. Raw Materials

The steel slag was provided by Oriental Runan Group Co. (Changzhou, China), processed by ball milling, and sieved through 20 mesh. Its chemical and mineral composition are shown in Table 1 and Figure 1. Steel slag mainly contains silica, calcium carbonate, RO phase (oxides of magnesium, iron, etc.), olivine, tricalcium silicate, dicalcium silicate, C_2F, $C_{12}A_7$, and so on. Figure 2 presents the FTIR pattern of the steel slag.

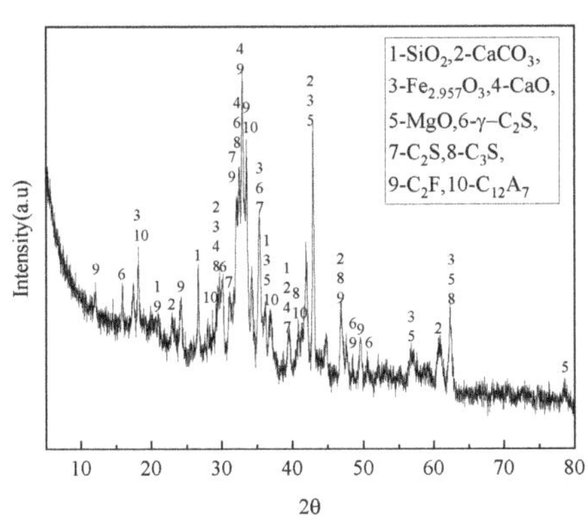

Figure 1. XRD pattern of steel slag.

Table 1. Chemical composition of steel slag.

Component	LOI	SiO_2	CaO	MgO	Al_2O_3	Fe_2O_3	SO_3
wt%	2.81	16.86	39.11	8.45	6.38	19.73	0.48

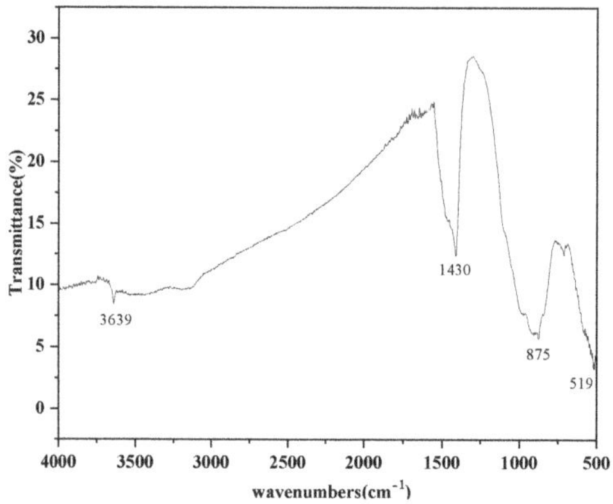

Figure 2. The FTIR patterns of steel slag.

The salted soil originating from Xinjiang was dried naturally and sieved through a 1 mm nylon sieve. Then, 20 g of salted soil was weighed into a conical flask, 100 mL of deionized water was added, and it was placed in a constant-temperature water bath oscillator and oscillated at a constant temperature of 25 °C for 2 min, and the salt washing wastewater was filtered immediately after the end of oscillation. The concentrations of Cl^- and SO_4^{2-} in the salt-washing wastewater were 12,486 mg/L and 6602 mg/L, respectively. The chemical reagents used in this paper are acetic acid (CH_3COOH), anhydrous ethanol, sodium hydroxide, and deionized water, all of which are analytically pure. Acetic acid was obtained from Sinopharm Group Chemical Reagent Co. (Shanghai, China) and NaOH was obtained from Xilong Science Co., Ltd. (Shantou, China); anhydrous ethanol was purchased from Wuxi Yasheng Chemical Co. (Wuxi, China).

2.2. Acidolysis of Steel Slag

Initially, the three-necked flask is prepared for the experiment by adding the acetic acid solution and inserting the thermometer. Once the thermometer reaches the desired temperature, steel slag is introduced to the flask. Stirring is ceased after a specific amount of time, and the resulting mixture is filtered by a pump to acquire the leaching solution of the steel slag.

The effect of acetic acid concentration on the selective leaching of target ions from steel slag was considered. Five different acetic acid concentrations of 3 mol/L, 4 mol/L, 5 mol/L, 6 mol/L, and 7 mol/L were selected during the experiment. To investigate the effect of liquid–solid ratio on ion leaching, five different liquid–solid ratios of 10 mL/g, 15 mL/g, 20 mL/g, 25 mL/g, and 30 mL/g were selected for the experiment. To investigate the effect of reaction time and reaction temperature on ion leaching, three different temperatures of 70 °C, 80 °C, and 90 °C were selected, respectively.

2.3. Preparation of Hydrotalcite

Hydrotalcite was synthesized hydrothermally using steel slag leach solution as the primary material. A set amount of said solution was added to a beaker, with a thermometer utilized to measure the temperature throughout the reaction. The beaker was placed in a magnetically stirred water bath for mixing, and the reaction temperature was maintained at <40 °C throughout the bath. A pH meter (Shanghai Lei Magnetics Co., Shanghai, China,

PHS-25) was inserted for pH measurement during the reaction; sodium hydroxide solution was added while stirring to adjust the pH to the desired value. The slurry obtained from the reaction is transferred to the hydrothermal reactor. Then, the hydrothermal reactor is put into the oven to be heated at a specific temperature for a particular time. After the reaction, the hydrothermal reactor was removed and cooled to room temperature, and the supernatant was poured off. The remaining material was washed numerous times with deionized water and anhydrous ethanol, and neutralized through centrifugation and filtration. The filter residue was dried to a constant weight in an oven at 80 °C, and then milled to obtain the Ca-Mg-Al-Fe hydrotalcite.

In order to investigate the effect of pH on the formation of phases, the pH was adjusted to 9, 10, 11, and 12 with sodium hydroxide. In order to investigate the effect of reaction temperature on the formation of phases in the hydrothermal reaction, four different reaction temperatures were chosen: 60 °C, 90 °C, 120 °C, and 150 °C. Four different reaction times, 12 h, 16 h, 20 h, and 24 h, were chosen to investigate the effect of reaction time on the formation of phases in the hydrothermal reaction.

2.4. Adsorption of Cl^- and SO_4^{2-} from Salt-Washing Wastewater by Hydrotalcite

Six 50 mL portions of salt-washing wastewater were taken, and 3 g of hydrotalcite was added to the solution. The solutions were placed on a thermostatic oscillator at a temperature of 25 °C and a speed of 150 rpm for 6 h. A portion of the solution was removed every 1 h filtered, and the concentrations of Cl^- and SO_4^{2-} in the remaining solution were measured. The above experiment was repeated several times to measure the concentrations of Cl^- and SO_4^{2-} in the remaining solution.

The adsorption of Cl^- and SO_4^{2-} in salt-washing wastewater is calculated as in Equation (1):

$$q = \frac{(c_0 - c_r) \times V}{m} \tag{1}$$

where:

q: Adsorption amount of hydrotalcite (mg/g);
C_0: Initial concentration of Cl^- and SO_4^{2-} (mg/L);
C_r: Ion concentration (mg/L) at the end of adsorption;
V: Volume of solution (L);
m: Amount of hydrotalcite added (g).

2.5. Testing Methods

The steel slag leachate samples were diluted 1000 times and analyzed by inductively coupled plasma optical emission spectroscopy (ICP-OES, Perkin Elmer, Waltham, MA, USA, Optima 7000DV) by analyzing the content of Ca^{2+}, Mg^{2+}, Al^{3+}, Fe^{3+}, and Si^{4+}. Ion chromatography ion chromatograph (IC, Metrohm, Herisau, Switzerland, ECO) was used to measure the concentrations of Cl^- and SO_4^{2-} in the salt-washing wastewater before and after hydrotalcite adsorption.

The prepared hydrotalcite was ground to a fine powder using an agate mortar, and the hydrotalcite was examined using X-ray diffraction (XRD, Rigaku, Tokyo, Japan, Smart Lab). The test conditions were the copper target, 30 Kv, 15 mA, scanning angle 10–90°, and scanning rate 10°/min.

The prepared hydrotalcite powders were analyzed using field emission scanning electron microscopy (FESEM, JEOL, Tokyo, Japan, JSM-6701F) and transmission electron microscopy (TEM, FEI, Hillsboro, OR, USA, FEI TF20) to investigate the morphology of hydrotalcite further.

The prepared hydrotalcite samples were analyzed using Fourier-transform infrared spectroscopy (FT-IR, Thermo Fisher Scientific, Waltham, MA, USA, Nicolet is50) to determine the type of bonds contained in the hydrotalcite structure, with a 400–4000 cm^{-1} test wave number.

Steel slag, prepared hydrotalcite samples, and adsorbed hydrotalcite samples were tested using a Surface Area and Porosimetry Analyzer (Gold APP, Beijing, China, V-Sorb 1800) to compare the size of the particular surface area of the hydrotalcite samples and to analyze the changes in their adsorption capacity.

3. Results

3.1. Determination of Steel Slag Acidolysis System

3.1.1. The Effect of Acetic Acid Concentration

Figure 3 illustrates the impact of acetic acid concentration on the selective leaching of Ca^{2+}, Mg^{2+}, Al^{3+}, Fe^{3+}, and Si^{4+} in steel slag. The results indicate that the concentrations of Ca^{2+} and Mg^{2+} in the leaching solution remain relatively constant, while the concentrations of Al^{3+} and Fe^{3+} gradually increase with the acetic acid concentration increases. Conversely, the concentration of Si^{4+} decreases as the acetic acid concentration increases. Consequently, it can be inferred that the concentration of acetic acid minimally impacts the selective leaching of Ca^{2+} and Mg^{2+} from the steel slag. However, it significantly affects the selective leaching of Al^{3+}, Fe^{3+}, and Si^{4+}. Additionally, it is worth noting that silicon in the steel slag tends to be leached as colloidal particles [30]. As the acetic acid concentration increases, the content of silicate gel may also rise, potentially hindering effective filtration. Therefore, it is crucial to strike a balance between attaining higher levels of Ca^{2+}, Mg^{2+}, Al^{3+}, and Fe^{3+} in the leaching solution and decreasing the concentration of Si^{4+}. Consequently, a 5 mol/L acetic acid solution was opted for, in order to accomplish these objectives.

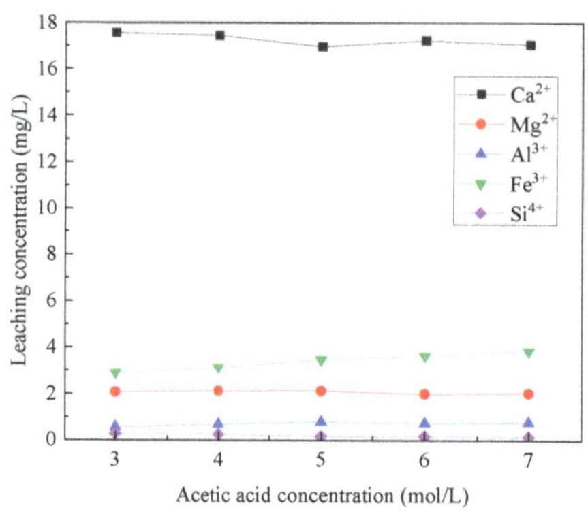

Figure 3. Effect of acetic acid concentration on selective leaching of steel slag.

3.1.2. The Effect of Liquid–Solid Ratio

The effects of different liquid–solid ratios on the selective leaching of Ca^{2+}, Mg^{2+}, Al^{3+}, Fe^{3+}, and Si^{4+} from steel slag are shown in Figures 4 and 5. Notably, the liquid–solid ratio plays a significant role in the selective leaching of these ions. Observing the figures, it is evident that the concentration of Ca^{2+}, Mg^{2+}, Al^{3+}, Fe^{3+}, and Si^{4+} in the leaching solution decreases with the increase of the liquid–solid ratio. The leach solution exhibits a higher concentration and decrease rate of calcium ions compared to other ions, which have a relatively lower concentration and slower decrease rate. The ratio of trivalent metal cations in the leaching solution to the sum of divalent and trivalent metal cations was calculated. The liquid–solid ratios of 10:1, 15:1, 20:1, 25:1, and 30:1 correspond to 0.148, 0.147, 0.151, 0.155, and 0.156, respectively. None of the ratios are between 0.2–0.33 [31], so it is necessary to choose the closer ones. It is also required to satisfy the relatively high content of Ca^{2+}, Mg^{2+}, Al^{3+}, and Fe^{3+} in the leaching solution and the relatively low content of Ca^{2+}, Mg^{2+}, Al^{3+}, Fe^{3+}, and other ions in the residual solution after the preparation of hydrotalcite. Therefore, the liquid–solid ratio of 25:1 was selected.

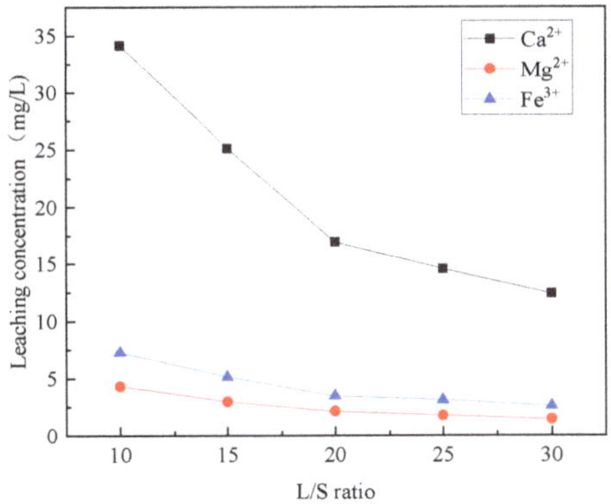

Figure 4. Effect of liquid–solid ratio on selective leaching of Ca^{2+}, Mg^{2+}, and Fe^{3+}.

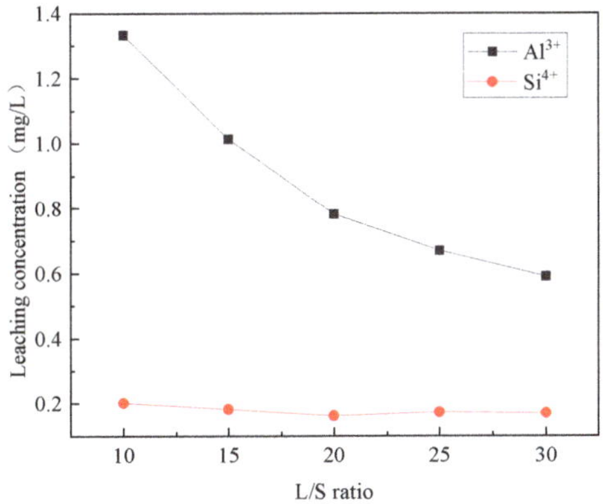

Figure 5. Effect of liquid–solid ratio on the selective leaching of Al^{3+} and Si^{4+}.

3.1.3. The Effect of Reaction Temperature

Figures 6–10 provide insights into the influence of different reaction temperatures and reaction times on the selective leaching of Ca^{2+}, Mg^{2+}, Al^{3+}, Fe^{3+}, and Si^{4+} from steel slag, respectively. Examining the figures, the concentration of Ca^{2+}, Mg^{2+}, Al^{3+}, and Fe^{3+} in the leaching solution increased continuously at first, before reaching a steady state, as the acidolysis of steel slag progressed. Conversely, the concentration of Si^{4+} underwent a decrease and then remained constant. The leaching of Ca^{2+} was completed within 120 min, whereas Mg^{2+} and Al^{3+} reached completion within 60 min. Fe^{3+} was leached rapidly in the initial stage of the reaction but slowly in the later stage. The overall leaching rate was relatively slow, and the leaching time was extended, taking about 180 min at 70 °C and 120 min at 80 °C and 90 °C. The Si^{4+} concentration in the leaching solution rapidly decreased in the first 20 min of the reaction and then slowly decreased after 20 min until the end. The figures further indicate that, with the increase in reaction temperature, the concentration of Ca^{2+}, Mg^{2+}, Al^{3+}, Fe^{3+}, and Si^{4+} in the leachate reached an extreme value, initially increased and then decreased. The leach solution at 80 °C had the highest concentration of Ca^{2+}, Mg^{2+}, Al^{3+}, Fe^{3+} and Si^{4+} at 15.88 mg/L, 1.762 mg/L, 0.671 mg/L, and 3.891 mg/L, respectively.

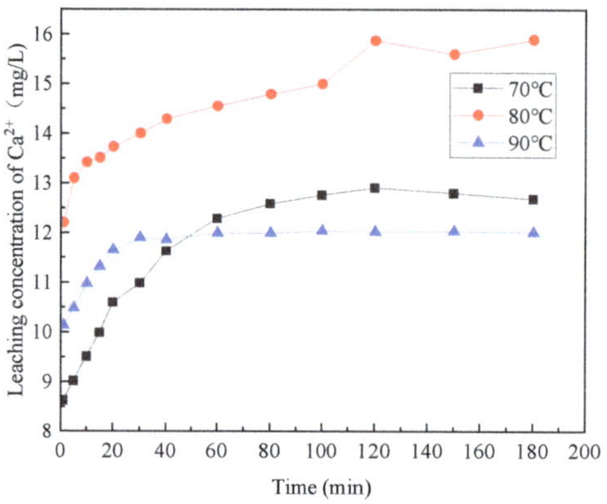

Figure 6. Effect of reaction temperature on the selective leaching of Ca^{2+}.

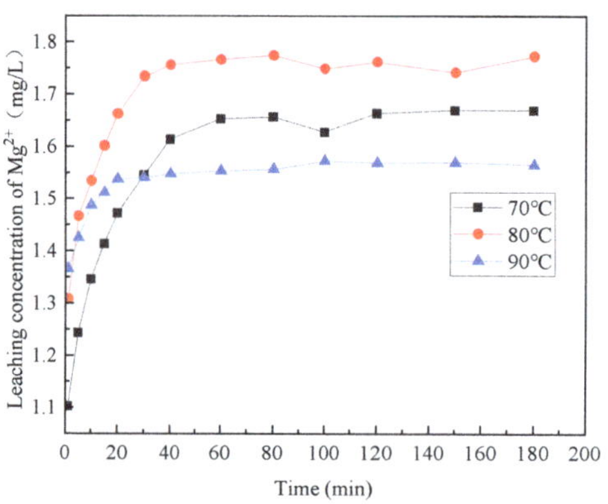

Figure 7. Effect of reaction temperature on the selective leaching of Mg^{2+}.

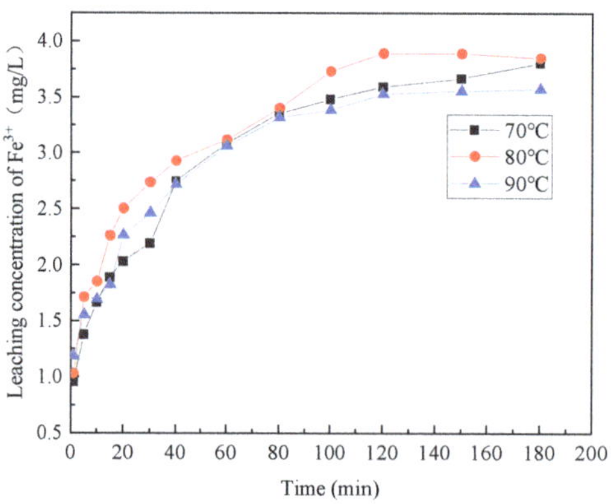

Figure 8. Effect of reaction temperature on the selective leaching of Fe^{3+}.

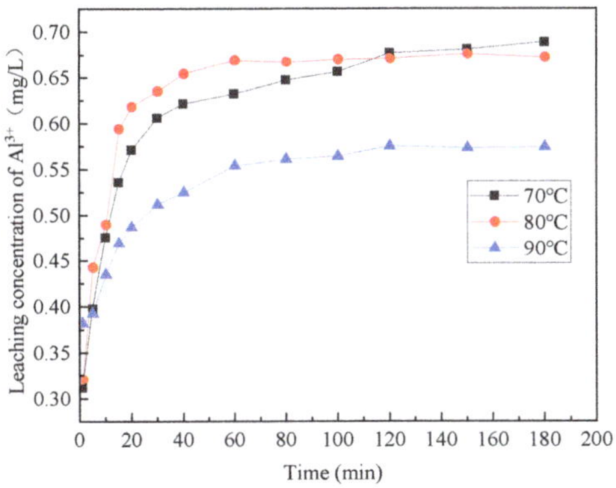

Figure 9. Effect of reaction temperature on the selective leaching of Al^{3+}.

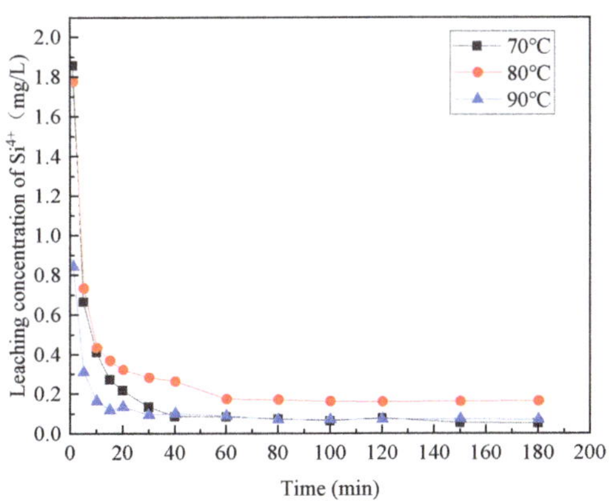

Figure 10. Effect of reaction temperature on the selective leaching of Si^{4+}.

Combining the above analyses, the highest concentration of ions was leached at the temperature of 80 °C, so 80 °C was chosen for the subsequent experiments. With the increase of reaction time, Ca^{2+}, Mg^{2+}, Al^{3+}, and Fe^{3+} leaching gradually increased, and Si^{4+} leaching gradually decreased until the ion concentration in the leach solution tends to be the constant at 2 h, so the reaction time was chosen as 2 h for the subsequent experiments.

Therefore, the optimal process conditions for the acidolysis of steel slag were an acetic acid concentration of 5 mol/L, liquid–solid ratio of 25:1, reaction temperature of 80 °C, and reaction time of 2 h. At this time, the concentrations of Ca^{2+}, Mg^{2+}, Al^{3+}, and Fe^{3+} in the leach solution were 15.88 mg/L, 1.762 mg/L, 0.671 mg/L, and 3.891 mg/L, respectively. The ratio of trivalent metal cations in the leaching solution to the sum of divalent and trivalent metal cations was 0.167, and the relatively high content of metal cations in the leaching solution allowed the preparation of more hydrotalcite.

3.2. Preparation of Steel-Slag-Based Hydrotalcite

3.2.1. The Effect of pH

The experiments utilized the aforementioned steel slag leach solution as the raw material, with pH values adjusted to 9, 10, 11, and 12, a temperature of 120 °C, and a duration of 16 h. Figure 11 showed the XRD patterns of the samples under different pH conditions. By comparing these patterns with the XRD patterns of standard magnesium–aluminum hydrotalcite, it was observed that the diffraction peaks of the samples indicated the presence of $CaCO_3$ and Fe_2O_3 with no evidence of hydrotalcite formation at pH 9. Conversely, when the pH values were 10, 11, and 12, the diffraction peaks of the hydrothermal reaction product aligned with those of the standard magnesium–aluminum hydrotalcite. The characteristic peaks of crystal surfaces (003), (006), and (009) were present, confirming the formation of hydrotalcite in the product. However, the intensity of the diffraction peaks for the prepared sample was lower compared to those of the standard Mg-Al hydrotalcite, indicating a lower degree of crystal growth along the crystal surfaces. Consequently, the crystallinity of the prepared sample was found to be inferior to that of the standard Mg-Al hydrotalcite.

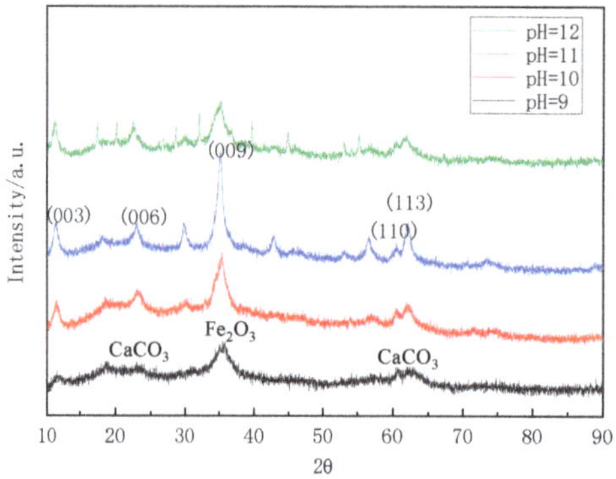

Figure 11. XRD plots of hydrotalcite under different pH conditions.

The XRD test data of samples under different pH conditions is illustrated in Table 2, which lists the corresponding angles and d values of the samples synthesized at pH 10, 11, and 12 on the (003), (006), and (009) crystal planes, as well as the half-peak widths on the (003) plane. It can be seen that the relationship $d_{(003)} = 2d_{(006)} = 3d_{(009)}$ was satisfied at pH 10 and 11, indicating that the hydrotalcite samples had a more regular lamellar structure [32]. Moreover, the half-peak width of 0.0119 rad on the (003) crystalline plane at pH 11 was smaller than that of 0.0142 rad at pH 10, which indicated that the diffraction peaks on the (003) crystalline plane of the sample were sharp and narrow. Thus, the crystallinity was better at pH 11. Therefore, the pH value of 11 was chosen for the subsequent study of the process conditions.

Table 2. XRD test data of hydrotalcite under different pH conditions.

Sample	(003)			(006)		(009)	
	$2\theta_{(003)}/°$	$d_{(003)}$/nm	$\beta_{(003)}$/rad	$2\theta_{(006)}/°$	$d_{(006)}$/nm	$2\theta_{(009)}/°$	$d_{(009)}$/nm
pH = 10	11.58	0.7635	0.0142	23.42	0.3795	35.46	0.2529
pH = 11	11.48	0.7704	0.0119	23.24	0.3824	35.18	0.2549
pH = 12	11.24	0.7867	0.0094	22.78	0.3900	35.36	0.2537

Figure 12 depicted the FT-IR spectra of hydrotalcite at various pH values. As depicted in the figure, the samples prepared at pH values of 10, 11, and 12 exhibited distinct peaks at 3458 cm^{-1}, 1559 cm^{-1}, 1365 cm^{-1}, and 590 cm^{-1}. The peak at 3458 cm^{-1} corresponded to the stretching vibrations of H-O-H on the laminar plate and the stretching vibrations of hydroxide bonding in water molecules within the interlayer. The peak at 1559 cm^{-1} represented the bending vibrations of the water of crystallization, while the peak at 1365 cm^{-1} signified the C-O stretching vibrations in CH_3COO^-. Additionally, the vibrational absorption peak at 590 cm^{-1} indicated the presence of metal–oxygen bonding. These findings served as further evidence of hydrotalcite synthesis.

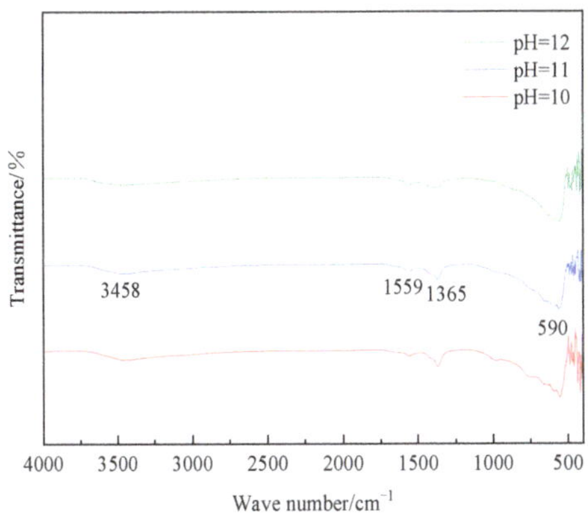

Figure 12. FT-IR plots of hydrotalcite under different pH conditions.

The SEM images of hydrotalcite at different pH values are given in Figure 13, from which it can be seen that hydrotalcite with hexagonal lamellar structure was formed at pH 10, 11, and 12, with a good dispersion of hydrotalcite crystals, uniform surface of the particles, and significant angularity, indicating that the synthesized hydrotalcite had an excellent homogeneity and regularity. The flake structure of hydrotalcite at pH 11 is the most obvious and the most regular, indicating that pH 11 is the optimal condition.

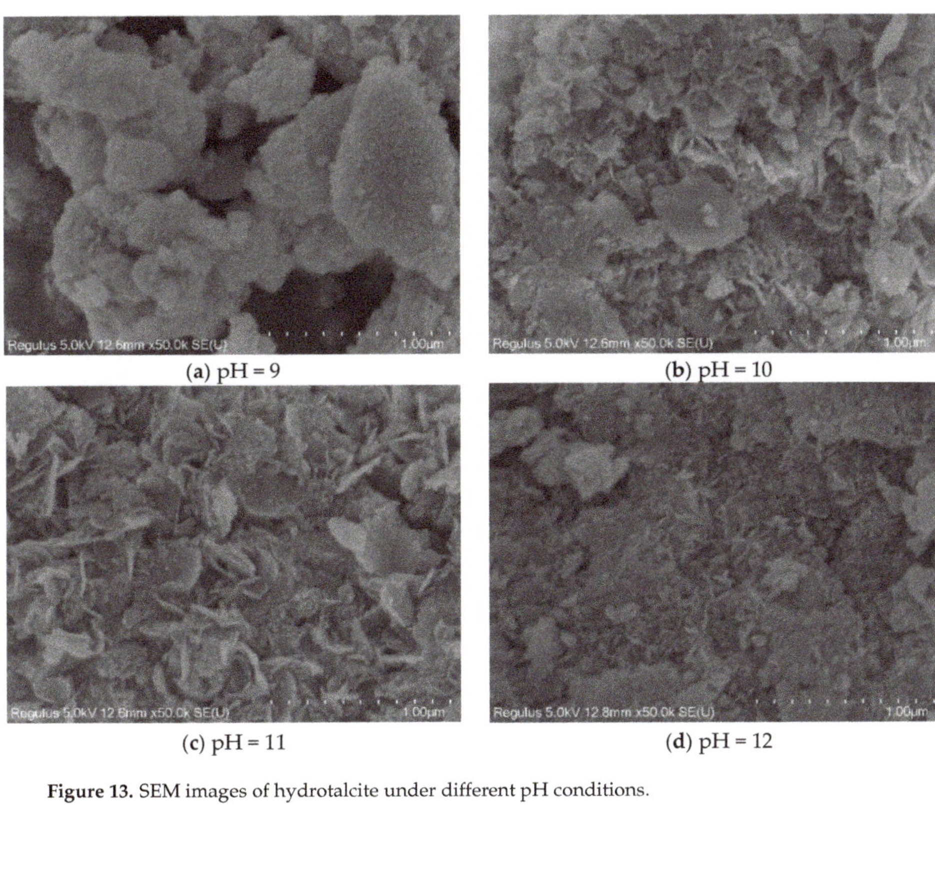

Figure 13. SEM images of hydrotalcite under different pH conditions.

3.2.2. The Effect of Reaction Temperature

Figure 14 provided the XRD patterns of hydrotalcite under different reaction temperatures. By comparing the XRD patterns with those of standard magnesium–aluminum hydrotalcite, it could be seen that the diffraction peaks of the hydrothermal reaction products were the same as those of the standard magnesium–aluminum hydrotalcite when the temperatures are 60 °C, 90 °C, 120 °C, and 150 °C, proving that the hydrotalcite was generated in the sample. Furthermore, Figure 15 demonstrates corresponding peaks at 3458 cm^{-1}, 1559 cm^{-1}, 1365 cm^{-1}, and 590 cm^{-1}, providing additional supporting evidence for the synthesis of hydrotalcite.

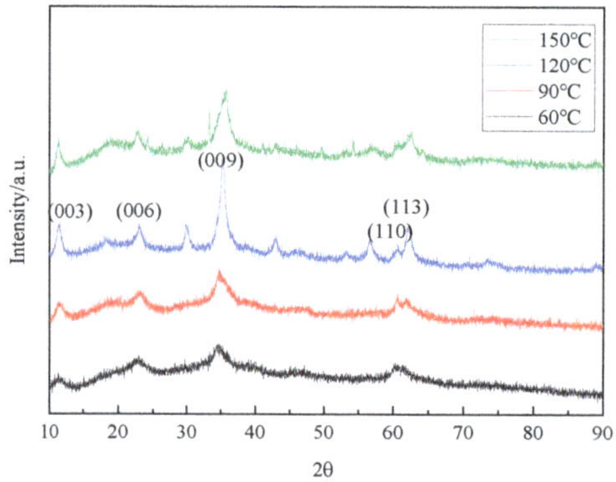

Figure 14. XRD patterns of hydrotalcite under different reaction temperature conditions.

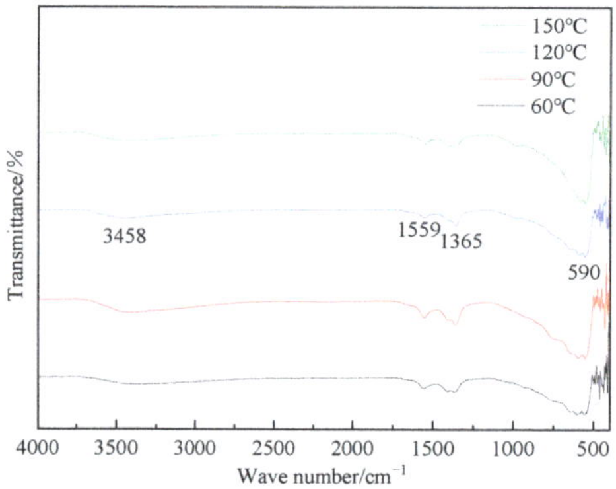

Figure 15. FT-IR plots of hydrotalcite under different reaction temperature conditions.

In Table 3, the XRD data of the synthesized hydrotalcite at various temperatures are presented. The angles and d values of hydrotalcite synthesized under reaction system temperatures of 60 °C, 90 °C, 120 °C, and 150 °C are listed, along with the half-peak widths of hydrotalcite at (003), (006) and (009) crystal planes. It was observed that only the temperature at 90 °C and 120 °C satisfied the relationship of $d_{(003)} = 2d_{(006)} = 3d_{(009)}$, indicating a more regular lamellar structure in the prepared hydrotalcite. The half-peak width of the hydrotalcite synthesized at 120 °C on the (003) crystal plane was 0.0119 rad smaller than that of 0.0124 rad at 90 °C. This suggested that the diffraction peaks on the (003) crystal plane of the hydrotalcite prepared at 120 °C were sharper and narrower, and exhibited improved crystallinity. Therefore, the optimum temperature for the preparation of hydrotalcite was 120 °C.

Table 3. XRD test data of hydrotalcite under different reaction temperature conditions.

Sample	(003)			(006)		(009)	
	$2\theta_{(003)}/^{\circ}$	$d_{(003)}$/nm	$\beta_{(003)}$/rad	$2\theta_{(006)}/^{\circ}$	$d_{(006)}$/nm	$2\theta_{(009)}/^{\circ}$	$d_{(009)}$/nm
60 °C	11.41	0.7751	0.0084	22.99	0.3865	34.40	0.2605
90 °C	11.32	0.7810	0.0124	22.72	0.3910	34.59	0.2591
120 °C	11.48	0.7704	0.0119	23.24	0.3824	35.18	0.2549
150 °C	11.30	0.7824	0.0102	22.66	0.3921	35.64	0.2517

3.2.3. The Effect of Reaction Time

The XRD patterns of hydrotalcite under different reaction time conditions were shown in Figure 16. A comparison between these patterns and those of standard magnesium–aluminum hydrotalcite revealed that the diffraction peaks on the XRD patterns of the hydrothermal reaction products at 12 h, 16 h, 20 h, and 24 h aligned with those of the standard hydrotalcite. These peaks corresponded to the characteristic crystalline surfaces (003), (006), and (009), confirming the presence of hydrotalcite in the products. Further supporting evidence for the synthesis of hydrotalcite was provided by the FT-IR plots shown in Figure 17.

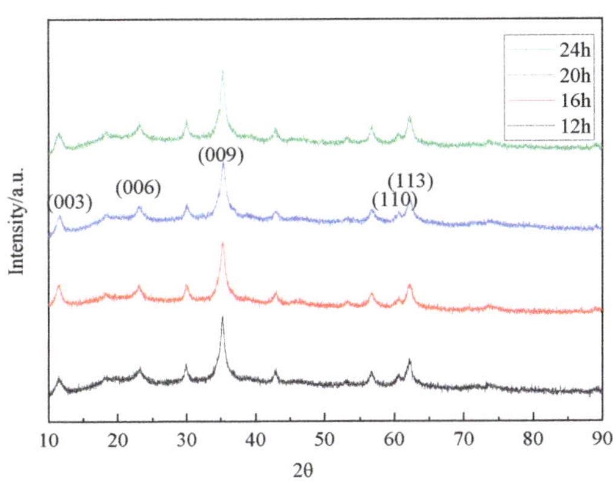

Figure 16. XRD pattern of hydrotalcite under different reaction time conditions.

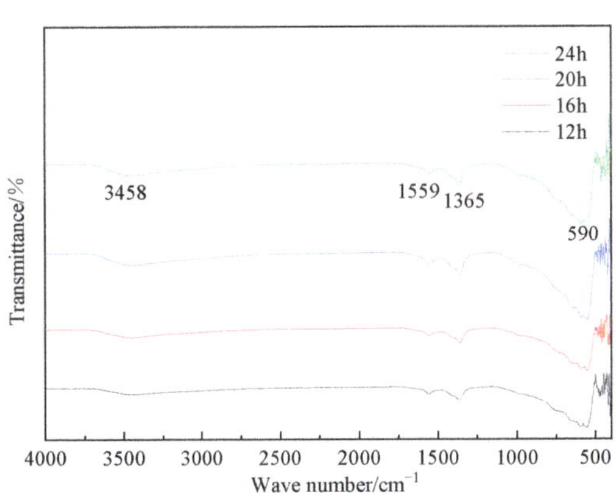

Figure 17. FT-IR plots of hydrotalcite under different reaction time conditions.

The XRD test data of hydrotalcite under different reaction time conditions were summarized in Table 4, which listed the angle and d value corresponding to the (003),

(006), and (009) crystal planes, as well as the half-peak width on the (003) crystal plane of the samples synthesized at reaction times of 12 h, 16 h, 20 h, and 24 h, respectively. It could be observed that, at 12 h, 16 h, 20 h, and 24 h, the relationship of $d_{(003)} = _{(006)} = 3d_{(009)}$ was satisfied, indicating a more regular lamellar structure. Notably, the half-peak width on the (003) crystalline surface was 0.0101 rad at a reaction time of 12 h, suggesting sharp and narrow diffraction peaks on this crystalline surface, indicative of better crystallinity. Combined with the above analysis and economic factors, 12 h was chosen for the subsequent process conditions.

Table 4. XRD test data of hydrotalcite under different reaction time.

Sample	(003)			(006)		(009)	
	$2\theta_{(003)}/°$	$d_{(003)}$/nm	$\beta_{(003)}$/rad	$2\theta_{(006)}/°$	$d_{(006)}$/nm	$2\theta_{(009)}/°$	$d_{(009)}$/nm
12 h	11.44	0.7727	0.0101	23.22	0.3827	35.26	0.2543
16 h	11.48	0.7704	0.0119	23.24	0.3824	35.18	0.2549
20 h	11.50	0.7688	0.0111	23.06	0.3854	35.24	0.2545
24 h	11.40	0.7756	0.0116	23.08	0.3850	35.24	0.2545

The investigation of process conditions for the hydrothermal preparation of hydrotalcite involves three factors: pH, reaction temperature, and reaction time. The hydrotalcite could be successfully prepared when the pH of the solution system ranges from 10 to 12. However, at pH 9, the main products of the hydrothermal reaction were found to be $CaCO_3$ and Fe_2O_3. The impact of the reaction temperature primarily influences the formation of the hydrotalcite phase in terms of its laminar structure and crystallinity. The effect of the reaction time was relatively small. The diffraction peaks of hydrotalcite generated at the reaction time of 12 h in the (003) crystal plane were sharp and narrow, and the crystallinity was better. In summary, the optimal process conditions for the hydrothermal preparation of hydrotalcite could be summarized as follows: a pH of 11, a reaction temperature of 120 °C, and a reaction time of 12 h.

3.3. Characterization of Steel-Slag-Based Hydrotalcite

3.3.1. Morphology of Hydrotalcite

The FESEM images of the hydrotalcite sample were shown in Figure 18. The hydrotalcite had a hexagonal lamellar structure; the particles of the crystals were more uniformly dispersed, with better dispersion and a flat surface. Figure 19 displayed the TEM of the hydrotalcite sample images. Figure 19a provided an overall view, while Figure 19b displayed a high-resolution image. Figiure 19a depicted the hydrotalcite as having a flat, block-like shape composed of stacked layers, as indicated by the red circle. The high-resolution image reveals the presence of diffraction fringes within the hydrotalcite. The bright fringes correspond to elements with lower atomic numbers, emitted primarily by the anions in the interlayer. Conversely, the dark fringes represent elements with higher atomic numbers, emitted predominantly by the cations forming the lamellae. The orderly arrangement of atoms confirms the regular morphology of the synthesized hydrotalcite.

3.3.2. Specific Surface Area Analysis of Hydrotalcite and Steel Slag

Figure 20a displayed the specific surface area and pore size of hydrotalcite, while Figure 20b showed the specific surface area and pore size of steel slag. The data revealed that the pore size distribution of hydrotalcite was primarily concentrated within the range of 2–20 nm. Conversely, in the case of steel slag, the pore size distribution was more widely spread, spanning from 10–80 nm.

Figure 18. FESEM of hydrotalcite samples.

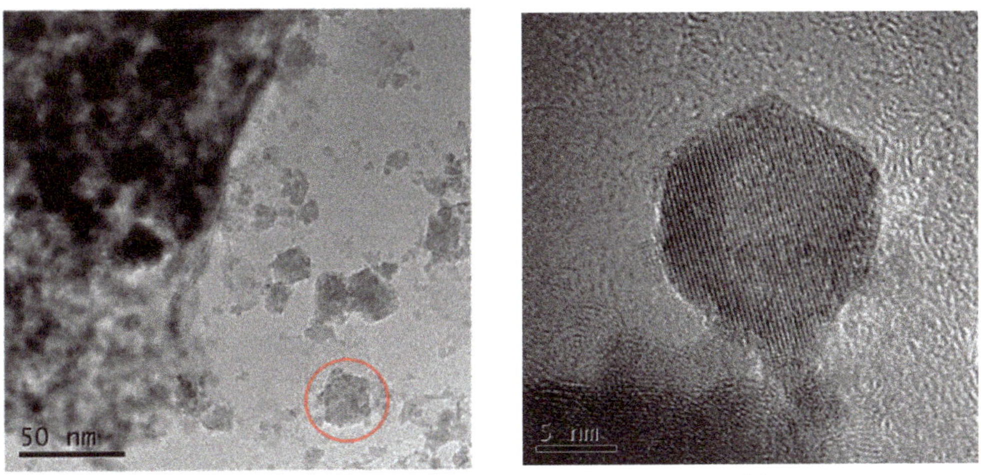

(**a**) (**b**)

Figure 19. TEM of hydrotalcite samples. (**a**) TEM of the whole; (**b**) TEM at high resolution.

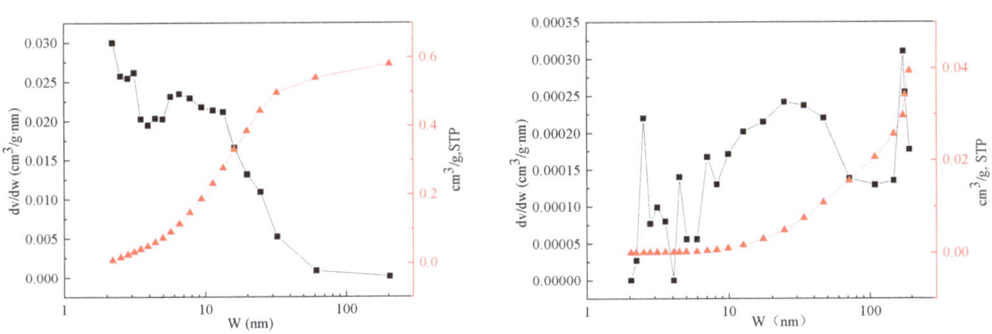

(**a**) (**b**)

Figure 20. Specific surface area analysis of hydrotalcite and steel slag. (**a**) Hydrotalcite; (**b**) Steel slag.

Table 5 presented the surface area, average pore size, and pore volume of hydrotalcite and steel slag. It demonstrated that hydrotalcite exhibits a specific surface area of 207.27 m^2/g, an average pore diameter of 10.89 nm, and a pore volume of 0.56 cm^3/g. In contrast, steel slag had a specific surface area of 3.88 m^2/g, an average pore diameter of 40.68 nm, and a pore volume of 0.04 cm^3/g. The pore diameter of hydrotalcite was approximately a quarter of that of steel slag, while its pore volume was significantly larger. Moreover, hydrotalcite possessed a substantially larger specific surface area compared to steel slag. A larger specific surface area typically corresponded to a more robust adsorption capacity. Accordingly, the larger specific surface area and pore volume of hydrotalcite made it a more favorable material for adsorption purposes when compared to steel slag.

Table 5. Specific surface area data for hydrotalcite and steel slag.

Samples	Specific Surface Area (m^2/g)	Average Pore Size (nm)	Pore Volume (cm^3/g)
hydrotalcite	207.27	10.89	0.56
steel slag	3.88	40.68	0.04

3.4. Adsorption of Cl^- and SO_4^{2-} from Salt-Washing Wastewater by Steel-Slag-Based Hydrotalcite

3.4.1. Effect of Adsorption Time

Figures 21 and 22 illustrate the Cl^- and SO_4^{2-} adsorption per unit mass of hydrotalcite and the corresponding removal rates from salt-washing wastewater. The figures demonstrate that, as the adsorption time increased within the first 2 h, both the Cl^- and SO_4^{2-} adsorption per unit mass of hydrotalcite and their removal rates from the wastewater increased. Between 2 and 4 h, only the adsorption of SO_4^{2-} per unit mass of hydrotalcite and the removal rate of SO_4^{2-} from the wastewater continued to rise. After 4 h, the amount of Cl^- and SO_4^{2-} adsorbed per unit mass of hydrotalcite, as well as the removal rates, reached a constant level. Consequently, the adsorption of hydrotalcite on salt-washing wastewater reached equilibrium at 4 h. At this stage, the amount of Cl^- adsorbed per unit mass of hydrotalcite was 13.3 mg/g, with a corresponding removal rate of 12.8% in the solution. Additionally, the amount of SO_4^{2-} adsorbed per unit mass of hydrotalcite was 20.9 mg/g, with a removal rate of 38% in the solution. The adsorption capacity of SO_4^{2-} was stronger than that of Cl^-, possibly because of the higher charge density of sulfates, which resulted in a greater affinity for intercalation by sulfates [33].

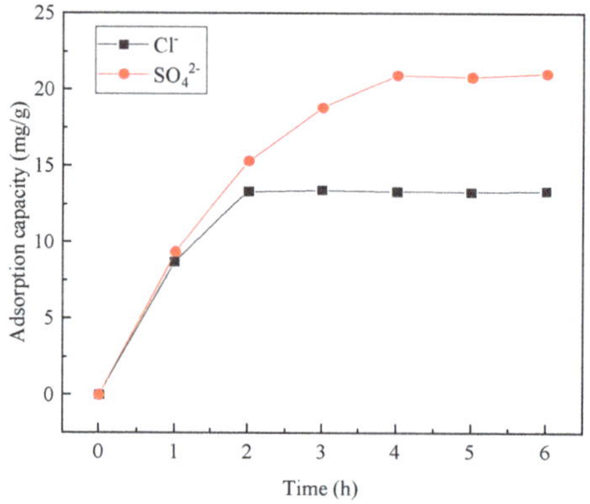

Figure 21. Adsorption of Cl^- and SO_4^{2-} per unit mass of hydrotalcite.

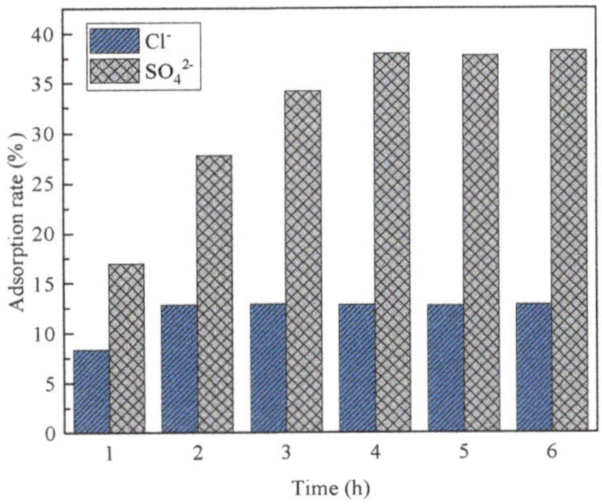

Figure 22. Removal rate of Cl^- and SO_4^{2-} in salt-washing wastewater.

3.4.2. Multiple Adsorption of Steel-Slag-Based Hydrotalcite on Salt-Washing Wastewater

Hydrotalcite was used to adsorb salt-washing wastewater multiple times, with subsequent measurements of Cl^- and SO_4^{2-} concentrations in the wastewater. The removal rate of Cl^- and SO_4^{2-} after repeated hydrotalcite adsorption was shown in Figure 23. It was evident from the figure that the amount of remaining Cl^- and SO_4^{2-} in the wastewater had decreased with multiple adsorptions. The fourth adsorption achieved a 96.4% removal rate of SO_4^{2-}, while the sixth adsorption resulted in a 98% removal rate of Cl^-, effectively treating Cl^- and SO_4^{2-} in salt-washing wastewater. Consequently, hydrotalcite could partially remove Cl^- and SO_4^{2-}, offering a novel avenue for salt-washing wastewater treatment research.

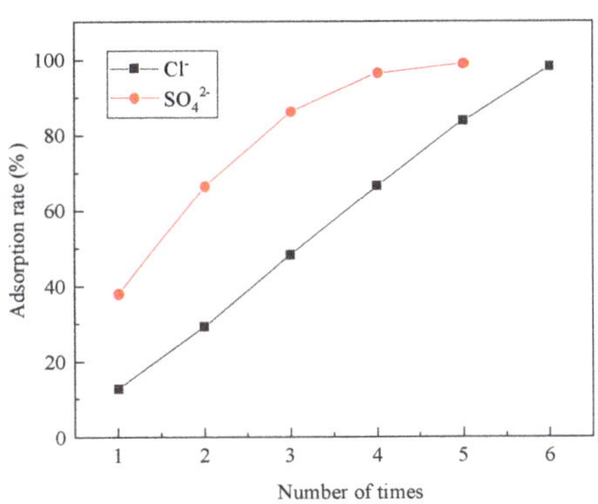

Figure 23. Removal rate of Cl^- and SO_4^{2-} in salt-washing wastewater after the multiple adsorption.

4. Conclusions

The study aimed to examine the leaching mechanism of metal from steel slag using acetic acid as the reaction medium and determine the optimal leaching process for preparing hydrotalcite. Subsequently, hydrotalcite was synthesized through hydrothermal synthesis using the leach solution obtained from steel slag, and its structure was characterized. Additionally, the adsorption properties of hydrotalcite derived from steel slag on Cl^- and SO_4^{2-} in salt-washing wastewater were investigated. The results are outlined below:

1. The acidolysis of steel slag exhibited the highest efficiency when conducted under specific conditions: an acetic acid concentration of 5 mol/L, liquid–solid ratio of 25:1, reaction temperature of 80 °C, and reaction time of 2 h. The ratio of trivalent metal cations to the total sum of divalent and trivalent metal cations in the leaching solution

was determined to be 0.167. The relatively high concentration of metal cations in the leaching solution proved advantageous for the production of a greater quantity of hydrotalcite.

2. The optimum process conditions for the preparation of hydrotalcite by the hydrothermal method were pH 11, a reaction temperature of 120 °C, and reaction time of 12 h. FT-IR, FESEM, and TEM were used to analyze the hydrotalcite samples under the optimum conditions. The analytical results showed that the prepared hydrotalcite had a hexagonal lamellar structure, regular crystal structure, large specific surface area, and layered structure with ordered atomic arrangement.
3. In the adsorption experiments of steel-slag-based hydrotalcite in salt-washing wastewater, the adsorption amount of Cl^- per unit mass of hydrotalcite was 13.3 mg/g, and the removal rate of Cl^- in salt-washing wastewater reached 12.8%; the adsorption amount of SO_4^{2-} per unit mass of hydrotalcite was 20.9 mg/g, and the removal rate of SO_4^{2-} reached 38%. The adsorption of salt-washing wastewater was carried out several times, and it was found that the removal rate of SO_4^{2-} reached 96.4% in the fourth adsorption, and the removal rate of Cl^- reached 98% in the sixth adsorption. These findings indicated that the adsorption process effectively eliminated Cl^- and SO_4^{2-} from the salt-washing wastewater, essentially eliminating their presence.

Author Contributions: Conceptualization, B.H.; data curation, Z.D. and T.Z.; writing—original draft preparation, Z.D.; writing—review and editing, B.H., N.L. and Z.M. All authors have read and agreed to the published version of the manuscript.

Funding: This work was supported by the Priority Academic Program Development of Jiangsu Higher Education Institutions (PAPD).

Institutional Review Board Statement: Not applicable.

Informed Consent Statement: Not applicable.

Data Availability Statement: The data are contained within the article.

Acknowledgments: The authors gratefully acknowledge the assistance from Zhongyang Mao and Na Liu from NJTECH, and the staff from State Key Laboratory of Materials-Oriented Chemical Engineering.

Conflicts of Interest: The authors declare no conflict of interest.

References

1. Kapourchal, S.A.; Homaee, M. Minimizing freshwater usage for quality improvement of saline soils in semi-arid regions. *Irrig. Drain.* **2022**, *71*, 1240–1253. [CrossRef]
2. Barnard, J.H.; Rensburg, L.D.; Bennie, A.T. Leaching irrigated saline sandy to sandy loam apedal soils with water of a constant salinity. *Irrig. Sci.* **2010**, *28*, 191–201. [CrossRef]
3. Huang, J.; Kong, Y.; Xu, Q.; Zhu, C.; Zhu, L.; Cao, X.; Hong, X.; Zhang, J. Progresses for Characteristics and Amelioration Measures of Saline Soil. *Soils* **2022**, *54*, 18–23.
4. Duan, W.; Li, G.; Wang, Z.; Wang, D.; Yu, Q.; Zhan, Y. Highly efficient production of hydrotalcite-like compounds from blast furnace slag. *Appl. Clay Sci.* **2022**, *219*, 106441. [CrossRef]
5. Guo, Q.; Yin, C.; Chen, Y.; Fang, Z.; Xiao, H.; He, J.; Huang, L.; Wu, G.; Zeng, Z.; Shen, F.; et al. Periodic pH regulation controls the phosphate uptake-release behavior and structural evolution of layered double hydroxides. *Chem. Eng. J.* **2023**, *459*, 141584. [CrossRef]
6. Hong, X.; Zhu, S.; Xia, M.; Du, P.; Wang, F. Investigation of the efficient adsorption performance and adsorption mechanism of 3D composite structure La nanosphere-coated Mn/Fe layered double hydrotalcite on phosphate. *J. Colloid Interface Sci.* **2022**, *614*, 478–488. [CrossRef]
7. Cai, H.; Ma, K.; Zhang, Y.; Li, X.; Wang, W.; Tong, S. Carbonizing hollow metal–organic framework/layered double hydroxide (MOF/LDH) nanocomposite with excellent adsorption capacity for removal of Pb(II) and organic dyes from wastewater. *Carbon Res.* **2023**, *2*, 23. [CrossRef]
8. Dai, X.; Yi, W.; Yin, C.; Li, K.; Feng, L.; Zhou, Q.; Yi, Z.; Zhang, X.; Wang, Y.; Yu, Y.; et al. 2D-3D magnetic NiFe layered double hydroxide decorated diatomite as multi-function material for anionic, cationic dyes, arsenate, and arsenite adsorption. *Appl. Clay Sci.* **2022**, *229*, 106664. [CrossRef]
9. Martin, G.P.; Robinson, A.J.; Worsley, D.A. Hydrotalcite mineral stabilization of titanium dioxide photocatalysed degradation of unplasticised polyvinylchloride. *Mater. Sci. Technol.* **2006**, *22*, 375–378. [CrossRef]

10. Cui, P.; Wu, S.; Xiao, Y.; Hu, R.; Yang, T. Environmental performance and functional analysis of chip seals with recycled basic oxygen furnace slag as aggregate. *J. Hazard. Mater.* **2021**, *405*, 124441. [CrossRef]
11. Ma, M.; Yang, H.; Bai, B.; Xie, G. Preparation and mechanism of silicate-based sintering material from large amount of steel slag. *Mater. Express* **2022**, *9*, 125503. [CrossRef]
12. Li, H.; Cui, C.; Sheng, Y.; Zhang, M.; Feng, Z.; Li, L.; Xue, H.; Zhang, Y. Application of composite steel slag as subgrade Filler: Performance evaluation and enhancement. *Constr. Build. Mater.* **2023**, *370*, 130448. [CrossRef]
13. Gao, B.; Yang, C.; Zou, Y.; Wang, F.; Zhou, X.; Barbieri, D.M.; Wu, S. Compaction Procedures and Associated Environmental Impacts Analysis for Application of Steel Slag in Road Base Layer. *Sustainability* **2021**, *13*, 4396. [CrossRef]
14. Zhao, G.; Li, Y.; Dai, W.; Cang, D. Preparation Mechanism of Glass-ceramics Based on Steel Slag Using One-step Sintering Process. *Bull. Chin. Ceram. Soc.* **2014**, *33*, 3288–3294.
15. Liu, L.; Liu, J.; Liu, T.; Zhang, Y.; Liang, H.; Ning, T.; Lin, X.; Bai, Z.; Lu, A. Preparation, crystallization kinetics and stabilization behavior of the heavy metal ions of all-solid waste-based glass-ceramics from steel slag and coal gangue. *J. Non-Cryst. Solids* **2022**, *592*, 121750. [CrossRef]
16. Gao, Y.; Li, Z.; Zhang, J.; Zhang, Q.; Wang, Y. Synergistic use of industrial solid wastes to prepare belite-rich sulphoaluminate cement and its feasibility use in repairing materials. *Constr. Build. Mater.* **2020**, *264*, 120201. [CrossRef]
17. Adolfsson, D.; Menad, N.; Viggh, E.; Björkman, B. Steelmaking slags as raw material for sulphoaluminate belite cement. *Adv. Cem. Res.* **2007**, *19*, 147–156. [CrossRef]
18. Wang, X.; Wang, K.; Li, J.; Wang, W.; Mao, Y.; Wu, S.; Yang, S. Heavy metals migration during the preparation and hydration of an eco-friendly steel slag-based cementitious material. *J. Clean. Prod.* **2021**, *329*, 129715. [CrossRef]
19. Huang, Y.; Pei, Y.; Qian, J.; Gao, X.; Liang, J.; Duan, G.; Zhao, P.; Lu, L.; Cheng, X. Bauxite free iron rich calcium sulfoaluminate cement: Preparation, hydration and properties. *Constr. Build. Mater.* **2020**, *249*, 118774. [CrossRef]
20. Yang, K.; Lu, D.; Yan, S.; Xu, Z. Mineral Composition and Microstructure of Portland Cement Clinker with Steel Slag in the Raw Meal. *Bull. Chin. Ceram. Soc.* **2013**, *32*, 1032–1036, 1047.
21. Zhao, D.; Zhang, D.; Shen, W.; Huang, J.; Tang, X.; Yang, Y.; Deng, Y.; Wang, Y. Investigation on industrial trial production of multi-phased clinker with crude granular steel slag. *J. Clean. Prod.* **2022**, *337*, 130467. [CrossRef]
22. Cao, L.; Shen, W.; Huang, J.; Yang, Y.; Zhang, D.; Huang, X.; Lv, Z.; Ji, X. Process to utilize crushed steel slag in cement industry directly: Multi-phased clinker sintering technology. *J. Clean. Prod.* **2019**, *217*, 520–529. [CrossRef]
23. Chen, M.; Feng, C.; Li, D. Research on Feasibility of Using Steel Slag as Mineral Admixture in Concrete. *Bull. Chin. Ceram. Soc.* **2011**, *30*, 751–754.
24. Liu, J.; Wang, D. Influence of steel slag-silica fume composite mineral admixture on the properties of concrete. *Powder Technol.* **2017**, *320*, 230–238. [CrossRef]
25. Liu, Z.; Huang, Z. Influence of Steel Slag-Superfine Blast Furnace Slag Composite Mineral Admixture on the Properties of Mortar and Concrete. *Adv. Civ. Eng.* **2021**, *2021*, 9983019. [CrossRef]
26. Roslan, N.H.; Ismail, M.; Abdul-Majid, Z.; Ghoreishiamiri, S.; Muhammad, B. Performance of steel slag and steel sludge in concrete. *Constr. Build. Mater.* **2016**, *104*, 16–24. [CrossRef]
27. Ma, S.; Song, C.; Zhang, Z.; Zhang, Y. Preparation of hydrotalcite flame retardant from steel slag and its application in EVA foam composites. *J. Mater. Eng.* **2021**, *49*, 151–157.
28. Liu, G.; Yang, J.; Xu, X. Synthesis of hydrotalcite-type mixed oxide catalysts from waste steel slag for transesterification of glycerol and dimethyl carbonate. *Sci. Rep.* **2020**, *10*, 10273. [CrossRef]
29. Song, Z.; Gao, H.; Liao, G.; Zhang, W.; Wang, D. A novel slag-based Ce/TiO_2@LDH catalyst for visible light driven degradation of tetracycline: Performance and mechanism. *J. Alloys Compd.* **2022**, *901*, 163525. [CrossRef]
30. Qin, W.; Li, W.; Lan, Z.; Qiu, G. Simulated small-scale pilot plant heap leaching of low-grade oxide zinc ore with integrated selective extraction of zinc. *Miner. Eng.* **2007**, *20*, 694–700. [CrossRef]
31. Cavani, F.; Trifiro, F.; Vaccari, A. hydrotalcite-type anionic clays: Preparation, properties and applications. *Catal. Today* **1991**, *11*, 173–301. [CrossRef]
32. Wang, C.; Chen, Y.; Shang, X.; Hou, X.; Li, H.; Guo, Z. Facile synthesis of Ca/Mg/Al/Fe layered double hydroxides using steelmaking slag as raw material. *Mater. Lett.* **2016**, *173*, 115–118. [CrossRef]
33. Eiby, S.H.J.; Tobler, D.J.; Nedel, S.; Bischoff, A.; Christiansen, B.C.; Hansen, A.S.; Kjaergaard, H.G.; Stipp, S.L.S. Competition between chloride and sulphate during the reformation of calcined hydrotalcite. *Appl. Clay Sci.* **2016**, *132–133*, 650–659. [CrossRef]

materials

MDPI

Article

Motion of Air Bubbles in a Cement Slurry

N'dri Arthur Konan [1,2], Eilis Rosenbaum [3] and Mehrdad Massoudi [3,*]

1 National Energy Technology Laboratory, 3610 Collins Ferry Road, Morgantown, WV 26507, USA; ndri.konan@netl.doe.gov
2 NETL Support Contractor, 3610 Collins Ferry Road, Morgantown, WV 26507, USA
3 National Energy Technology Laboratory, 626 Cochran Mill Road, Pittsburgh, PA 15236, USA; eilis.rosenbaum@netl.doe.gov
* Correspondence: mehrdad.massoudi@netl.doe.gov

Abstract: The dynamics of air (gas) bubbles in a column of cement slurry is examined numerically. The air injected at the bottom of a laboratory-scale column through a porous distributor plate spatially distributes and migrates as a swarm of bubbles throughout the slurry toward the freeboard. The two-phase system of the cement slurry and the air bubbles is modeled using the conservation equations of mass and linear momentum in the framework of the volume-of-fluid (VOF) approach. The cement slurry is modeled using the Herschel–Bulkley and Bingham fluid models. Results show that the mean Sauter diameter and the mean rise velocity of the bubbles decrease with the gas flow rate. Meanwhile, it is found that the rising of the bubbles is controlled by breakup events, along with relatively weak path instabilities of the bubbles resulting in relatively straight trajectories, independent of the gas flow rate. The extent of the yielded region appears larger for the Herschel–Bulkley model compared to the Bingham fluid model (by approximately 10%).

Keywords: yield stress fluid; cement slurry; gas migration; Herschel–Bulkley fluid; Bingham fluid; VOF; bubble

1. Introduction

Environmental risks associated with the integrity of wellbores remain an active concern for oil and gas industries (see, e.g., Vidic et al., 2013 [1]). Survey-based analyses of thousands of reports of conventional and unconventional oil/gas wells in Pennsylvania/USA by Ingraffea et al. (2014) [2] suggest that these integrity issues may be correlated to cementing operation failures (e.g., cement density, premature gelation, high permeability in the cement slurry, etc.). It is known that cemented Aeolian sand-fly ash backfill material can be used to fill the mining area to help with the damage caused by underground coal mining (Yang et al., 2023 [3]). Bonett and Pafitis (1996) [4] discuss circumstances that may impair the cementing of the wellbores on the basis of three states of the cement: (1) a slurry that is a granular-type fluid with an evolving yield stress that further exhibits a vulnerability to the invasion of the gas from the surrounding formation because of pressure differentials between these two media as the cement hardens; (2) a fully self-supporting two-phase material marked by continued setting and accelerated hydration that result in an internal shrinkage, a weak solid structure, subject to viscoelastic fractures caused by the influx of gas; and (3) an elastic and brittle material where gas migration can no longer occur, although gas flows through interfacial channels can take place. These authors further discuss four gas migration patterns (i.e., bubbling, slug, interface, and rising plume) through the cement slurry with a dependence on the gas flow rate (typically within the range of 10^{-9} to 10^{-7} m^3/s). Wellbore cementing challenges are many, and great technical efforts are put forward to prevent compromising the integrity of wellbore cementing (see, e.g., Renteria et al., 2022 [5]; Maleki and Frigaard, 2019 [6]). It is known that air bubbles can have detrimental effect on concrete and cement, reducing the compressive strength of hardened concrete during the later stages of cement hydration process. For a review of important issues related to gas migration in cement, see Tao et al. (2021) [7].

Several studies point toward the design of cements with controlled and enhanced properties that could prevent the migration (see, e.g., Velayati et al., 2015 [8]; Ahmed et al., 2020 [9]; Moradi and Nikolaev, 2016 [10]; Teodoriu and Asamba, 2015 [11]). Velayati et al. [8] use additives in the cement slurries of a laboratory scale and report significant reductions in the migration. However, following Bonett and Pafitis (1996) [4], one would expect a bubbling migration pattern for the lower flow rates reported by Velayati et al. (2015) [8]. Given that the cement slurry could still show vulnerabilities despite improved cement designs, additional insights into the process of gas migration should help better understand the strength of the wellbore and improve and anticipate mitigations. The following questions need to be considered. (1) How does the migrating gas spatially distribute throughout the slurry column? (2) What are the size and aspect ratio distributions of the gas bubbles? (3) When the bubbles are not motionless, how fast do they rise in the slurry?

Cement slurries, in general, behave as complex nonlinear fluids that may exhibit viscoelasticity, thixotropy, yield stress, shear-thinning effects, etc. (see the review by Tao et al., 2020 [12]; Banfil, 2006 [13], 1993 [14]; and Rodrigues and de Souza Mendes, 2019 [15]). As indicated by Yuan, et al. (2022) [16], the thixotropy of cement becomes crucial when a cement slurry is used for mine-out area backfilling, where it should be pumped easily but should also become a solid-like material at very small velocities. An extensive literature discussing fundamental aspects of air bubble injection, detachment, growth, and rising mechanisms in different types of non-Newtonian fluids exists. These studies mostly have focused on single or a few bubbles with controlled injection points.

Ghosh and Ulbrecht (1989) [17] examine experimentally the formation and detachment of a single bubble in stagnant shear-thinning and viscoelastic fluids characterized by the power-law and the two-parameter Oldroyd models, respectively. For the sake of comparative analysis, they also look at a viscous Newtonian fluid. Gosh and Ulbrecht [17] observe a delayed detachment and slower growth rate of the bubbles injected in the viscoelastic fluid compared to both the shear-thinning and the viscous Newtonian fluids. They also report that while the bubble volumes are found to increase with the air flow rate, higher bubble volumes are obtained with fluids of higher apparent viscosities. They observe that bubbles elongate in the direction of their vertical axis at higher gas flow rates for Newtonian and low-viscosity fluids, whereas for the viscoelastic fluid, elongation phenomenon occurs even at low gas flow rates, and bubble elongation increases with the gas flow rate. Within a column filled with yield stress Herschel–Bulkley fluid, Terasaka and Tsuge (2001) [18] similarly investigate experimentally the formation and the growth of single gas bubbles injected through a nozzle. The authors report a strong correlation between the rheological properties of the fluids and the bubble volumes. Indeed, their observations show that, at a fixed gas injection flow rate, the larger the yield stress is, the larger the bubble volume.

Dubash and Frigaard (2004) [19] study the permanent entrapment and moving conditions of single air bubbles in a viscoplastic fluid (modeled as a Herschel–Bulkley fluid). They derived two stopping conditions by relying on variational-based minimizations of the symmetric part of the velocity gradient and the stress fields, respectively. These conditions, which can be identified as realizability criteria, stipulate that above a critical yield stress (or a critical Bingham number), the bubble will not rise in the fluid. Dubash and Frigaard [19] show that these stopping conditions are independent of the fluid constitutive parameters, namely, the consistency index and the power-law exponent; they explain that until a bubble begins to move, it only sees the yield stress of the fluid. Indeed, the bubble rises when the force caused by the buoyancy of the bubble is sufficient to overcome the yield stress. In a follow-up experimental study, Dubash and Frigaard (2007) [20] discuss the validity of those conditions against measurements of the bubble rise velocities by assuming that bubbles have axisymmetric shapes for the sake of simplicity. The experiments consist of injecting single bubbles of several different sizes through a nozzle, where the column is filled with a fluid assumed to behave as a Herschel–Bulkley fluid. The experiment shows that the yield stress parameters are beyond the range of the stopping conditions previously derived by Dubash and Frigaard (2004) [19]. They also show from an energy-based analysis that the bubble buoyancy is balanced by both the yield and the surface tension stresses; they report that the surface tension effects appear to be significant and more important for smaller bubbles. Dubash and Frigaard [20] also observe a rapid increase in the bubble terminal velocity as the bubble-to-column radius ratio increases, while above a certain bubble-to-column radius ratio they report a maximum terminal velocity.

Stein and Buggisch (2000) [21] investigate the possibility of initially static small bubbles to rise when subjected to pulsations, i.e., an additional mechanism beyond the buoyancy allowing the bubbles to overcome the yield stress. Their experiments show that there exists a pulsating amplitude threshold with a dependence on the bubble size below which the bubble does not rise. In addition, they observe that larger bubbles rise faster than small ones for a given pulsation amplitude. Similarly, Iwata et al. (2008) [22] investigated experimentally the enhancement of a single bubble rising velocity by controlling the fluid pressure through a sine-wave-based mechanical vibration. Their measurements are for a shear-thinning Carreau–Yasuda fluid (i.e., not a yield stress fluid) and show that higher oscillation frequency results in higher shear rates, along with a substantial decrease in the fluid effective viscosity around the bubble. The authors observe that the bubble's rising velocity increases by two orders of magnitude, which they partly attribute to the reduction in the apparent viscosity in the vicinity of the bubble. In a recent work, De Corato et al. (2019) [23] investigated numerically Iwata et al.'s experiments and reported that the bubble's radial motion reduces the viscosity of the surrounding fluid and subsequently may greatly reduce the friction coefficient of the bubble. This mechanism can be used to control the enhancement of the bubble's rising velocity.

For the flow of interest in this work (which could also be applicable to geothermal applications), the gas is mostly expected to invade the cement slurry through the interface between the external periphery of the wellbore annulus of the stagnant cement slurry and the surrounding porous medium formation. This implies that the gas is injected over a surface, and therefore, multiple bubbles can be expected to detach (either at once or successively). In addition, the pressure fluctuations along with the change in the influx rate of the gas percolating the porous formation could generate flow pulsations at any of the gas invasion locations. Such a flow is understood to be adequately approximated by the investigation of a column of cement slurry subject to a continuous air (gas) injection at the bottom through a porous distributor plate.

In this paper, we look at the motion of bubbles as they rise in a cement slurry column, where the cement is modeled as a Herschel–Bulkley fluid. Quantities such as the geometrical dimensions of the bubbles, along with their rise angle, position, and velocities, are discussed as the flow rate is varied. To gain further insight and to check our formulation and numerical scheme, we simulate the experiment of Terasaka and Tsuge (2001) [18] on the formation and growth of a single bubble in a yield stress fluid, an aqueous solution of xanthan gum. In the next section, the conservation equations of the mass and the linear momentum within the framework of the volume-of-fluid (VOF) approach is presented. An overview of the numerical methods and the simulation conditions are followed by thorough discussions of the results. Finally, some concluding remarks are drawn, and possible future works are discussed.

2. Mathematical Framework

Cement is a multicomponent complex material with microstructure, undergoing chemical and physical changes, and its constitutive modeling can be a daunting task. A cement slurry is composed of particles, water, air, etc. The presence of bubbles in the cement slurry adds further complexity to the problem at hand. The mixture, as a suspension when flowing, behaves as a nonlinear fluid with shear-thinning effect, thixotropy, yield stress, etc. In the paste form, cement can behave as a visco-plastic or visco-elastic material, and after it has hardened, it can behave as a nonlinear poro-elastic material. The most common rheological models used are the Bingham model and the Herschel–Bulkley model (Konan et al., 2022 [24]; Plancher et al., 2022 [25]). To mathematically model a cement slurry, in general, we can use (1) a suspension rheology approach where a single constitutive relation is provided for the Cauchy stress tensor $\boldsymbol{T}$ (see Tao, et al., 2020 [12]), or (2) use a two-phase (multi-phase) approach where constitutive relations are needed for the two stress tensors $\boldsymbol{T}_1$ and $\boldsymbol{T}_2$ for cement particles and water, respectively (with ϕ_1 and ϕ_2 as the volume fractions of the two phases), and an interaction force $\boldsymbol{f}_I$ (see Massoudi, 2003 [26]), or (3) use a modified suspension approach where a stress tensor $\boldsymbol{T}$ for the suspension and a flux vector $\boldsymbol{N}$ are required for the whole system (see Tao et al., 2019 [27]). The volume-of-fluid (VOF) approach is similar to the modified suspension approach, and it presents a computationally feasible method for certain multiphase applications. In this paper, we use this method.

2.1. Volume-of-Fluid (VOF) Method

Here, we provide a brief description of the traditional VOF method, and later, we generalize this to the case of a cement slurry with air bubbles. Let us consider the physical domain, Ω, of two immiscible fluids (i.e., the cement slurry and the gas). Following Hirt and Nichols (1981) [28] and Nichols et al. (1980) [29], let us introduce a volume fraction, $\phi(\mathbf{x}, t)$, whose value of unity at any point ($\mathbf{x} \in \Omega$) indicates the full presence of a given fluid (i.e., cement slurry) and zero otherwise. A value of the volume fraction ϕ between 0 and 1 signifies the presence of an interface between the two fluids. At any point $\mathbf{x} \in \Omega$ in time $t(> 0)$, the two-fluid system is considered as "one-fluid" with a total velocity $\mathbf{u}(\mathbf{x}, t)$ and a density $\rho(\mathbf{x}, t)$. This "one-fluid" is assumed to be incompressible:

$$\text{div}\ \mathbf{u} = 0 \tag{1}$$

where "div" stands for the divergence operator. The density of the "one-fluid" is defined using the volume fraction $\phi(\mathbf{x}, t)$, such that:

$$\rho(\mathbf{x}, t) = \phi\rho_1 + (1 - \phi)\rho_2 \tag{2}$$

where ρ_1 and ρ_2 are the constant densities of the fluids labeled 1 and 2, i.e., the cement slurry and the gas, respectively.

The linear momentum of the "one-fluid" is assumed to be described by:

$$\frac{\partial}{\partial t}(\rho\mathbf{u}) + \text{div}(\rho\mathbf{u}\mathbf{u}) = div\mathbf{T} + \rho\mathbf{b} + [[\boldsymbol{\Sigma}]] \tag{3}$$

where $\mathbf{T}$ is the Cauchy stress tensor, $\mathbf{b}$ is the body force vector, and $[[\boldsymbol{\Sigma}]]$ is the source term because of the interfacial interaction between the two fluids; this accounts for the stress jump at the interface and the immiscibility assumption.

The stress for the traditional "one-fluid" VOF approach is described through a single pressure (p) and a single viscosity (μ) by relying upon the volume fraction, such that:

$$\mu(\mathbf{x}, t) = \phi\mu_1 + (1 - \phi)\mu_2 \tag{4}$$

where μ_1 and μ_2 are the viscosities of the fluids labelled 1 and 2, respectively. At any point $\mathbf{x} \in \Omega$ in time $t(> 0)$, we assume that the Cauchy stress tensor for the "one-fluid" is given by:

$$\mathbf{T}(\mathbf{x}, t) = -p(\mathbf{x}, t)\boldsymbol{I} + \mu(\mathbf{x}, t)\boldsymbol{A}_1 \tag{5}$$

where $\boldsymbol{I}$ is the identity tensor, and $\boldsymbol{A}_1 = \text{grad}\ \mathbf{u} + (\text{grad}\ \mathbf{u})^T$ is the kinematical tensor, related to $\boldsymbol{D}$, the symmetric part of the velocity gradient [$\boldsymbol{A}_1 = 2\boldsymbol{D}$].

The volume fraction ϕ, which serves as a flag identifying the presence of a given fluid, is governed by a convection–diffusion type equation of the "one-fluid", such that:

$$\frac{\partial \phi}{\partial t} + \mathbf{u} \cdot \nabla \phi = 0 \tag{6}$$

where ∇ is the gradient operator.

2.2. Constitutive Relations

2.2.1. Interfacial Interaction between the Two Fluids

The difference in the physical properties such as density or viscosity of the two immiscible fluids induce interfacial interactions. The net tensile force acting on the interface of curvature κ, separating the two fluids can be modeled as (Brackbill et al., 1992 [30]):

$$[[\boldsymbol{\Sigma}]] = \int_{\mathcal{V}(t)} \sigma \kappa(\mathbf{x}) \mathbf{n}(\mathbf{x}) \delta[\mathbf{n}(\mathbf{x}_s) \cdot (\mathbf{x} - \mathbf{x}_s)] d\mathbf{x} \tag{7}$$

$$[[\boldsymbol{\Sigma}]] \cong \sigma \kappa(\mathbf{x}_s) \mathbf{n}(\mathbf{x}_s) \tag{8}$$

where σ is the surface tension, $\mathbf{n}$ is the unit normal vector to the interface, and δ is the Dirac function.

The curvature κ of a surface at a point $\mathbf{x}_s$, along the unit vector $\hat{\mathbf{n}}$ to the interface (within a given cell volume), are assumed to be given by:

$$\kappa = -\text{div}\, \hat{\mathbf{n}} \tag{9}$$

$$\hat{\mathbf{n}} = \frac{\nabla \phi}{|\nabla \phi|} \tag{10}$$

where ∇ is the gradient operator. Brackbill et al. (1992) [30] argue, based on practical computational challenges, that the normal vector in $\mathbf{n}$ (Equation (8)) can be replaced by $-\nabla \phi$, so that the source term because of the interfacial interaction between the two fluids is given by:

$$[[\boldsymbol{\Sigma}]] = \sigma \left[\text{div} \left(\frac{\nabla \phi}{|\nabla \phi|} \right) \right] \nabla \phi \tag{11}$$

This interaction term can appear in two-phase flows, as one of the many interaction forces, in the form of:

$$[[\boldsymbol{\Sigma}]] = \omega \nabla \phi$$

along with other forces such as drag, lift, etc. (see Massoudi 2003 [26]).

2.2.2. Stress Tensor

For our problem, we assume that air (fluid 2) is injected into cement slurry (fluid 1). The densities ρ_1 and ρ_2 of the cement slurry and the air (gas), respectively, are considered constant, as indicated by Equation (2).

In general, as shown by Tao et. al. (2019 [27], 2020 [12]), a cement slurry may exhibit yield stress, thixotropy, shear-dependent viscosity, concentration-dependent viscosity, etc. In this paper, we assume that the cement slurry can be modeled as a Herschel–Bulkley fluid, and we ignore thixotropy and concentration-dependent viscosity. Thus, we assume that the stress in the "one-fluid" of the VOF method is given by:

$$\mathbf{T}(\mathbf{x}, t) = -p(\mathbf{x}, t)\boldsymbol{I} + \boldsymbol{\tau}(\mathbf{x}, t) = -p(\mathbf{x}, t)\boldsymbol{I} + \boldsymbol{\tau}_1(\mathbf{x}, t) + \boldsymbol{\tau}_2(\mathbf{x}, t) \tag{12}$$

where $\boldsymbol{\tau}_1$ and $\boldsymbol{\tau}_2$ are the stress tensors associated with the cement slurry and the air, respectively.

For the cement slurry, we assume it behaves as a Herschel–Bulkley fluid, where $\boldsymbol{\tau}_1$ (in its proper tensorial form as given in Macosko, 1994 [31]) is given by:

$$\boldsymbol{\tau}_1(\mathbf{x}, t) = \phi \left[k|\mathrm{II}_{2\boldsymbol{D}}|^{\frac{n-1}{2}} + \frac{\tau_0}{|\mathrm{II}_{2\boldsymbol{D}}|^{1/2}} \right] (2\boldsymbol{D}) \quad \text{for } \mathrm{II}_{\tau_1}^{1/2} > \tau_0 \tag{13}$$

$$\boldsymbol{D} = 0 \quad \text{for } \mathrm{II}_{\tau_1}^{1/2} \leq \tau_0 \tag{14}$$

where τ_0 is the yield stress, k is the consistency index, and n is the power-law exponent, a measure of the nonlinearity of the fluid, related to the shear-thinning effects (when $n < 0$) or shear-thickening effects (when $n > 0$). II_{τ_1} and $\mathrm{II}_{2\boldsymbol{D}}$ are the second invariants of the stress tensor $\boldsymbol{\tau}_1$ and of the symmetric part of the velocity gradient tensor $2\boldsymbol{D} = (\text{grad } \mathbf{u}) + (\text{grad } \mathbf{u})^T$, respectively. Note that the Bingham fluid is obtained from the above Equation (13) when the power-law exponent $n = 1$, i.e., the viscosity is now independent of the shear rate.

The stress tensor $\boldsymbol{\tau}_2$ for the air (gas) is given by:

$$\boldsymbol{\tau}_2(\mathbf{x}, t) = (1 - \phi)\mu_2(2\boldsymbol{D}) \tag{15}$$

where μ_2 is the viscosity of the air. For the sake of simplicity, the flow is assumed to be laminar. We should note here that in the VOF method, as opposed to the general case of two-phase flow formulation, we only have one velocity $\mathbf{u}$ and only one velocity gradient.

3. Numerical Approach

The computational fluid dynamics (CFD) finite volume method from the open-source toolbox/library, OpenFOAM, is used on unstructured grids to solve the mass and the linear momentum conservation equations, using the solver "*multiphaseInterFoam*" customized with non-Newtonian viscosity libraries. To avoid the numerical implementation challenges associated with the discontinuity (singularity) in the stress tensor field $\boldsymbol{\tau}_1$ of the cement slurry fluid between the unyielded and the yielded regions, the regularization method is used, and the stress tensor of the cement slurry $\boldsymbol{\tau}_1$ is replaced with an ϵ-dependent small parameter, such that:

$$\boldsymbol{\tau}_{1,\epsilon}(\mathbf{x}, t) = \phi\mu_{1,\epsilon}\left(|II_{2\mathbf{D}}|^{1/2}\right)(2\mathbf{D}) \tag{16}$$

where the ϵ-dependent viscosity $\mu_{1,\epsilon}$ is approximated according to Papanastasiou (1987) [32] by:

$$\mu_{1,\epsilon}\left(|II_{2D}|^{1/2}\right) = k|II_{2D}|^{(n-1)/2} + \frac{\tau_0}{|II_{2D}|^{1/2}}\left[1 - \exp\left(-\frac{|II_{2D}|^{1/2}}{\epsilon}\right)\right] \tag{17}$$

This approach has recently been used by Konan et al. (2022) [24] in their study of a cement slurry, as well as by Espinoza et al. (2022) [33] in a similar context of the Herschel–Bulkley fluid coupled with the VOF approach to investigate the primary cementing processes of an oil well. From a numerical standpoint, the advection of the volume fraction ϕ through (Equation (6)) may undesirably lead to the smearing of ϕ and, thus, losing the sharp interface separating the two fluids. To preserve this discontinuity nature of the volume fraction with the transition region as narrow as possible, the so-called "interface compression" method is introduced through an additional convective term in the advection equation of ϕ (see, e.g., Rusche, 2003 [34]; Berberović et al., 2009 [35]; Hoang et al., 2013 [36]; Cifani et al., 2016 [37]). Indeed, the advection of the volume fraction in each fluid, advected by $\mathbf{u}_1$ and $\mathbf{u}_2$, are:

$$\frac{\partial \phi}{\partial t} + \text{div}(\phi \mathbf{u}_1) = 0 \tag{18}$$

$$\frac{\partial}{\partial t}(1 - \phi) + \text{div}((1 - \phi)\mathbf{u}_2) = 0 \tag{19}$$

Adding Equations (18) and (19), we have:

$$\mathrm{div}(\phi\mathbf{u}_1 + (1-\phi)\mathbf{u}_2) = \mathrm{div}\mathbf{u} = 0 \tag{20}$$

where the total velocity defined as $\mathbf{u} = \phi\mathbf{u}_1 + (1-\phi)\mathbf{u}_2$ would result in $\mathbf{u} = \mathbf{u}_1$, when the interface is strictly sharp. Introducing a relative velocity $\mathbf{u_r}$ (to "compress" the interface) as:

$$\mathbf{u_r} = \mathbf{u}_1 - \mathbf{u}_2 \tag{21}$$

we have:

$$\phi\mathbf{u}_1 = \phi\mathbf{u} + \phi(1-\phi)\mathbf{u_r} \tag{22}$$

Substituting Equation (22) in Equation (18), we obtain:

$$\frac{\partial\phi}{\partial t} + \mathrm{div}(\phi\mathbf{u}) + \mathrm{div}(\phi(1-\phi)\mathbf{u_r}) = 0 \tag{23}$$

where the additional convective term is only active within the narrow interface region restricted by $\phi(1-\phi)$. This convective term aims at avoiding the use of a special discretization scheme for the convection in the advection equation of ϕ, which could allow a better resolution for sharper interfaces. The relative (or the interface compression) velocity is assumed to be given by:

$$\mathbf{u_r} = c_\phi|\mathbf{u}|\mathbf{n} \tag{24}$$

where c_ϕ is an adjustable compression factor, and $\mathbf{n}$ is the unit normal vector to the interface.

Using Equation (10) for the unit vector, the relative velocity becomes:

$$\mathbf{u_r} = c_\phi|\mathbf{u}|\frac{\nabla\phi}{|\nabla\phi|} \tag{25}$$

Now, substituting Equations (4)–(25) in Equation (3), along with the incompressibility condition of the "one-fluid", the basic equations that need to be solved numerically are:

$$\mathrm{div}\ \mathbf{u} = 0 \tag{26}$$

$$\begin{aligned}\tfrac{\partial}{\partial t}(\rho\mathbf{u}) + \mathrm{div}(\rho\mathbf{u}\mathbf{u}) = & -\nabla p + \mathrm{div}\left\{\left[\phi\mu_{1,\epsilon}\left(|II_{2\mathbf{D}}|^{1/2}\right) + (1-\phi)\mu_2\right](2\boldsymbol{D})\right\} \\ & +\rho\mathbf{g} + \sigma\left[\mathrm{div}\left(\tfrac{\nabla\phi}{|\nabla\phi|}\right)\right]\nabla\phi\end{aligned} \tag{27}$$

$$\rho(\mathbf{x},t) = \phi\rho_1 + (1-\phi)\rho_2 \tag{28}$$

$$\frac{\partial\phi}{\partial t} + \mathbf{u}\cdot\nabla\phi + \mathrm{div}\left[c_\phi|\mathbf{u}|\phi(1-\phi)\frac{\nabla\phi}{|\nabla\phi|}\right] = 0 \tag{29}$$

In the context of the experiments of gas migration into a column of cement slurry as described in the experimental section of this paper, the boundary conditions can be formulated as follows.

At the distributor plate:

$$\mathbf{u} = \mathbf{U}_g \tag{30}$$

$$\mathbf{n}\cdot\nabla p = 0 \tag{31}$$

where $\mathbf{U}_g$ is the velocity of the gas injected at the distributor plate.

At the wall:

$$\mathbf{u} = 0 \tag{32}$$

$$\mathbf{n}\cdot\nabla p = 0 \tag{33}$$

At the outlet of the column:

$$\mathbf{n} \cdot \nabla \mathbf{u} = 0 \tag{34}$$

$$p = p_{out} \tag{35}$$

where p_{out} is the outlet pressure, set as the atmospheric pressure.

Initially, the column of the slurry is at rest; thus:

$$\mathbf{u}(\mathbf{x}, t) = 0 \tag{36}$$

$$p(\mathbf{x}, t) = 0 \tag{37}$$

4. Parameters Used in the Simulation

The slurry was prepared using the preconditioned class H cement (Headrick et al., 2023 [38]). The rheological properties from the fitting of the data to the Herschel–Bulkley and the Bingham models are summarized in Table 1. The density of the slurry is 1970 kg/m^3. The surface tension is not measured; however, for the sake of simulations, σ = 0.07 N/m is used.

Table 1. Summary of the slurry properties for fitting the models (Headrick et al., 2023 [38]).

Model	τ_0 [Pa]	η_p [Pa·s]	k [Pa·s^n]	n [-]
Herschel–Bulkley	12.87	-	1.96	0.69
Bingham	25.97	0.27	-	-

The column is filled up with 0.68 kg of slurry, at an initial height of 0.17 m above the distributor plate. Summarized in Table 2 are the nine flow rates used in this work, which covers a broad range of typical gas influx rates in a wellbore (see, e.g., Bonett and Pafitis, 1996 [4]). Equations (26)–(37) are solved numerically on unstructured hexa-dominated grids (see Figures 1–3) that are generated using the open-source code, cfMesh. Given in Table 3 is the summary of the mesh densities along with the grid spacings (distributed over the flow domain) that are used for the calculation. A variable time step with the maximum $\delta t = 2.5 \times 10^{-4}$ s is used to advance the solutions; this ensures that the Courant–Friedrichs–Lewy (CFL) remains below 0.2 and is allowed to resolve the fastest capillary waves in the domain (and subsequently to avoid undesirable oscillation phenomena of the interfaces caused by surface tension), as the time step remains smaller than the capillary time (see, e.g., Galusinski and Vigneaux, 2008 [39]; Brackbill et al., 1992 [30]).

Although Papanastasiou's ϵ-dependent viscosity $\mu_{1,\epsilon}$ regularization technique is employed in this work, a parametric study is not carried out. ϵ is selected as small as 10^{-6} in order to limit plausible spurious effects on the bubble shapes (see, e.g., Dimakopoulos et al., 2013 [40]). The convective terms in the momentum equations, as well as the transport equation of the volume fraction, are discretized using the second order "linear" scheme. Spatial gradients are also discretized using the second order "linear" scheme (central differences with linear interpolation). The unsteady terms are discretized with a backward Euler scheme. Table 4 summarizes the descriptions of the quantities used throughout the manuscript.

Table 2. Summary of the gas flow rate, along with the superficial gas velocity.

Gas Flow Rate (mL/min)	2	4	8	15	20	30	60	90	120
Superficial velocity ($\times 10^{-2}$) (mm/s)	1.65	3.29	6.58	12.34	16.45	24.68	49.35	74.03	98.70

Table 3. Summary of the computational grids.

Mesh	Cell Count	Δ_{min} [mm]	Δ_{max} [mm]
1	338,424	0.32	1.31
2	650,840	0.24	1.05
3	1,517,008	0.17	0.79

Table 4. Summary of the descriptions of the quantities discussed throughout the manuscript. V_b and S_b are the volume and the surface of the bubble. $z_b^{(n)}$ denotes the vertical position of the bubble at a given time ($n\delta t$), δt being the time step. $U_{g,s}$ is the gas superficial velocity.

Quantity	Description
Bubble Sauter diameter, d_{32}	$d_{32} = 6V_b/S_b$
Bubble aspect ratio, AR	$AR = \frac{\text{max.horizontal length}}{\text{vertical max.length}}$
Bubble rise velocity, u_b	$u_b = \frac{z_b^{(n+1)} - z_b^{(n)}}{\delta t}$
Scaling velocity, U_0	$U_0 = \frac{\rho g d_{32}^2}{4\mu_0}$
Scaling viscosity, μ_0	$\mu_0 = k\left(\frac{U_{g,s}}{d_{32}}\right)^{n-1}$
Bingham number, Bi	$\text{Bi} = \frac{2\tau_0}{\rho g d_{32}}$
Bond number, Bo	$\text{Bo} = \frac{\rho g d_{32}^2}{4\sigma}$

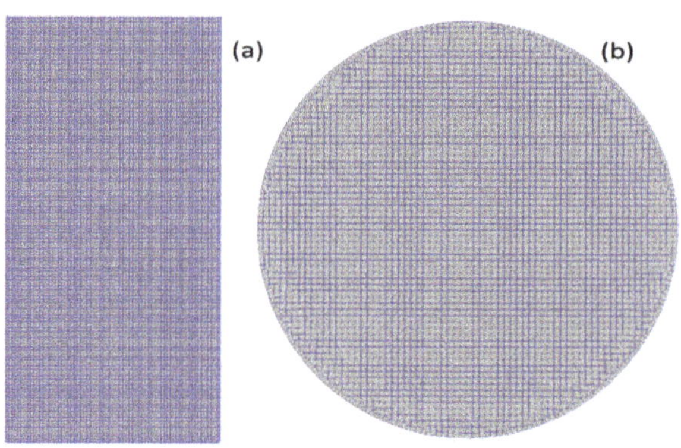

Figure 1. Computational grid of the column: (**a**,**b**) the longitudinal and the transversal plans, respectively.

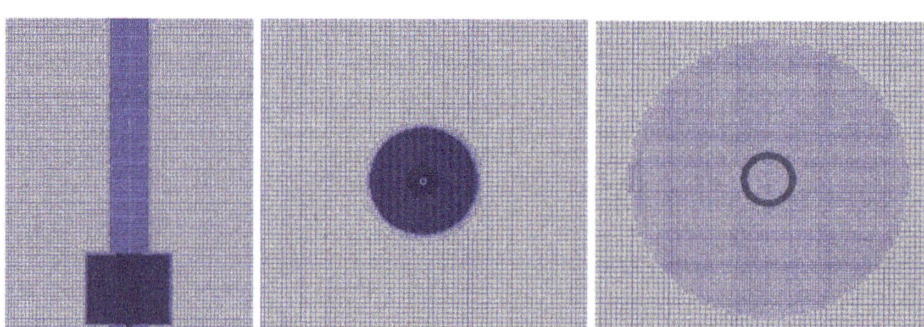

Figure 2. Computational grid of the single bubble injection domain.

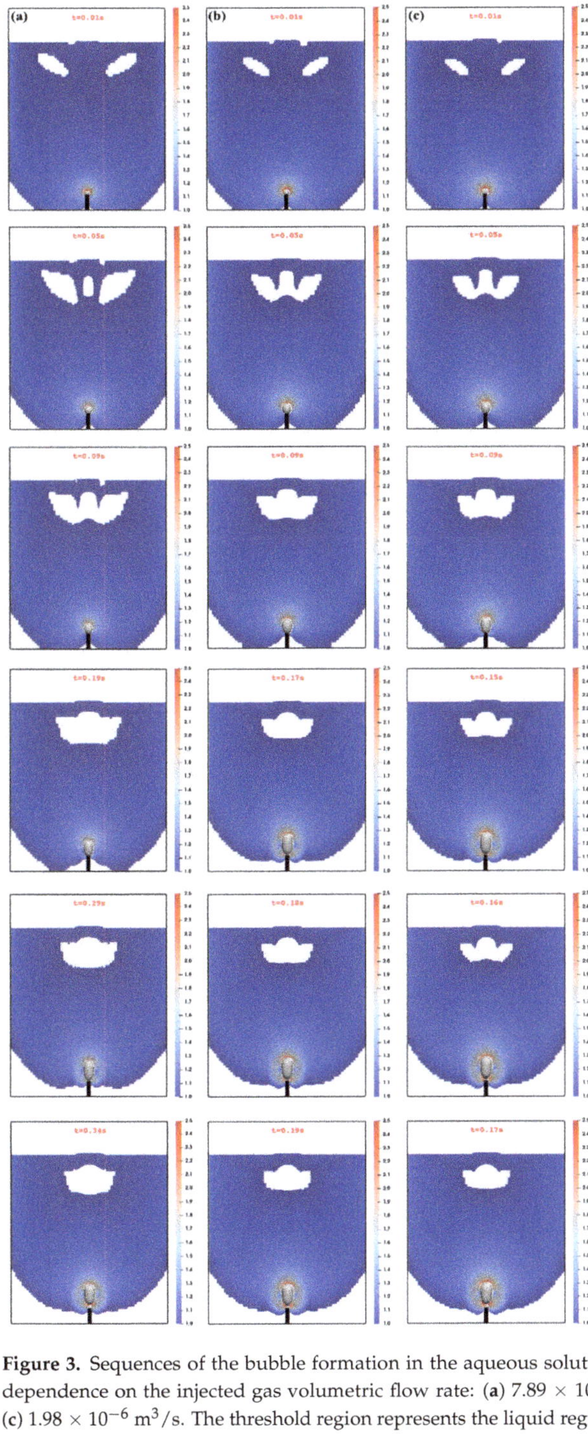

Figure 3. Sequences of the bubble formation in the aqueous solution of the xanthan gum with a dependence on the injected gas volumetric flow rate: (**a**) 7.89 × 10^{-7}, (**b**) 1.60 × 10^{-6} m^3/s, and (**c**) 1.98 × 10^{-6} m^3/s. The threshold region represents the liquid region colored with the ratio of the second invariant of the viscous tensor to the yield stress of the xanthan gum (i.e., $|II_{\tau_1}|^{\frac{1}{2}}/\tau_0$).

5. Results

5.1. Formation of a Single Bubble and Its Growth

Prior to investigating the motion of bubbles in the cement slurry, the experimental work of Terasaka and Tsuge (2001) [18] on a single bubble formation and growth in a stagnant yield stress fluid is used to verify the results from the numerical simulations of the equations presented in the previous section. Terasaka and Tsuge (2001) [18] inject nitrogen gas through a cylindrical nozzle centered at the bottom of a square cross-section column, which is filled with an aqueous solution of xanthan gum; it is assumed that the fluid behaves as a yield stress fluid whose rheological data are fitted with the Herschel–Bulkley model. The column cross-section is 10×10 cm^2, while the diameter of the nozzle is 1.5 mm. The properties of the fluid are summarized in Table 5. The depth of the aqueous solution above the top of the nozzle is 10 cm. Three different gas flow rates, namely 7.89×10^{-7}, 1.60×10^{-6} m^3/s, and 1.98×10^{-6} m^3/s corresponding to gas orifice velocities of about 0.45, 0.91, and 1.12 m/s, respectively, are used, and the bubble volumes at the detachment are compared against the measurements from Terasaka and Tsuge [18].

Table 5. Rheological and physical properties of the xanthan gum aqueous solution.

Density (ρ_1)	Surface Tension (σ)	Yield Stress (τ_0)	Consistency (k)	Power-Law Exponent (n)	Temperature (T)
975 kg/m^3	0.0551 N/m	10.2 Pa	4.06 Pa.s^n	0.307	297

The calculations are performed on an unstructured hexa-dominated computational mesh, with the smallest cells concentrated around the injection nozzle and along the axial region of the column (see Figure 2) to ensure an adequate resolution of the formation and growth of the bubble as the gas penetrates the fluid, causing strong velocity gradients. Thus, the cross-section of the gas injection nozzle is clustered with 16 cells such that the grid size in the injection nozzle and surroundings is $\Delta \simeq 0.094$ mm. The largest cells (i.e., with spacing $\Delta \simeq 1.5$ mm) are distributed within the outer region along the column. The generated grid consists of 6,707,927 cells. The numerical schemes and the regularization outlined above remain the same for this case study.

The sequential snapshots of the bubble shown in Figure 3 exhibit markedly different stages during the bubble formation, which is consistent with the experimental observations (see, e.g., Terasaka and Tsuge, 2001 [18]), along with the flow region where the second invariant of the viscous tensor (i.e., $|II_{\tau_1}|^{\frac{1}{2}}$) exceeds the yield stress (τ_0). Indeed, for each of these flow rates, the bubble appears first to increase in volume by expanding both radially and axially. Figure 4a, which illustrates the time evolution of the interface between the bubble and the fluid, further quantifies the growth of the bubble and the change in its aspect ratio (along with the range of shape). Along the second stage, which can be associated with the onset of a conical-like shaping of the bubble, the base diameter is progressively reduced (relative to the first stage) as a result of the upward stretching and elongations while the gas continues to be injected (see, e.g., Figure 4b for the quantitative change). Figure 4c shows that the third and the last stage of the formation of the bubble is essentially marked by the formation of the neck, which results in the detachment.

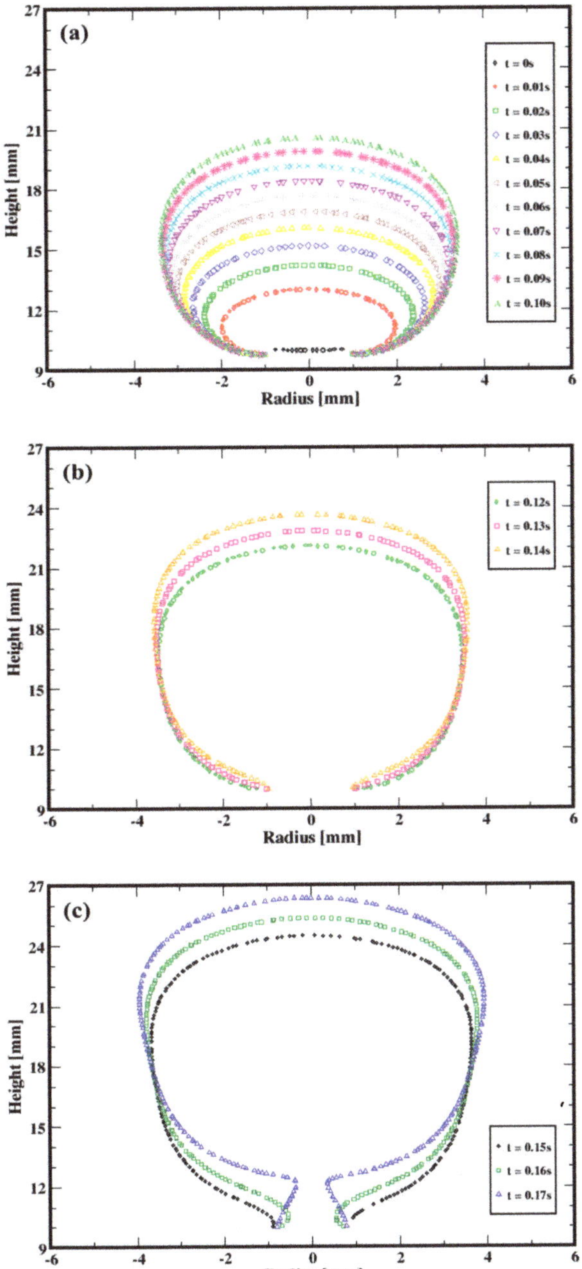

Figure 4. Sequential evolution of the interface between the injected gas and the aqeous solution of xanthan for the gas volumetric flow rate of 1.98×10^{-6} m^3/s. (**a**) First stage (**b**) Second stage (**c**) Third stage.

Figure 5 shows the bubble growth curves for the three flow rates. The growth rate increases with the gas injection rate, while the bubble formation period decreases.

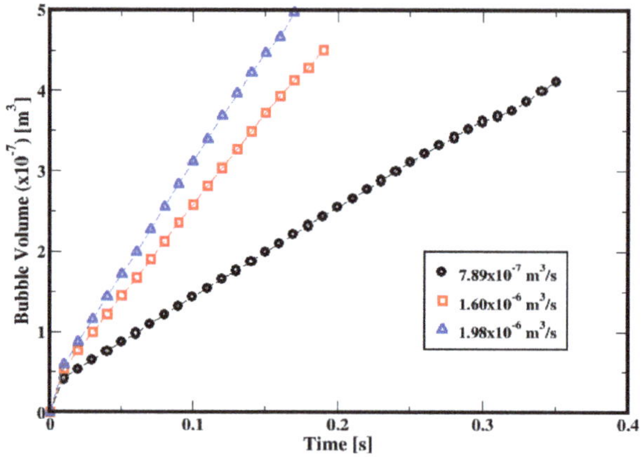

Figure 5. Dependence of bubble growth on the gas volumetric flow rate.

Figure 6 compares the predicted volume of the bubbles at the time of the detachment against the measurements from Terasaka and Tsuge (2001) [18]. The deviations in the predictions associated with the flow rates 7.89×10^{-7}, 1.60×10^{-6}, and 1.98×10^{-6} m^3/s are −3.93%, −1.34%, and 7.63%, respectively. Although the uncertainties in the measurements are not available, these deviations, which remain within ±10% margins relative to the measurements, can be regarded as relatively small considering that the mathematical models along with the numerical approach reasonably approximate the flow dynamics.

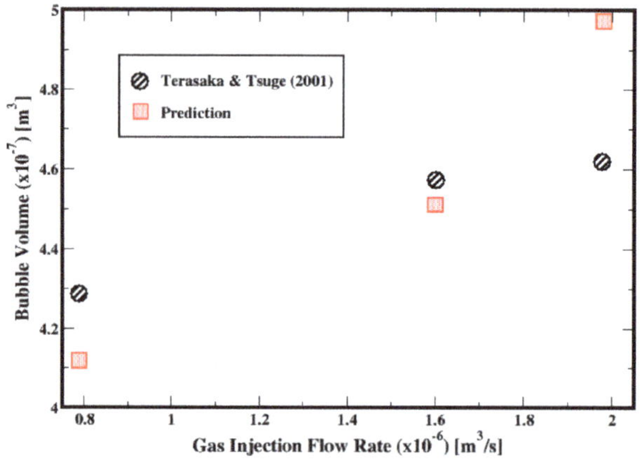

Figure 6. Prediction of bubble volume with respect to the gas volumetric flow rate [18].

5.2. Multiple-Bubble Cement Slurry Column

5.2.1. Mesh Refinement

The sensitivity of the solution on the grid resolution is based on the three grid densities summarized in Table 3. The generation of the finer grid (i.e., mesh #3), considered hereafter as the reference, is guided by the reasonable accuracy of the single bubble injection predictions. Indeed, the current "fine" mesh exhibits larger cells almost twice as small, while smaller cells are about twice as large, compared to single bubble injection. Given the considerable difference in the superficial velocity of the injected gas between the two systems (i.e., 2 to 4 order of magnitudes smaller with the cement column), the flow gradients can be assumed to be less strong within the column than for the single bubble formation and detachment. Therefore, the reference "fine" mesh is considered fine enough to capture the flow. The simulations are performed using the Herschel–Bulkley fluid model with a gas flow rate of 20 mL/min.

Table 6 compares the quantities of interest in this work, using these three grids. The extent of the yielded material, as well as the mean rise radial position and the rise angle, appears to be somewhat the same. The most significant deviations occur with the mean Sauter diameter and the mean rise velocity for which the deviations relative to the grid density #3 (i.e., the finer mesh) are quite large compared with the grid density #1 (i.e., the coarser mesh). In general, the deviations with the grid density #2 relative to the finer mesh are below 10%, which suggests that the details of the flow are adequately captured with the grid density #2.

Table 6. Summary of mesh refinement. The deviations are calculated with respect to the finer mesh (i.e., mesh #3).

Mesh	#1		#2		#3	
	Mean	**Deviation (%)**	**Mean**	**Deviation (%)**	**Mean**	**Deviation (%)**
Yielded fraction (-)	0.61	−3.17	0.64	1.59	0.63	-
Sauter diameter (mm)	4.73	42.90	3.14	−5.14	3.31	-
Aspect ratio (-)	0.82	−18.81	1.11	9.90	1.01	-
Rise position, r/R (-)	0.51	−13.56	0.60	1.69	0.59	-
Rise velocity (mm/s)	3.65	31.77	2.50	−9.75	2.77	-
Rise angle (deg)	84	0	84	0	84	-

5.2.2. Flow Patterns

Shown in Figures 7–9 are the snapshots of the distribution of the volume fraction of the cement slurry and the bubbles in the cylindrical domain for different flow rates (see Table 2), along with the iso-surfaces (i.e., a three-dimensional surface passing through data of the same value) of 80% of the gas presence, which is considered at the external boundary of the gas bubbles. As we can see, different flow patterns can readily be identified for different lengths, sizes, and shapes of these iso-surface bubbles as the gas injection flow rate changes. Indeed, for low injection rates, relatively small bubbles rise toward the freeboard. With moderate flow rates, small to large bubbles appear to rise, while smaller bubbles often accompany the larger ones in their wakes. This is most likely a signature of the breakup of large bubbles as they rise. For high flow rates, the iso-surfaces exhibit stretched and elongated shapes toward the vertical axis as channeling paths for the gas; these bubbles appear to rise vertically in the column. Furthermore, these flow patterns remain consistent for both the Herschel–Bulkley and Bingham models used in this paper.

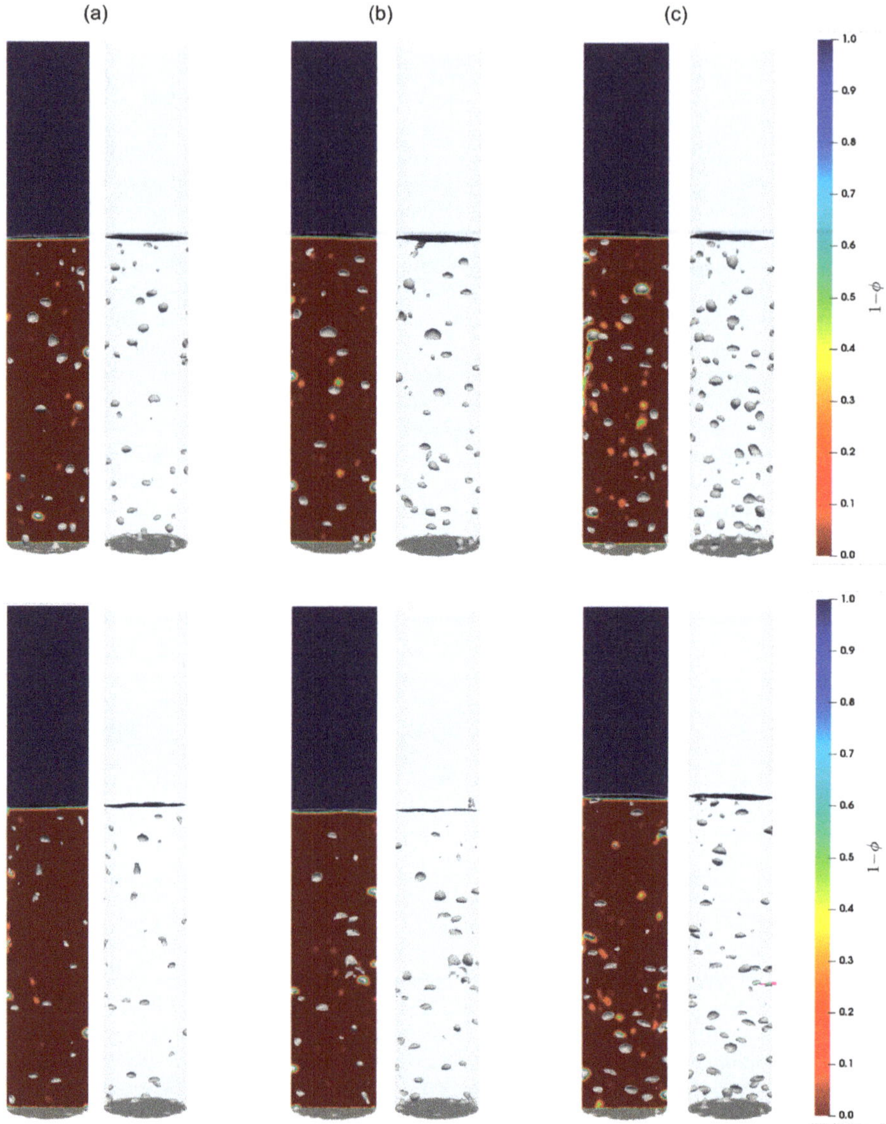

Figure 7. Contours of the cement slurry along with the air bubbles in the column; (**a**–**c**) correspond to the flow rates 2 mL/min, 4 mL/min, and 8 mL/min, respectively. Top and bottom parcels show the contours for the Herschel–Bulkley and the Bingham fluids, respectively.

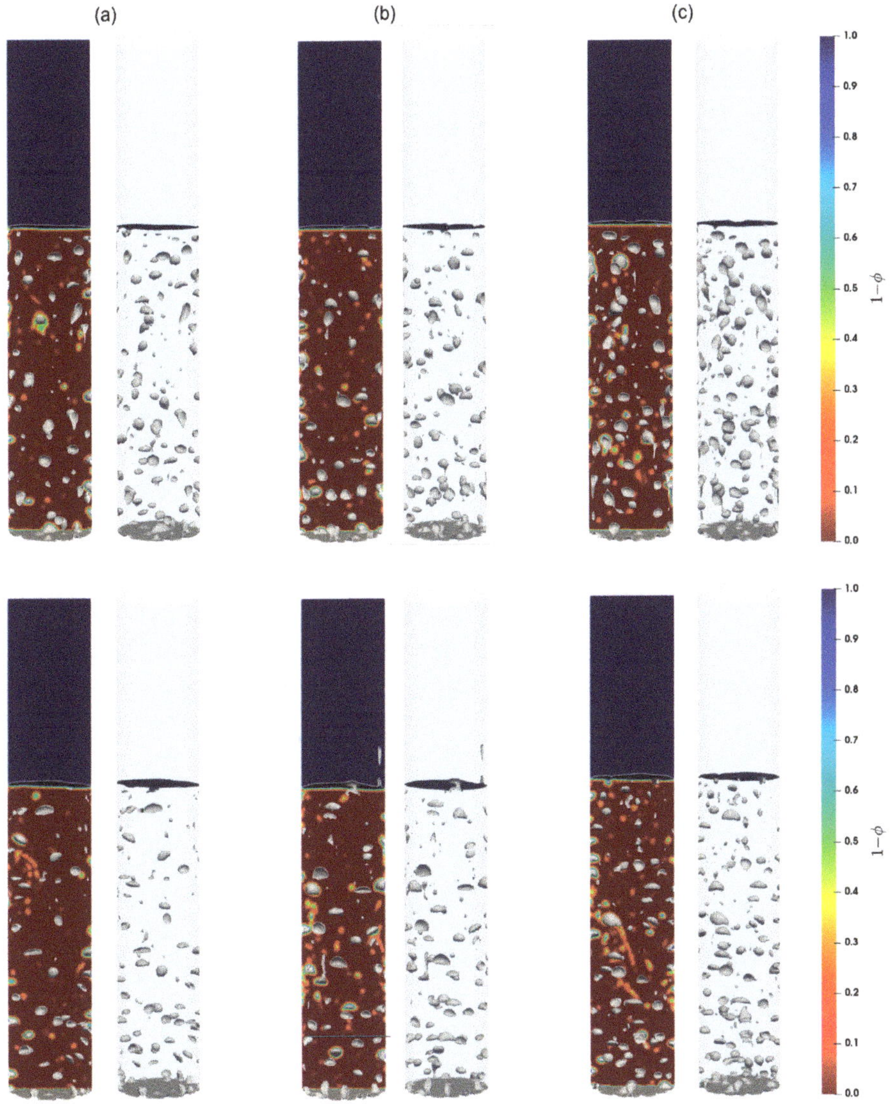

Figure 8. Contours of the cement slurry, along with air bubbles in the column; (**a**–**c**) correspond to the flow rates 15 mL/min, 20 mL/min, and 30 mL/min, respectively. Top and bottom parcels show the contours for the Herschel–Bulkley and the Bingham fluids, respectively.

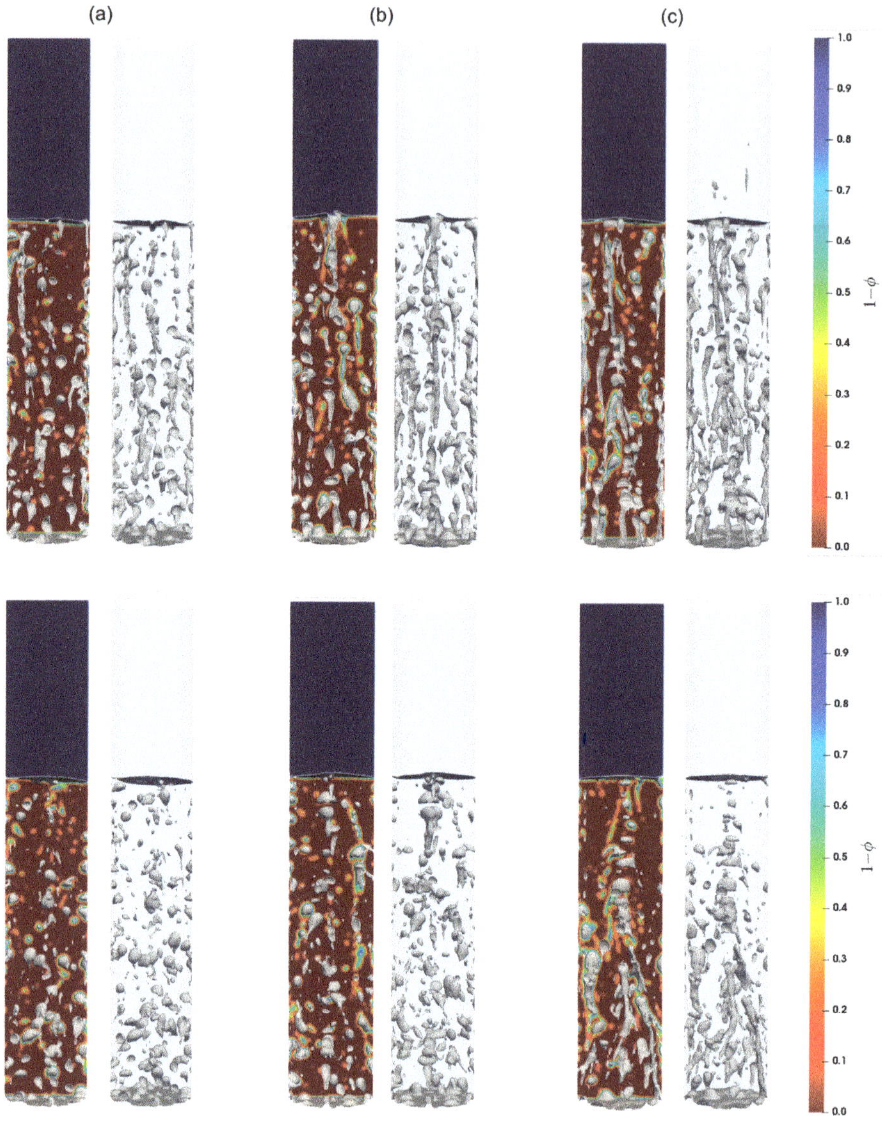

Figure 9. Contours of the cement slurry, along with air bubbles in the column; (**a**–**c**) correspond to the flow rates 60 mL/min, 90 mL/min, and 120 mL/min, respectively. Top and bottom parcels show the contours for the Herschel–Bulkley and the Bingham fluids, respectively.

5.2.3. Flow Behavior

It is known that an object moving through a yield stress fluid exhibits a surrounding envelope in which the material is yielded (see, e.g., Oldroyd, 1947 [41]; Beris et al., 1985 [42]; Piau, 2002 [43]; Tsamopoulos et al., 2008 [44]). In the context of a single freely rising bubble in a stagnant Herschel–Bulkley fluid, Dimakopoulos et al. (2013) [40] discuss the extent of the surrounding yielded region, as well as the unyielded areas around the bubble with respect to the Bingham number (for given Bond and Archimedes numbers), merging the circumstances of the unyielded material that may result in the entrapment of the bubble.

Clearly, the evolution and the dynamics of the yielded (or unyielded) material can have strong implications on the dynamics of the rising bubbles.

In this work, several bubbles rise in the column, and in order to quantify the fraction of the yielded material within a domain Ω, we introduce the conditional average of the material that yields within the domain:

$$\left\langle \phi \middle| II_{\tau_1}^{1/2} > \tau_0 \right\rangle = \frac{\int_\Omega \phi H\left(II_{\tau_1}^{1/2} - \tau_0\right) dV}{\int_\Omega \phi dV} \tag{38}$$

where H is the Heaviside function (i.e., equal to 1 when $II_{\tau_1}^{1/2} > \tau_0$ and is zero otherwise); ϕ is the volume fraction of the cement slurry. $II_{\tau_1}\left(= \frac{1}{2}\boldsymbol{\tau}_1 : \boldsymbol{\tau}_1\right)$ is the second invariant of the stress tensor $\boldsymbol{\tau}_1$, and τ_0 is the yield stress of the material.

Shown in Figure 10 is the dependence of the yielded cement slurry in the column for both the Bingham and the Herschel–Bulkley fluids. The increase in the gas flow rate seems to indicate that the fraction of the yielded material is virtually independent of the flow rates investigated here, since the differences appear marginal. However, it can be observed that approximately more than 10% of the material is yielded when using the Herschel–Bulkley model than the Bingham fluid. This difference in the yielded material may be related to the yield stress τ_0 of the Bingham fluid (for this slurry), which is approximately twice for the Herschel–Bulkley model (see Table 1). Note that in this paper, we focus on the behavior of the bubble, its rise velocity and rise angle, etc. This figure also shows the change in the conditional average of the material that yields in the column with respect to the flow rate. Following the previous studies (also shown in the single bubble study in the current work), the rising of a bubble causes the material to locally yield. With the swarm of rising bubbles, the yielding behavior of the material would somewhat exhibit a dependence on the number of the bubbles, if one assumes an additive influence. However, it appears that such an effect (if any) is not shown, given that the flow rate increase results in marginal differences. In our next study, we will present detailed analysis of the flow field for the velocity, pressure fields, etc.

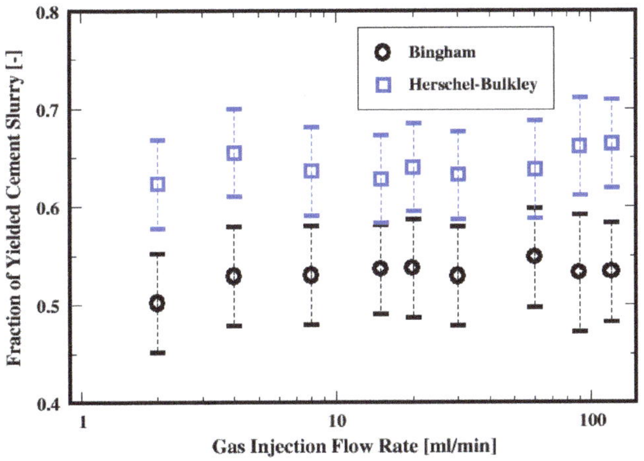

Figure 10. Dependence of the yielded material on the flow rate for the two fluid models.

5.2.4. Bubble Statistics

Size Distribution

The distribution of the air bubble Sauter diameters (i.e., volume-to-surface ratio equivalent diameter) are plotted in Figure 11, Figure 13 and Figure 15 for different gas flow rates. For low flow rates (identified here as 2 mL/min and 4 mL/min), the bubble size distributions mainly exhibit more than one mode (which is defined as a commonly found value of the distribution) for both the Herschel–Bulkley and the Bingham fluid models (see Figure 11). Both large and small size bubbles appear regardless of the fluid model, although a third mode is also apparent for the Bingham fluid. Furthermore, large bubbles are systematically larger for the Herschel–Bulkley model compared to Bingham. Figure 12a,b also show the average Sauter diameter along the column where the bubble sizes decrease as they rise toward the freeboard, regardless of the fluid model.

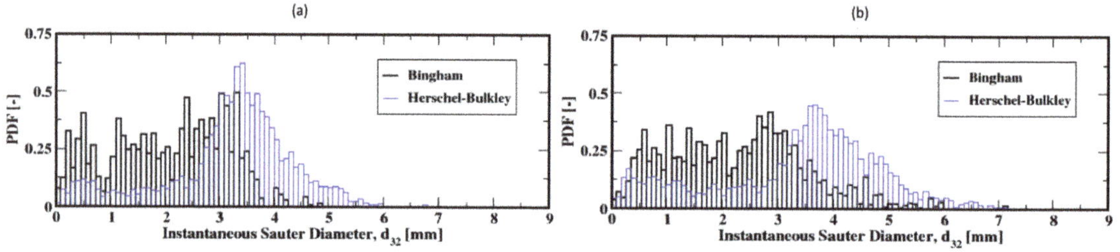

Figure 11. Distribution of the instantaneous Sauter diameter of the air bubbles in the column; (**a**,**b**) correspond to the flow rates 2 mL/min and 4 mL/min, respectively.

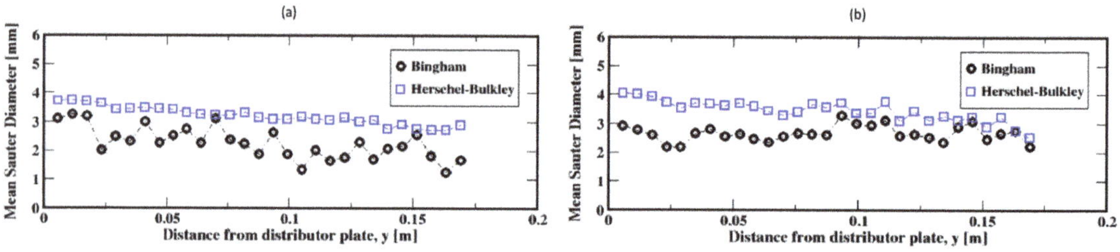

Figure 12. Mean Sauter diameter of the air bubbles in the column; (**a**,**b**) correspond to the flow rates 2 mL/min and 4 mL/min, respectively.

The bubble size distributions shown in Figure 13a–d, associated with gas flow rates of 8, 15, 20, and 30 mL/min, exhibit an interesting bimodal trend of large and small bubble sizes marked with a progressive decrease in the frequency of large bubbles to the benefit of the small ones. This trend is independent of the fluid model used in this work. The apparent difference is the slightly larger bubbles for the Herschel–Bulkley model. Plotted in Figure 14a–d are the average Sauter diameters of the bubbles along the column. The exhibited unique pattern of decreasing bubble sizes with the distance above the distributor plate is indicative of a rising dynamics mostly controlled by breakup events in the column.

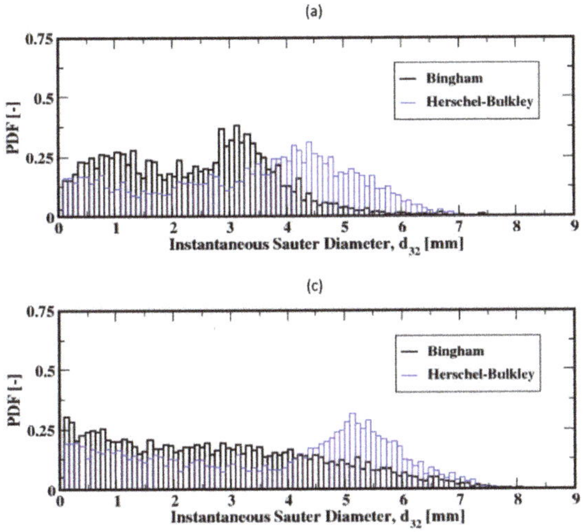

Figure 13. Distribution of the instantaneous Sauter diameter of the air bubbles in the column; (**a**–**d**) correspond to the flow rates 8 mL/min, 15 mL/min, 20 mL/min, and 30 mL/min, respectively.

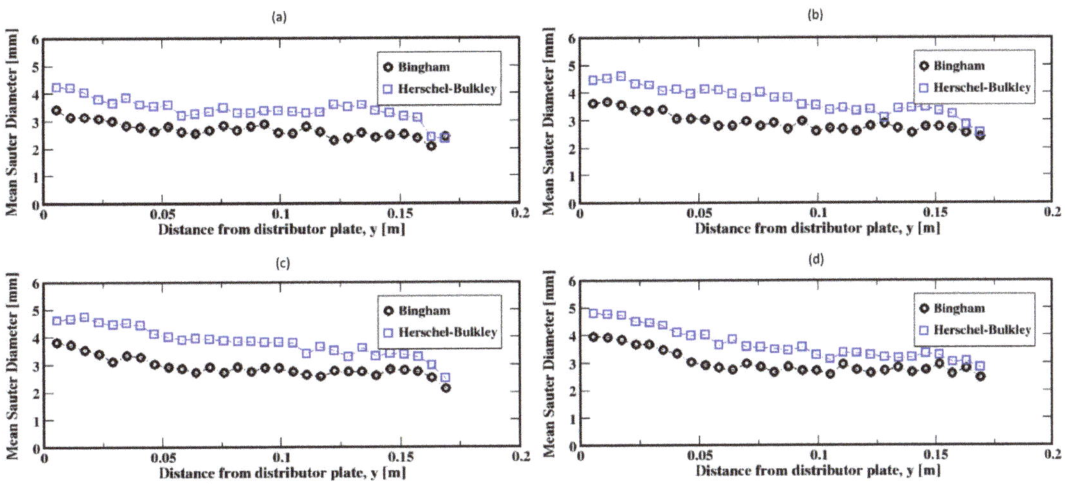

Figure 14. Mean Sauter diameter of the air bubbles in the column; (**a**–**d**) correspond to the flow rates 8 mL/min, 15 mL/min, 20 mL/min, and 30 mL/min, respectively.

Further increases in the gas flow rate (e.g., to 60, 90, and 120 mL/min) result in exponential-like size distributions of the Sauter diameters of the bubbles, both for the Herschel–Bulkley and the Bingham models (see Figure 15a–c). This indicates that small Sauter diameter bubbles populate the column, regardless of the fluid rheology. However, the slow decrease in the average size of the bubbles along the column, as shown in Figure 16a–c, suggests that the small Sauter size bubbles are distributed throughout the column following the breakup events of the bubbles as they rise toward the freeboard.

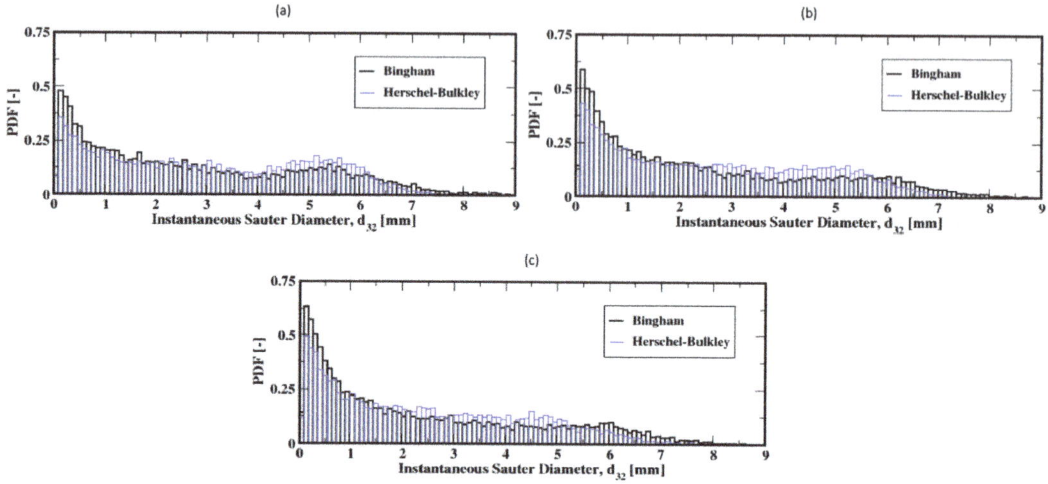

Figure 15. Distribution of the instantaneous Sauter diameter of the air bubbles int the column; (**a**–**c**) correspond to the flow rates 60 mL/min, 90 mL/min, and 120 mL/min, respectively.

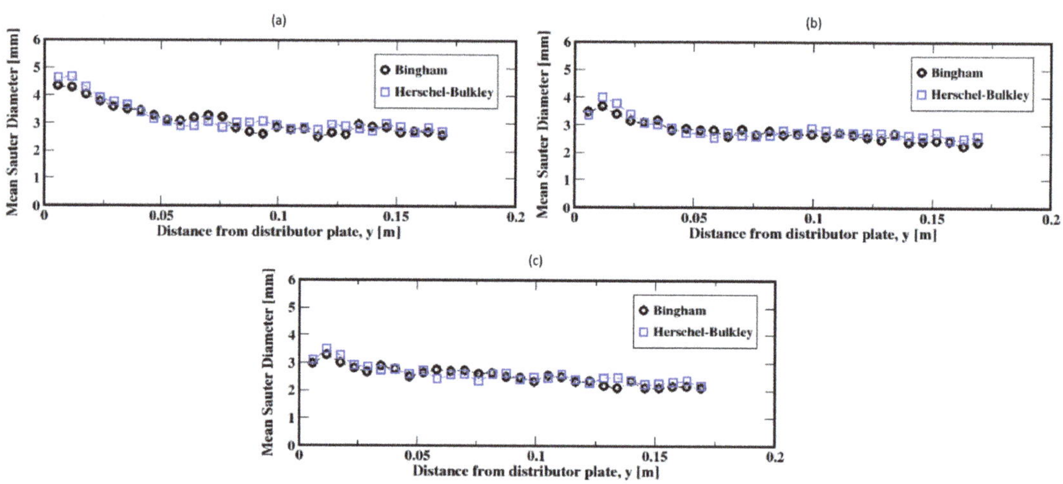

Figure 16. Mean Sauter diameter of the air bubbles in the column; (**a**–**c**) correspond to the flow rates 60 mL/min, 90 mL/min, and 120 mL/min, respectively.

Both the influence of gas flow rate and the cement properties on the average Sauter diameters of the bubbles are summarized in Figure 17. It is apparent that regardless of which model is used, the average bubble Sauter diameter decreases with the increase in the gas flow rate. Furthermore, relatively large dispersions are shown around these average diameters, thus indicating a broad range of bubbles evolving in the column. The difference in the fluid properties results in smaller average Sauter diameters for the Bingham fluid relative to the Herschel–Bulkley model. This is more pronounced for the moderate gas flow rates.

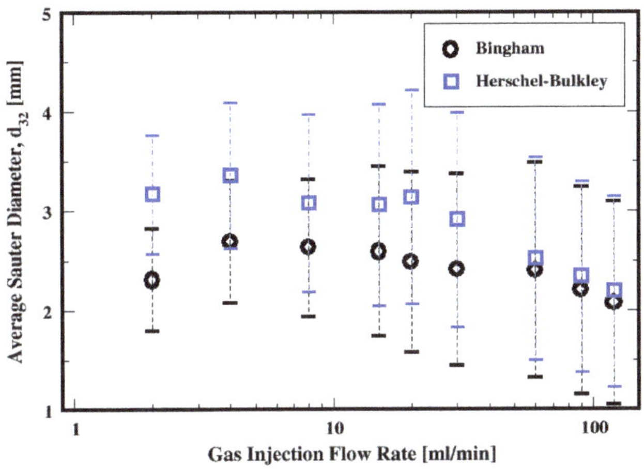

Figure 17. Dependence of the mean Sauter diameter on the flow rate and the fluid model.

Aspect Ratio Distribution

The shapes of the rising bubbles are studied by looking at the distributions of the instantaneous bubble aspect ratio (i.e., ratio of the maximum horizontal length to the maximum vertical length) for different gas flow rates and fluid models.

The distributions of the aspect ratios at low flow rates, which do not exhibit a unique shape, are plotted in Figure 18a,b. They show that the bubbles are most frequently wide/oblate and especially wider for the Bingham fluid compared to the Herschel–Bulkley fluid model. For these low flow rates, the average aspect ratio is presented in Figure 19a,b; they show that on average the bubbles tend to rise with moderately stable wide shapes for the Herschel–Bulkley fluid. Additionally, the bubbles appear to detach from the distributor plate with wide shapes regardless of the fluid model used.

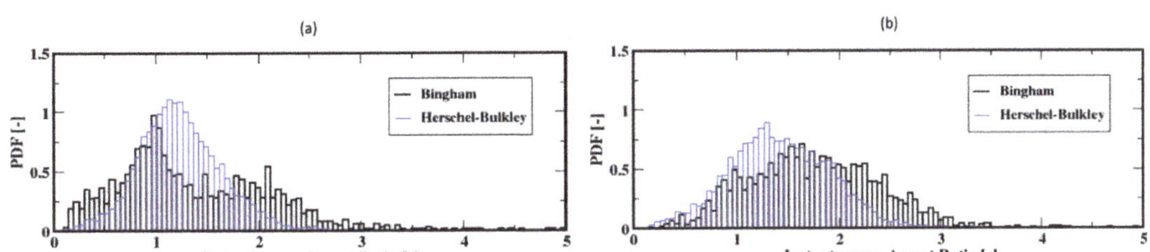

Figure 18. Distribution of the instantaneous aspect ratio of the air bubbles; (**a**,**b**) correspond to the flow rates 2 mL/min and 4 mL/min, respectively.

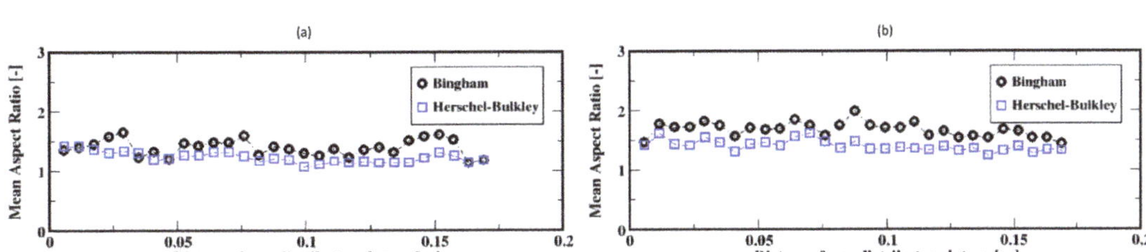

Figure 19. Mean aspect ratio of the air bubbles; (**a**,**b**) correspond to the flow rates 2 mL/min and 4 mL/min, respectively.

As the gas flow rate increases, the aspect ratio distributions, Figure 20a–d, exhibit a shift toward the smaller ratios that peak around unity. This suggests that the bubbles are mostly of spherical shape and also that dependence of the shape of bubbles on the fluid model appears to fade out. Figure 21a–d present the average aspect ratios of the bubbles, which exhibit quite stable aspect ratios despite noticeable decreases. This indicates that the bubbles undergo smaller deformations (i.e., change in shape) after their detachment from the plate up to the freeboard. This trend is independent of the fluid model. However, the bubbles are wider for the Bingham fluid when compared to the case of the Herschel–Bulkley fluid for each of these moderate gas flow rates.

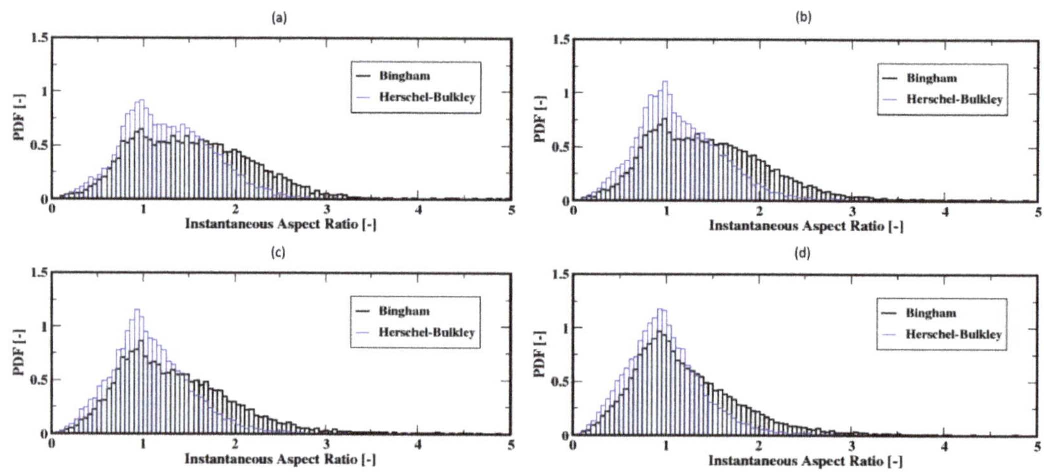

Figure 20. Distribution of the instantaneous aspect ratio of the air bubbles; (**a**–**d**) correspond to the flow rates 8 mL/min, 15 mL/min, 20 mL/min, and 30 mL/min, respectively.

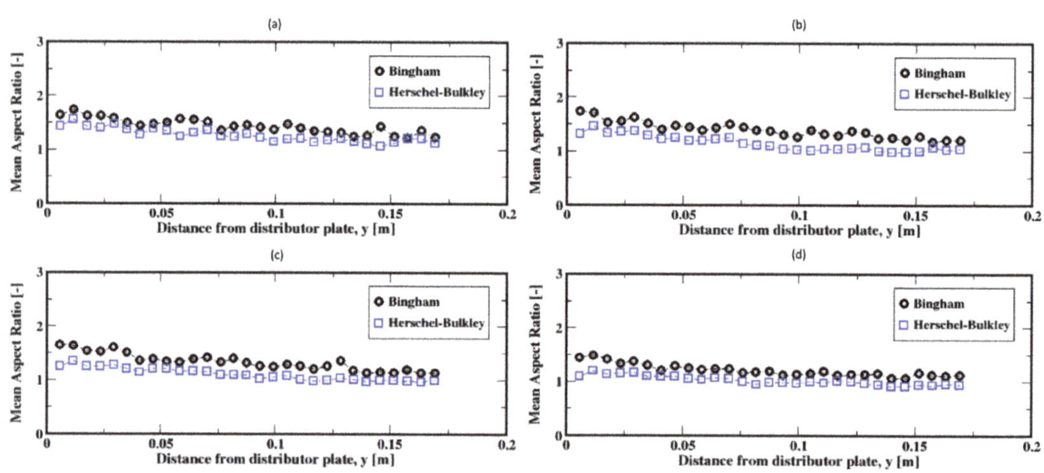

Figure 21. Mean aspect ratio of the air bubbles; (**a**–**d**) correspond to the flow rates 8 mL/min, 15 mL/min, 20 mL/min, and 30 mL/min, respectively.

For the higher flow rates, the distributions mostly shift toward the range of smaller ratios with most frequent values below the unity for the aspect ratio for both the Herschel–Bulkley and the Bingham fluids (see Figure 22a–c). That is, the bubbles are mostly elongated in the vertical direction, regardless of the fluid model used. Furthermore, the average aspect ratios along the column show that the bubbles tend to rise with stable elongated shapes but also that the Herschel–Bulkley fluid results in more elongated shapes compared to the Bingham fluid case (see Figure 23a,b).

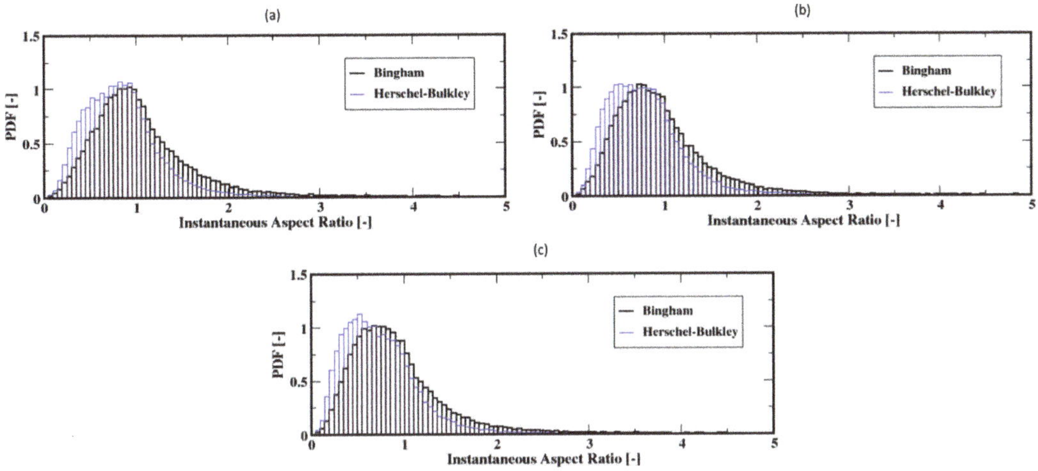

Figure 22. Distribution of the instantaneous aspect ratio of the air bubbles; (**a**–**c**) correspond to the flow rates 60 mL/min, 90 mL/min, and 120 mL/min, respectively.

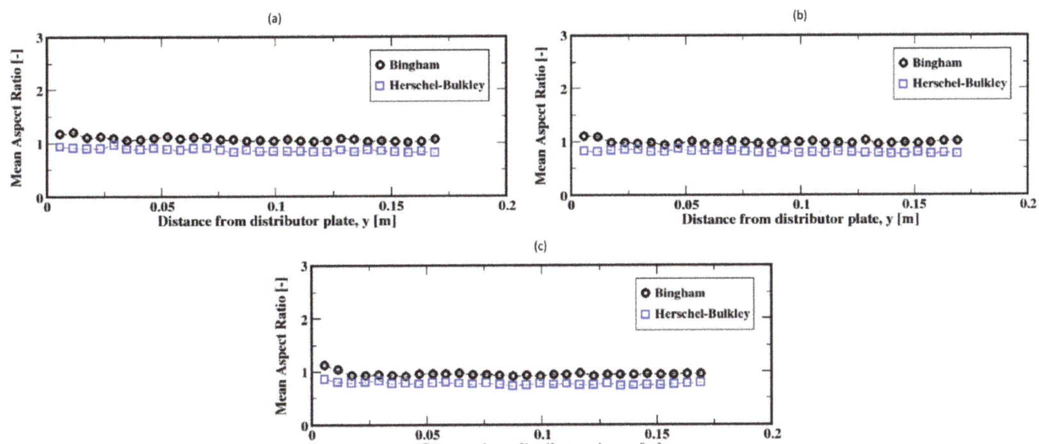

Figure 23. Mean aspect ratio of the air bubbles; (**a**–**c**) correspond to the flow rates 60 mL/min, 90 mL/min, and 120 mL/min, respectively.

Shown in Figure 24 is the dependence of the average aspect ratio on the gas flow rate, the fluid model, and the dispersion associated with each distribution. These average aspect ratios monotonically decrease with increasing gas flow rate, regardless of the fluid model. This indicates a change in the shapes from wide (or "kidney-like") shapes to elongated (or "finger-like") shapes with increasing gas flow rate.

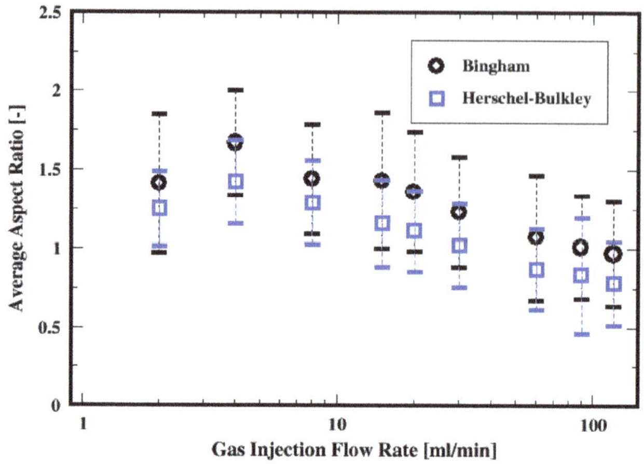

Figure 24. Dependence of the mean aspect ratio of the air bubbles on the flow rate and the fluid models.

Bubble Rise Velocity

Figure 25 shows the dependence of the cross-sectional average bubble rise velocity on the gas flow rate. It is apparent that, on average, the rise velocities of the bubbles decrease with the gas flow rate, regardless of the fluid model. This is not surprising since the bubble size decreases as well with the gas flow rate (as shown in Figure 17), and the smaller bubbles rise more slowly than the larger ones. This dependence of the rise velocity on the bubble size (or volume) is well established and has been extensively discussed for a single bubble rising in non-Newtonian fluids (see, e.g., De Kee et al., 1990 [45]; Tsamopoulos et al., 2008 [44]; Amirnia et al., 2013 [46]; Ravisankar et al., 2022 [47]). Tsamopoulos et al. (2008) [44] discuss such an increase in the rise velocity with the bubble diameter; they attribute this to the decrease in the Bingham number. Moreover, they find that the rise velocity decreases with the Bingham number because of the increase in the drag coefficient.

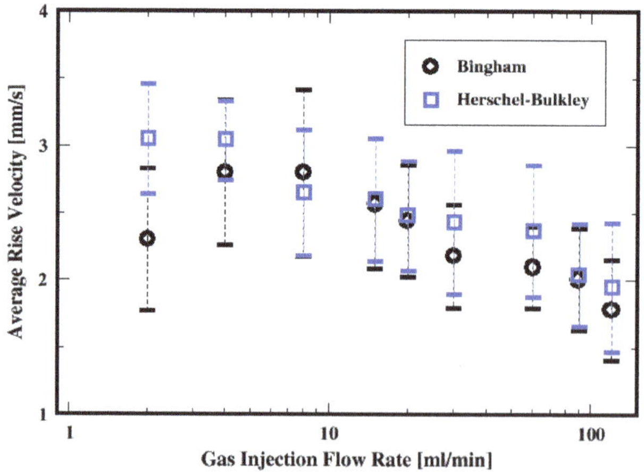

Figure 25. Dependence of the mean rise velocity of the air bubbles on the flow rate and the fluid model.

The mean rise velocity of the bubbles (which is defined as the arithmetic average of the bubble velocities calculated from the difference of consecutive axial centroid coordinates of the bubbles and divided by the time interval) is plotted as a function of the mean diameter in Figure 26a. Regardless of the fluid model used, the mean rise velocity increases with the mean diameter of the bubbles. However, the distributions of the rise velocities and the bubble diameters suggest that the description of the average rise dynamics must account for these dispersions and to a certain extent for additional mechanisms such as the breakup and coalescence of the bubbles. Buchholz et al. (1978) [48] report that bubbles in swarm have higher rising velocities than single bubbles (for a power-law Ostwald–de Waele fluid). With similar results, Yuan et al. (2021) [49] report that the group rising velocity increases with the number of bubbles, and also, the vertical chains become unstable for multiple bubbles because of the distinct oscillation of the uppermost ones. Vélez-Cordero et al. (2011) [50] discuss the dependence of the drag coefficient and the rise velocity on the interactions (e.g., drafting and kissing but not tumbling) between rising bubbles in shear-thinning fluid. The mean rise velocity plotted as a function of the mean bubble diameter (see Figure 26a) shows that the mean rising velocity is independent of the fluid properties for the smaller mean diameters, whereas the dependence becomes more apparent for larger mean diameters of the bubbles, where higher rise velocity occurs for the Bingham fluid. Similar mean rise velocity is noticed for small mean diameters for both the Herschel–Bulkley and the Bingham fluid descriptions of the cement slurry. This might have been controlled by the oscillating pressure at the inlet of the column (because of the zero gradient inlet pressure).

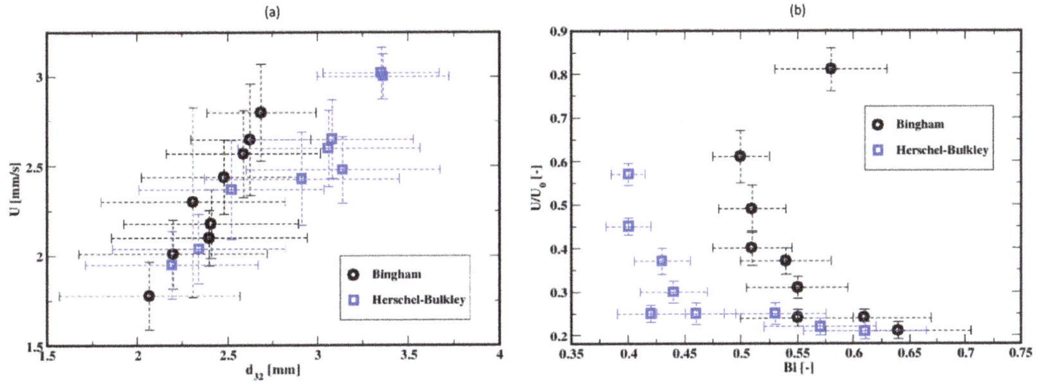

Figure 26. (**a**) Mean rise velocity of the air bubbles as a function of the mean Sauter diameter and (**b**) dimensionless mean rise velocity versus the Bingham number.

Shown in Figure 26b is the dependence of the dimensionless mean rise velocity (scaled by balancing the buoyancy and the viscous forces, i.e., $U_0 = \rho g d_{32}^2 / 4\mu_0$ with $\mu_0 = k(U_{g,s}/d_{32})^{n-1}$, where $U_{g,s}$ is the gas superficial velocity) on the Bingham number (defined as $\mathrm{Bi} = 2\tau_0/\rho g d_{32}$). Following Tsamopoulos et al. (2008) [44], such a scaling aims at highlighting the influence of the bubble size. The Bingham number increases with the superficial gas velocity since the mean diameter of the bubbles decrease (see Figure 17), while the mean bubble rise velocity decreases. Thus, with the increase in the gas superficial velocity (or the gas volumetric flow rate), the dimensionless mean rise velocity of the bubbles decreases with the Bingham number, as shown in Figure 26b. Clearly, this is attributable to the increase in the drag as indicated in previous studies (see, e.g., Tabuteau et al., 2007 [51]; Tsamopoulos et al., 2008 [44]; Dimakopoulos et al., 2013 [40]; Mirzaagha et al., 2017 [52]).

Rise Angle

The average rise angle of the bubbles, which is defined as the arithmetic average of angles formed between the vectors of consecutive centroid coordinates of a generic bubble and the unit vector in the plane of the distributor plate, is plotted as a function of gas flow rate in Figure 27 for both the Herschel–Bulkley and Bingham fluids. The plots show that the bubbles rise throughout the slurry with average angles of approximately 81° to 84° relative to the distributor plate and with variances ranging between 6° and 9°. In other words, they mostly rise in almost vertical paths while exhibiting some oscillations (as suggested by the variances) toward the freeboard. Only the conditions of the low flow rates exhibit an average rising angle of approximately 66° with large variances of approximately 20° when using the Bingham model to describe the cement slurry. This suggests that the bubbles rise with more pronounced zigzagging trajectories. This behavior is consistent for these low flow rate conditions for the Bingham fluid, where the bubbles experience strong yield stress effects resulting in the breakup of the bubbles away from the distributor. Furthermore, it is worth noticing that instabilities in the trajectories of bubbles rising through non-Newtonian fluids are observed and discussed in the literature; they are thought to be associated with deformation and vortex shedding in the wake of the bubbles (see, e.g., Premlata et al., 2017 [53]; Sharaf et al., 2017 [54]). Premlata et al. [53] also report that, using quiescent shear-thinning liquids, increasing the shear-thinning behavior promotes path instabilities. In the present work, in addition to the shear-thinning behavior of the cement slurry, the gas injection introduces additional perturbations to the system that might interact with the vortex shedding in the wake of the bubbles.

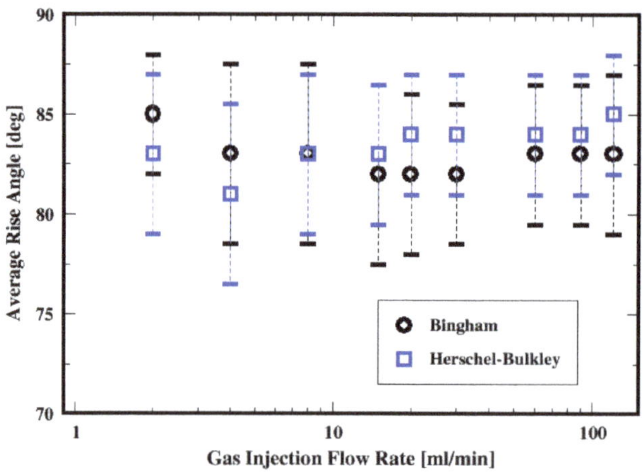

Figure 27. Dependence of the mean rise angle of the air bubbles on the flow rate and the two fluid models.

Rising Position

Figure 28 shows the dependence of the average radial positions of the bubbles as they rise. First, this shows that the bubbles tend to rise throughout the peripherical regions of the column. Second, increasing the gas flow rate results in the broadening of the rising regions to the vicinity of the walls as illustrated by the monotonically decreasing of their positions toward the center of the column; this seems to be independent of the fluid models used here.

Furthermore, for a given gas flow rate, the rising position behavior can be correlated to the resulting sizes of the bubbles. Indeed, it can be noticed that for a given flow rate, the larger the bubbles are, the farther the center of the bubbles are relative to the wall by comparison to the positions of the smaller ones. That is, when using the Herschel–Bulkley fluid, the bubbles are farther away from the wall, as opposed to the Bingham fluid case, since the bubbles are larger on average with the Herschel–Bulkley fluid (see Figure 17).

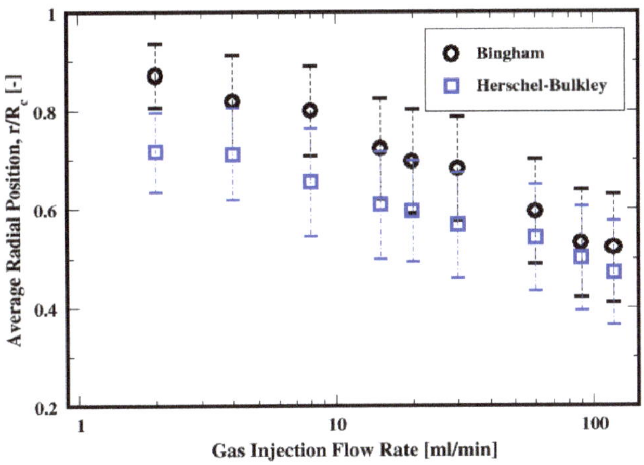

Figure 28. Dependence of the mean rise radial position of the air bubbles on the flow rate and the two fluid models. R_c denotes the radius of the column.

5.2.5. Influence of the Surface Tension

The influence of the surface tension on the bubbles' geometrical characteristics and their rising dynamics are examined in the framework of the Herschel–Bulkley fluid model for gas flow rates in low and moderate ranges, namely 2 mL/min and 20 mL/min. We use three different values for the surface tension, σ = 0.025, 0.05, and 0.1 N/m, in addition to the previous value, σ = 0.07 N/m.

Figure 29a illustrates the dependence of the bubble sizes on the surface tension, along with the Bond number defined as $\text{Bo} = \rho_0 g d_{32}^2 / 4\sigma$ and plotted in Figure 29b. Figure 29a shows that the mean Sauter diameter of the bubbles monotonically increases with the surface tension. This is mainly due to the smaller bubbles detaching from the distributor plate as shown in Figure 30. In other words, the surface tension controls the initial sizes of the bubbles as they rise. The Bond number, plotted in Figure 29b as a function of the surface tension, remains on average below unity over the range of the surface tension investigated in this work. This suggests that the bubbles are mostly in regimes dominated by the surface tension.

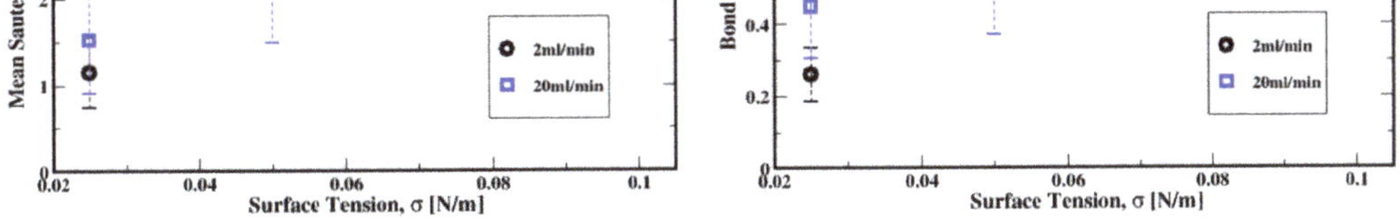

Figure 29. Dependence of the mean Sauter diameter of the air bubbles and the Bond number for gas flow rates of 2 mL/min and 20 mL/min. (**a**) the dependence of the bubble size on surface tension; (**b**) Dependence of the Bond number as a function of surface tension.

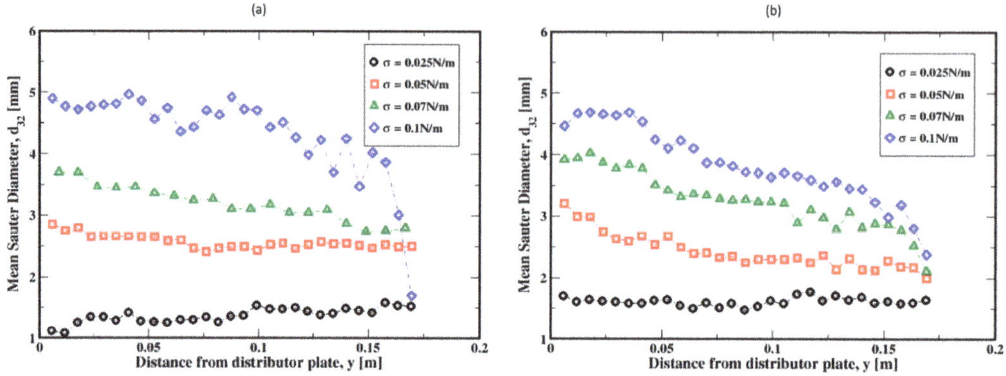

Figure 30. Dependence of the mean Sauter diameter of the air bubbles on the surface tension along the height of the column; (**a**,**b**) refer to the flow rates 2 mL/min and 20 mL/min, respectively.

Meanwhile, the aspect ratio presented in Figure 31 shows that the increase in the surface tension results in substantial deformations of the bubbles at a low gas flow rate. Indeed, it is apparent for this low flow rate that the bubble shapes vary from an elongation-like shape in their vertical axes (more like a "finger") to more oblate/flattened (or "kidney-like") shapes from small to high surface tension. However, for the moderate gas flow rate, the dependence of the bubble aspect ratio on the surface tension appears weak and the bubbles remain quasi-spherical.

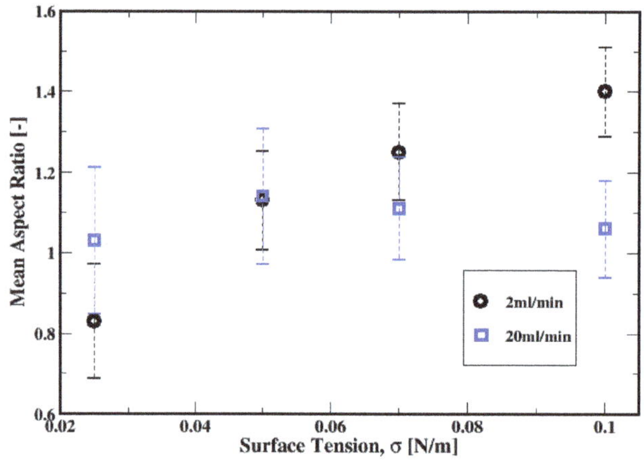

Figure 31. Dependence of the mean aspect ratio of the air bubbles on the surface tension for gas flow rates of 2 mL/min and 20 mL/min.

The average rise velocity plotted in Figure 32 exhibits a weak dependence on the surface tension for the low and the moderate flow rates. The range of the Bond number (i.e., 0.2 to 1.1) used here for different surface tensions does not substantially change the rise velocity. Dimakopoulos et al. (2013) [40] report a non-monotonic change in the rise velocity of a single bubble rising in a stagnant Herschel–Bulkley fluid for Bond numbers ranging from 0.01 to 50. In our study, the weak dependence of the rise velocity on the surface tension may not only be related to the narrow range of Bond numbers used here but also a consequence of the inlet fluctuations of the pressure perturbating the slurry and the interactions between the swarm of the rising bubbles.

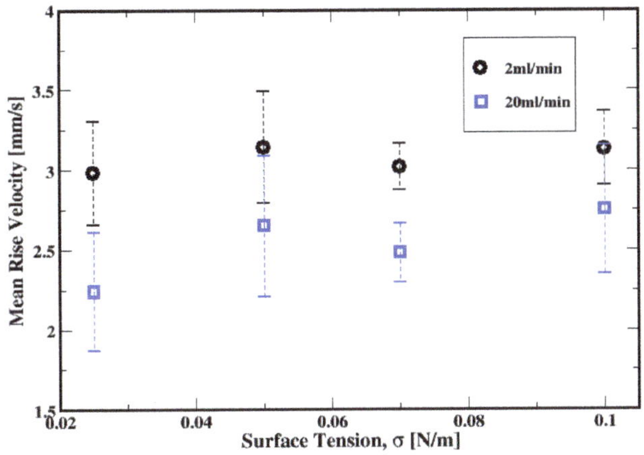

Figure 32. Dependence of the mean rise velocity of the air bubbles on the surface tension for gas flow rates of 2 mL/min and 20 mL/min.

Shown in Figure 33 is the mean rise angle, which also depicts a relatively weak dependence on the surface tension, especially at low flow rates. For the moderate flow rates, both the average and the extent of the variance of the rise angles indicate that zigzagging trajectories of the bubbles are more frequent with smaller values of the surface tensions.

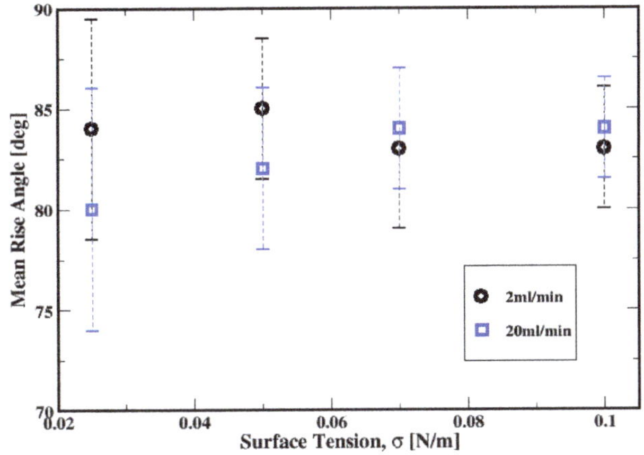

Figure 33. Dependence of the mean rise angle of the air bubbles on the surface tension for the gas flow rates of 2 mL/min and 20 mL/min.

6. Concluding Remarks

The motion of bubbles in a column of cement slurry in a freshly cemented wellbore is studied numerically. The two-phase system of the cement slurry and the air bubbles is described by using conservation equations of the mass and the linear momentum and solved in the framework of the volume-of-fluid (VOF) approach. Predictions of a single bubble formation and its growth in a yield stress fluid, namely, an aqueous solution of xanthan gum (Terasaka and Tsuge, 2001 [18]), shows good comparison with the experimental measurements of the bubble volume. The cement slurry is modeled as a Herschel–Bulkley fluid as well as a Bingham fluid for the sake of comparison. The simulations are performed over a wide range of gas flow rates, typical of gas influx rates in the wellbore.

The results show that increasing the gas flow rate results in mainly three identifiable flow patterns with varying lengths, sizes, and shapes of the air bubbles. Regardless of the fluid model, the mean bubble Sauter diameter decreases with increasing the gas flow rate. A unique pattern of decreasing bubble sizes at a distance above the distributor plate is noticed independently of the gas flow rate. The bubble shape changes from the wide-dominated case to vertically elongated shapes throughout the column when the gas flow rate increases. These results are similar for both the Herschel–Bulkley fluid and the Bingham fluid models.

The fraction of the yielded region appears virtually independent of the gas flow rate, but the extent of the yielded fluid is approximately 10% more with the Herschel–Bulkley model compared to the Bingham fluid. One plausible explanation of the non-dependence of the yielded fluid region on the gas flow rate could be due to the fluctuating nature of the pressure at the bottom of the column. The mean bubble rise velocities decrease with the gas flow rate, which can be attributed to the decrease in the bubble sizes with the gas flow rate, regardless of the fluid model. The dimensionless mean rise velocity decreases with the Bingham number, which is consistent with the results from previous works (see, e.g., Tabuteau et al., 2007 [51]; Dimakopoulos et al., 2013 [40]). The bubbles appear to rise more slowly in the Bingham fluid, probably because of the smaller bubble sizes or possibly because of the smaller extent of the yielded region found in the Bingham fluid.

Path instabilities of the bubbles, which are the deviations of the centroid motions of the bubbles from the vertical motion (i.e., zigzagging, or spiral paths), are found to be relatively weak and virtually independent of the gas flow rate. On average, the bubbles appear to rise with a relatively straight trajectory regardless of the fluid model used in this work.

The influence of the surface tension shows that the surface tension dominates the buoyancy forces. However, the average rise velocity exhibits a weak dependence on the surface tension for the low and moderate flow rates.

Finally, the current predictions appear to match well with the experimental observations and exhibit the dependence of the bubbles motion on the gas flow rate. Furthermore, both the Herschel–Bulkley and Bingham models mostly exhibit similar trends. Moreover, a comparative analysis against the limiting case for Newtonian fluids warrants further investigations.

The influence of the microstructure, i.e., inhomogeneity of the cement slurry that is due to the cement particles in the suspension, expected to locally affect the yield stress, and the effective viscosity are planned for future work.

Author Contributions: All authors jointly conceived the idea to do this study. M.M. did some of the theoretical work, and N.A.K. did some of the theoretical work, the literature review, and all the computational work. M.M. and E.R. contributed to the formal analysis, methodology, and interpretations of the results. All authors have read and agreed to the published version of the manuscript.

Funding: This project was funded by the U.S. Department of Energy, National Energy Technology Laboratory, in part, through a site support contract.

Institutional Review Board Statement: Not applicable.

Informed Consent Statement: Not applicable.

Data Availability Statement: Not applicable.

Conflicts of Interest: The authors declare no conflict of interest. The funders had no role in the design of the study, in the collection, analyses, or interpretation of data, in the writing of the manuscript, or in the decision to publish the results.

Nomenclature

Symbol	Description
$\boldsymbol{T}$	Cauchy stress tensor of the "one-fluid"
$\boldsymbol{A}_1$	kinematical tensor
$\boldsymbol{D}$	symmetric part of the velocity gradient tensor
$\boldsymbol{I}$	identity tensor
$\boldsymbol{\tau}_1$	cement slurry phase stress tensor
$\boldsymbol{\tau}_2$	gas (air) phase stress tensor
$[[\boldsymbol{\Sigma}]]$	interfacial interaction between the two phases
$II_{2\mathbf{D}}$	sec ond invariant of the tensor $2\boldsymbol{D}$
II_{τ_1}	second invariant of the stress tensor $\boldsymbol{\tau}_1$
p	pressure of the "one-fluid"
$\mathbf{u}$	velocity of the "one-fluid"
$\mathbf{u}_1$	velocity of the cement slurry phase
$\mathbf{u}_2$	velocity of the gas (air) phase
$\mathbf{u}_\mathbf{r}$	relative velocity between the two phases
$\mathbf{n}$	unit normal vector
k	consistency of the cement slurry (for Herschel–Bulkley fluid)
n	power-law exponent of the cement slurry (for Herschel–Bulkley fluid)
ϕ	volume fraction of the cement slurry
σ	surface tension
κ	curvature of the interface between the two phases
δ	Dirac function
τ_0	yield stress of the cement slurry
η_p	plastic viscosity of the cement slurry (for Bingham fluid)
μ	viscosity of the "one-fluid"
μ_1	viscosity of the cement slurry phase
μ_2	viscosity of the gas (air) phase
ρ	density of the "one-fluid"
ρ_1	density of the cement slurry phase
ρ_2	density of the gas (air) phase
ϵ	regularization parameter
δt	time step
Δ	computational cell size
∇	gradient operator
c_ϕ	compression factor

References

1. Vidic, R.D.; Brantley, S.L.; Vandenbossche, J.M.; Yoxtheimer, D.; Abad, J.D. Impact of shale gas development on regional water quality. *Science* **2013**, *340*, 1235009. [CrossRef]
2. Ingraffea, A.R.; Wells, M.T.; Santoro, R.L.; Shonkoff, S.B.C. Assessment and risk analysis of casing and cement impairment in oil and gas wells in Pennsylvania, 2000–2012. *Proc. Natl. Acad. Sci. USA* **2014**, *111*, 10955–10960. [CrossRef]
3. Yang, B.; Zheng, Z.; Jin, J.; Wang, X. Time-Dependent Rheological Properties of Cemented Aeolian Sand-Fly Ash Backfill Vary with Particles Size and Plasticizer. *Materials* **2023**, *16*, 5295. [CrossRef] [PubMed]
4. Bonett, A.; Pafitis, D. Getting to the root of gas migration. *Oilfield Rev.* **1996**, *8*, 36–49.
5. Renteria, A.; Sarmadi, P.; Thompson, C.; Frigaard, I.A. Effects of wellbore irregularity on primary cementing of horizontal wells, Part 1: Large scale effects. *J. Pet. Sci. Eng.* **2022**, *208*, 109581. [CrossRef]
6. Maleki, A.; Frigaard, I.A. Comparing laminar and turbulent primary cementing flows. *J. Pet. Sci. Eng.* **2019**, *177*, 808–821. [CrossRef]
7. Tao, C.; Rosenbaum, E.; Kutchko, B.G.; Massoudi, M. A brief review of gas migration in oilwell cement slurries. *Energies* **2021**, *14*, 2369. [CrossRef]

8. Velayati, A.; Kazemzadeh, E.; Soltanian, H.; Tokhmechi, B. Gas migration through cement slurries analysis: A comparative laboratory study. *Int. J. Min. Geo-Eng.* **2015**, *49*, 281–288.
9. Ahmed, A.; Elkatatny, S.; Adjei, S.; Al-Majed, A. Influence of weighting materials on the properties of oil-well cement. *ACS Omega* **2020**, *5*, 27618–27625. [CrossRef]
10. Moradi, S.S.T.; Nikolaev, N.I. Considerations of well cementing materials in high-pressure, high-temperature conditions. *Int. J. Eng. Trans. C Asp.* **2016**, *29*, 1214–1218.
11. Teodoriu, C.; Asamba, P. Experimental study of salt content effect on class G cement properties with application to well integrity. *J. Nat. Gas Sci. Eng.* **2015**, *24*, 324–329. [CrossRef]
12. Tao, C.; Kutchko, B.G.; Rosenbaum, E.; Massoudi, M. A review of rheological modeling of cement slurry in oil well applications. *Energies* **2020**, *13*, 570. [CrossRef]
13. Banfill, P.F.G. Rheology of fresh cement and concrete. In *Rheology Reviews 2006*; British Society of Rheology: Lincoln, UK, 2006; pp. 61–130.
14. Banfill, P.F.G. Structure and rheology of cement-based systems. *Mat. Res. Soc. Symp. Proc.* **1993**, *289*, 149–160. [CrossRef]
15. Rodrigues, E.C.; de Souza Mendes, P.R. Predicting the time-dependent irreversible rheological behavior of oil well cement slurries. *J. Pet. Sci. Eng.* **2019**, *178*, 805–813. [CrossRef]
16. Yuan, K.; Lu, Y.; Li, W.; Yu, H.; Gao, S. Rheological Characterization and Accumulation Tests for Strong Thixotropic Engineering Slurry. *Materials* **2022**, *15*, 6891. [CrossRef]
17. Ghosh, A.K.; Ulbrecht, J.J. Bubble formation from a sparger in polymer solutions—I. Stagnant liquid. *Chem. Eng. Sci.* **1989**, *44*, 957–968. [CrossRef]
18. Terasaka, K.; Tsuge, H. Bubble formation at a nozzle submerged in viscous liquids having yield stress. *Chem. Eng. Sci.* **2001**, *56*, 3237–3245. [CrossRef]
19. Dubash, N.; Frigaard, I. Conditions for static bubbles in viscoplastic fluids. *Phys. Fluids* **2004**, *16*, 4319–4330. [CrossRef]
20. Dubash, N.; Frigaard, I.A. Propagation and stopping of air bubbles in Carbopol solutions. *J. Non-Newton. Fluid Mech.* **2007**, *142*, 123–134. [CrossRef]
21. Stein, S.; Buggisch, H. Rise of pulsating bubbles in fluids with yield stress. *Z. Angew. Math. Mech.* **2000**, *80*, 827–834. [CrossRef]
22. Iwata, S.; Yamada, Y.; Takashima, T.; Mori, H. Pressure-oscillation defoaming for viscoelastic fluid. *J. Non-Newton. Fluid Mech.* **2008**, *151*, 30–37. [CrossRef]
23. De Corato, M.; Dimakopoulos, Y.; Tsamopoulos, J. The rising velocity of a slowly pulsating bubble in a shear-thinning fluid. *Phys. Fluids* **2019**, *31*, 083103. [CrossRef]
24. Konan, N.A.; Rosenbaum, E.; Massoudi, M. On the response of a Herschel–Bulkley fluid due to a moving plate. *Polymers* **2022**, *14*, 3890. [CrossRef]
25. Plancher, L.; Pierre, A.; Nguyen, G.T.M.; Hébert, R.L.; Ledésert, B.A.; Di Martino, P.; Mélinge, Y. Rheological Behaviour of Cementitious Materials Incorporating Solid–Solid Phase Change Materials. *Materials* **2022**, *15*, 20. [CrossRef] [PubMed]
26. Massoudi, M. Constitutive relations for the interaction force in multicomponent particulate flows. *Int. J. Non-Linear Mech.* **2003**, *38*, 313–336. [CrossRef]
27. Tao, C.; Kutchko, B.G.; Rosenbaum, E.; Wu, W.T.; Massoudi, M. Steady flow of a cement slurry. *Energies* **2019**, *12*, 2604. [CrossRef]
28. Hirt, C.W.; Nichols, B.D. Volume of fluid (VOF) method for dynamics of free boundaries. *J. Comput. Phys.* **1981**, *39*, 201–225. [CrossRef]
29. Nichols, B.D.; Hirt, C.W.; Hotchkiss, R.S. *SOLA-VOF: A Solution Algorithm for Transient Fluid Flow with Multiple Free Boundaries*; Technical Report (TRN: 80-017293); Los Alamos National Lab. (LANL): Los Alamos, NM, USA, 1980. [CrossRef]
30. Brackbill, J.U.; Kothe, D.B.; Zemach, C. A continuum method for modeling surface tension. *J. Comput. Phys.* **1992**, *100*, 335–354. [CrossRef]
31. Macosko, C.W. *Rheology principles. Measurements and Applications*; Wiley Publications: New York, NY, USA, 1994.
32. Papanastasiou, T.C. Flows of materials with yield. *J. Rheol.* **1987**, *31*, 385–404. [CrossRef]
33. Espinoza, P.J.T.; Varges, P.R.; Rodrigues, E.C.; Naccache, M.F.; de Souza Mendes, P.R. Displacement flow of yield stress materials in annular spaces of variable cross section. *J. Pet. Sci. Eng.* **2022**, *208*, 109614. [CrossRef]
34. Rusche, H. Computational Fluid Dynamics of Dispersed Two-Phase Flows at High Phase Fractions. Ph.D. Thesis, Imperial College of Science, Technology and Medicine, London, UK, 2002.
35. Berberović, E.; van Hinsberg, N.P.; Jakirlić, S.; Roisman, I.V.; Tropea, C. Drop impact onto a liquid layer of finite thickness: Dynamics of the cavity evolution. *Phys. Rev. E* **2009**, *79*, 036306. [CrossRef]
36. Hoang, D.A.; van Steijn, V.; Portela, L.M.; Kreutzer, M.T.; Kleijn, C.R. Benchmark numerical simulations of segmented two-phase flows in microchannels using the Volume of Fluid method. *Comput. Fluids* **2013**, *86*, 28–36. [CrossRef]
37. Cifani, P.; Michalek, W.R.; Priems, G.J.M.; Kuerten, J.G.M.; van der Geld, C.W.M.; Geurts, B.J. A comparison between the surface compression method and an interface reconstruction method for the VOF approach. *Comput. Fluids* **2016**, *136*, 421–435. [CrossRef]
38. Headrick, E.; Spaulding, R.; Rosenbaum, E.; Massoudi, M.; Kutchko, B. *The Effects of Conditioning and Additives on the Viscosity Measurement of Cement Slurries*; DOE.NETL-2022.3352; NETL Technical Report Series; U.S. Department of Energy, National Energy Technology Laboratory: Pittsburgh, PA, USA, 2023; p. 32. [CrossRef]
39. Galusinski, C.; Vigneaux, P. On stability condition for bifluid flows with surface tension: Application to microfluidics. *J. Comput. Phys.* **2008**, *227*, 6140–6164. [CrossRef]

40. Dimakopoulos, Y.; Pavlidis, M.; Tsamopoulos, J. Steady bubble rise in Herschel–Bulkley fluids and comparison of predictions via the Augmented Lagrangian Method with those via the Papanastasiou model. *J. Non-Newton. Fluid Mech.* **2013**, *200*, 34–51. [CrossRef]
41. Oldroyd, J.G. Two-dimensional plastic flow of a Bingham solid. A plastic boundary-layer theory for slow motion. *Proc. Camb. Philos. Soc. Math. Phys. Sci.* **1947**, *43*, 383–395. [CrossRef]
42. Beris, A.N.; Tsamopoulos, J.A.; Armstrong, R.C.; Brown, R.A. Creeping motion of a sphere through a Bingham plastic. *J. Fluid Mech.* **1985**, *158*, 219–244. [CrossRef]
43. Piau, J.-M. Viscoplastic boundary layer. *J. Non-Newton. Fluid Mech.* **2002**, *102*, 193–218. [CrossRef]
44. Tsamopoulos, J.; Dimakopoulos, Y.; Chatzidai, N.; Karapetsas, G.; Pavlidis, M. Steady bubble rise and deformation in Newtonian and viscoplastic fluids and conditions for bubble entrapment. *J. Fluid Mech.* **2008**, *601*, 123–164. [CrossRef]
45. De Kee, D.; Chhabra, R.P.; Dajan, A. Motion and coalescence of gas bubbles in non-Newtonian polymer solutions. *J. Non-Newton. Fluid Mech.* **1990**, *37*, 1–18. [CrossRef]
46. Amirnia, S.; de Bruyn, J.R.; Bergougnou, M.A.; Margaritis, A. Continuous rise velocity of air bubbles in non-Newtonian biopolymer solutions. *Chem. Eng. Sci.* **2013**, *94*, 60–68. [CrossRef]
47. Ravisankar, M.; Correa, A.G.; Su, Y.; Zenit, R. Hydrodynamic interaction of a bubble pair in viscoelastic shear-thinning fluids. *J. Non-Newton. Fluid Mech.* **2022**, *309*, 104912. [CrossRef]
48. Buchholz, H.; Buchholz, R.; Lucke, J.; Schugerl, K. Bubble swarm behaviour and gas absorption in non-Newtonian fluids in sparged columns. *Chem. Eng. Sci.* **1978**, *33*, 1061–1070. [CrossRef]
49. Yuan, W.; Zhang, M.; Khoo, B.C.; Phan-Thien, N. On a vertical chain of small bubbles ascending in a viscoelastic fluid. *Phys. Fluids* **2021**, *33*, 101704. [CrossRef]
50. Velez-Cordero, J.R.; Samano, D.; Yue, P.; Feng, J.J.; Zenit, R. Hydrodynamic interaction between a pair of bubbles ascending in shear-thinning inelastic fluids. *J. Non-Newton. Fluid Mech.* **2011**, *166*, 118–132. [CrossRef]
51. Tabuteau, H.; Coussot, P.; de Bruyn, J.R. Drag force on a sphere in steady motion through a yield-stress fluid. *J. Rheol.* **2007**, *51*, 125–137. [CrossRef]
52. Mirzaagha, S.; Pasquino, R.; Iuliano, E.; D'Avino, G.; Zonfrilli, F.; Guida, V.; Grizzuti, N. The rising motion of spheres in structured fluids with yield stress. *Phys. Fluids* **2017**, *29*, 093101. [CrossRef]
53. Premlata, A.R.; Tripathi, M.K.; Karri, B.; Sahu, K.C. Dynamics of an air bubble rising in a non-Newtonian liquid in the axisymmetric regime. *J. Non-Newton. Fluid Mech.* **2017**, *239*, 53–61. [CrossRef]
54. Sharaf, D.M.; Premlata, A.R.; Tripathi, M.K.; Karri, B.; Sahu, K.C. Shapes and paths of an air bubble rising in quiescent liquids. *Phys. Fluids* **2017**, *29*, 122104. [CrossRef]

materials

MDPI

Article

Recycling Local Waste Glass Bottles into Cement Paste: Effect on Hydration, Microstructure, and CO_2 Emission

Fengming Ren [1], Xiwen Zhang [1], Mingxin Lin [2], Qing Wang [1,*] and Jing Sun [1,*]

1 School of Civil Engineering, Guangzhou University, Guangzhou 510006, China; rfm@gzhu.edu.cn (F.R.); zxw@e.gzhu.edu.cn (X.Z.)
2 Panyu Dashi Construction Engineering Limited Company, Guangzhou 511430, China; djc1988@djc1988.com
* Correspondence: wangq@gzhu.edu.cn (Q.W.); jingslinks@gzhu.edu.cn (J.S.)

Abstract: Large amounts of waste glass are generated along with the manufacturing of glass products, causing detrimental effects on the environment. Through crushing and ball-milling, waste glass powder (WGP) can be acquired from glass bottles and has been suggested in cementitious systems due to its potential pozzolanic activity. To better understand the impact of WGP on cementitious composites, experimental tests of rheology, heat of hydration, and strength development were conducted on cement pastes with and without WGP. Results show that the rheological performance of cement paste is improved when WGP with particles passing through 80 μm sieves is incorporated. The retarding effect and pozzolanic reaction were observed through X-ray diffraction patterns and thermo-gravimetric parameter analyses. A calcium hydroxide (CH) content calculation further confirms the secondary reactivity of WGP in cement pastes. Compared with the samples without WGP, the normalized CH content of binder per unit mass containing 35% WGP decreased by 21.01%, 24.94%, and 27.41% at the ages of 1, 28, and 90 days, respectively, which contributes to late-age strength development of pastes. At the same time, the hydration per unit of cement was increased by 21.53%, 15.48%, and 11.68%, which improved the cement efficiency. In addition, WGP particles provide nuclei for hydration products, facilitating the subsequent growth of C-S-H and strength development in late ages. Based on value engineering analysis, WGP was found to reduce the impact of Portland cement on the environment by 34.9% in terms of carbon dioxide emissions, indicating a bright prospect for WGP in the cement industry.

Keywords: waste glass powder; cement paste; hydration; microstructure; CO_2 emission

1. Introduction

Glass materials are widely used in daily life and industrial applications due to their stable chemical properties and aesthetic appearance. Large amounts of waste glass are generated along with the manufacturing and application of various glass products, such as waste glass bottles. The annual production of waste glass is about 21.23 million tons in China, whereas the recycling rate is merely 30.4%, causing occupation of land resources and pollution of adjacent environments due to its unbiodegradable nature [1]. Thus, rational utilization of waste glass has become an important issue for glass factories and waste glass collectors.

The main components of waste glass are generally acknowledged as amorphous SiO_2, accounting for 70% or more [1,2]. This attracts researchers and engineers to take advantage of the pozzolanic activity in producing greener cementitious materials due to the fact that cement production is ecologically detrimental in terms of carbon dioxide (CO_2) emissions. For instance, in 2020, cement clinker production reached 1.579 billion tons on the Chinese mainland alone, and the emissions of CO_2 accompanying it reached over 1.23 billion tons, taking up 12.43% of the total CO_2 emissions in all industries. Aiming at mitigating the increasingly tense greenhouse effect, specific schemes for the strategy of carbon neutrality and peak emissions have been deployed, advocating new patterns for high-quality/sustainable cement development. The most effective way is to reduce the amount of silicate cement clinker by incorporating a high volume of supplementary cementitious materials (SCMs) [3,4], such as fly ash [5], furnace slag [6], and high-reactivity metakaolin [7]. Herein, waste glass has also been proposed to partially substitute cement in concrete after mechanical grinding or chemical activation [8,9]. Utilizing waste glass in cementitious composites would definitely decrease the consumption of cement and lower the cost of relevant products, contributing to solid waste utilization and carbon emission mitigation [10–13].

A decrease in the particle size of WGP was reported to significantly increase the pozzolanic activity [14]; 75 μm was considered to be the largest diameter [15,16]. Before application, waste glass was mechanically ground into powder with different particle sizes according to specific requirements. WGP would participate in the secondary reaction of hydration thanks to its high content of amorphous silica. No particular difference was detected in the secondary calcium silicate hydrates (C-S-H) in the micromorphologies of concrete containing WGP [14,17] and concrete with fly ash [18] or silica fume [14] based on scanning electron microscope (SEM) observations. Shao et al. [2] reported that concrete containing WGP exhibited higher strength in both early and late ages than that with fly ash. Schwarz and Neithalath [19] compared the non-evaporated water content of cementitious composites with fine WGP and fly ash, indicating high pozzolanic activity of WGP in the system. Yusuf et al. [20] investigated the synergistic effect of WGP and silica fume and showed that the synergistic action of WGP and SF enhances the reorganization of silicates, making the binder more amorphous and reducing hydroxyl compounds, but morphologically causing heterogeneity in the microstructure. The authors also studied the residual strength and density of glass-adulterated concrete at elevated temperatures (550 °C). It was found that glass enhances the workability of concrete, shortens the setting time, reduces the density, and increases the residual strength and density [21]. In addition, feasibility studies have been conducted on the optimization of cementitious composites with various amounts of WGP to obtain high performance and high durability [10,22,23].

Thus, recycling waste glass bottles into WGP to produce a greener cementitious composite seems practicable. However, the effect of WGP from local waste glass bottles on the hydration and microstructure of cement paste has not been fully understood, especially for WGP with similar particle sizes to cement. Whether the rheological and mechanical properties of pastes containing different amounts of local WGP could meet relevant standards will be critically investigated for further application. Therefore, it is necessary and of great significance to study the effects of recycled WGP from waste glass bottles on the fresh and hardened properties of cementitious composites. Experimental tests of WGP content on the performance variation of cement pastes were carried out accordingly. Mineral compositions from an X-ray diffractometer (XRD) and microstructure morphologies under scanning electronic microscopy (SEM) were obtained to analyze the impact of WGP in substitution with cement. The ecological effect, focusing on CO_2 emissions, was quantified correspondingly.

2. Experimental Program

2.1. Materials

Common white-type waste glass bottles were provided by a local waste collection station in Guangzhou, China. Those bottles were first cleaned and dried, then crushed and ground in a ball mill to powder, as schematically shown in Figure 1. After sieving, WGP with a similar size to P.O. 42.5 Portland cement (as per Chinese Standard GB175 [24]) was prepared. The particle size distributions of cement and WGP are displayed in Figure 2, which were analyzed by a laser particle size analyzer MS2000 from Malvern Panalytical, UK. They share the same size range of 0.1 μm to 110 μm, facilitating the substitution of WGP for cement without drastically changing the grading of grains. The diameter of WGP (D10, D50, and D90 as 6.66 μm, 29.29 μm, and 96.04 μm) is slightly larger than that of cement (D10, D50, and D90 as 2.97 μm, 18.43 μm, and 52.31 μm). The apparent densities of cement and WGP are 3.10 g/cm^3 and 2.51 g/cm^3, respectively.

Figure 1. Preparation of WGP from local waste glass bottles.

The chemical compositions of both materials are listed in Table 1, tested by X-ray fluorescence (XRF) via PANalytical Axios XRF-1800. The main oxides of WGP are SiO_2 (64.90 wt.%), CaO (19.62 wt.%), and Na_2O (10.06 wt.%), implying a typical soda-lime silica glass and an alkaline glass material. Such a high content of SiO_2 (mostly amorphous silica) would reveal pozzolanic activity based on previous research and related standards [25–27]. Figure 3 displays the micromorphology of WGP with its angular lumpy shape and irregular coarse surface, which might contribute to adequate contact with cement hydration products.

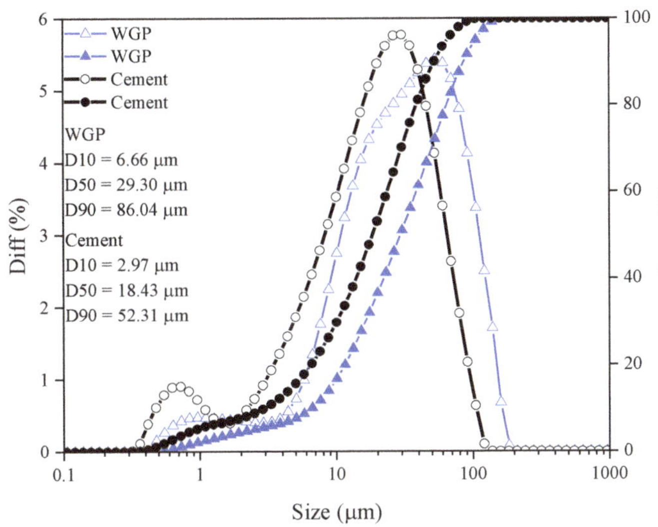

Figure 2. Particle size distribution of WGP and cement.

Table 1. Chemical compositions and physical properties of WGP and cement used (wt.%).

Types	CaO	SiO_2	Al_2O_3	Fe_2O_3	SO_3	MgO	K_2O	TiO_2	SrO	MnO	Na_2O	P_2O_5	Cl	ZnO	Rb_2O	ZrO_2	Density (g/cm^3)	D50 (μm)
Cement	68.52	14.03	5.98	3.80	3.08	2.55	0.91	0.58	0.16	0.14	0.13	0.06	0.02	0.02	0.01	-	3.10	18.43
WGP	18.62	64.90	3.17	0.26	0.16	1.54	0.88	0.12	0.10	-	10.06	0.04	0.05	-	0.04	0.08	2.51	29.29

Figure 3. Micromorphology of WGP used (×1000).

2.2. Mix Design and Specimen Preparation

Different substitution ratios of WGP for cement are proposed by British standard BS EN 196-1 [28], international standard ASTM C311 [29], and Chinese standard GB/T 12957 [30], 20%, 25%, and 30% in weight, respectively. In addition, BS EN 197-1 [4] also requires that the upper limit of pozzolanic material dosage be 35 wt.%. In terms of activity index, which is defined as the compressive strength ratio of samples with SCM to the control group, BS EN 450-1 [31] requires a value larger than 75% after curing for 28 days and 85% for 90 days. While in GB/T 2847 [32], the ratio becomes 65% at 28 days, and in ASTM C311 [29], it becomes 75% at 7 and 28 days.

To better illustrate the effect of local WGP on the properties of cement paste, the above weight percentages of 20%, 25%, and 30% were selected. In addition, 15 wt.% and 35 wt.% WGP replacements for cement were designed as contrasts. Detailed mix proportions with a constant water-to-cement ratio of 0.4 are listed in Table 2. The mixes in all dosages (even the 35% WGP content) showed good flow and no water secretion or segregation. The mixing procedure was consistent with Chinese standard GB/T 17671 [33]. Specimens with a size of 40 × 40 × 40 mm^3 were prepared for compressive strength evaluation at the ages of 7, 28, and 90 days. All specimens were demolded after film covering for 24 h, then moved to a standard curing chamber at 20 ± 2 °C and ≥98% relative humidity until the test ages. After the compression tests were conducted at an MTS hydraulic universal testing machine, samples were carefully prepared from the core of the cubes for further microstructural analysis.

Table 2. Mix proportions of cement pastes with and without WGP (kg/m^3).

Groups	Content		
	Cement	**WGP**	**Water**
Plain paste	1367.00	0.00	546.80
15% WGP	1161.95	205.05	546.80
20% WGP	1093.60	273.40	546.80
25% WGP	1025.25	341.75	546.80
30% WGP	956.90	410.10	546.80
35% WGP	888.55	478.45	546.80

2.3. *Test Methods*

2.3.1. Rheology

A rheometer with a co-axial cylindrical rheometer was applied to investigate the rheological parameters of all pastes. The shear speed was first increased from 5 to 100 revolutions per minute (RPM) in a step of 5 RPM, maintained for 15 s till the peak. Then, the shear speed was decreased at the same interval [34]. The Bingham model [35] was adopted to determine the rheological parameters.

2.3.2. Heat of Hydration

The heat evolution of all mixtures was obtained via the TAM air isothermal calorimeter. According to the equilibrium principle of specific heat capacity, Equation (1) was proposed to calculate the amount of cement and free water at a constant quartz sand quantity (m_S) of 15 g.

$$m_S c_S = m_C c_C + m_W c_W \tag{1}$$

where m_C denotes the mass of cement and m_W denotes the mass of water. Herein, the specific heat capacity of quartz sand c_S is 0.76 J/gK, cement c_c is 0.75 J/gK, and water c_W is 4.18 J/gK. To analyze the impact of WGP, 15% and 30% WGP were applied to replace Portland cement by weight. Before testing, the calorimeter was first turned on and calibrated for three days. The samples were preweighed and mixed inside a standard ampoule for 2 min. Thereafter, the ampoules were placed inside the isothermal calorimeter and sealed with metal taps. The temperature of the instrument was set constantly to 25 °C, and real-time heat flows were detected and recorded for 7 whole days (168 h).

2.3.3. Microstructure Analysis

XRD analysis was performed to study the physical phase composition of cement pastes with and without WGP. Operating conditions were set as 40 kV, 100 mA, and Cu target in the diffractometer of the Japan Ultima type VI. The continuously scanning range was from 5° to 60°, with an average speed of 10°/min.

Thermo-gravimetric and differential thermal analysis (TG-DTA) experiments were carried out on the NETZSCH STA 2500. The temperature range was 30 °C to 1000 °C at a rate of 10 °C/min under a dynamic N_2 atmosphere (20 mL/min).

The micromorphologies of hydration products were photographed by an SNE-4500 M Plus electron microscope. The instrument voltage was 15 kV, and thin layers of gold were sprayed on the surface of paste samples in advance to guarantee good conductivity for concrete specimens.

3. Results and Discussions

3.1. Rheological Properties

Shear stress-rate curves of fresh cement pastes with 15 wt.% WGP and 30 wt.% WGP are displayed in Figure 4 together with the control group. As expected, flowability varies in the presence of WGP, which is characterized by 90% grains smaller than 86μm and a relatively coarse surface. Shear stress of plain paste is higher than that with WGP addition, indicating a positive effect of WGP on workability, as also observed in [36]. When WGP content increases, shear stress decreases at the same shear rate, implying better flowability in paste with 30 wt.% WGP. It is attributed to the reduction in water demand because of the decrease in cement amounts when WGP is incorporated. Thus, more free water is released, facilitating the grain particles easy flow in the cementitious system. Another possible reason for the improvement in rheological performance lies in the appropriate grading of WGP particles. As reported by Lu et al. [37], large WGP particles (D50 $\geq$ 204 μm) show an adverse effect on fluidity due to the high friction of individual irregularly shaped particles, while small WGP particles (diameter less than 45 μm) also fail to improve flowability because of their high specific surface area and effective water demand [14,38]. Thus, a suitable WGP size range or proper ball-milling parameters for waste glass bottles should be duly considered in the design of cementitious composites with favorable flowability. WGP particles with sizes similar to cement grains were experimentally proven to be effective in improving the rheological performance in the present study.

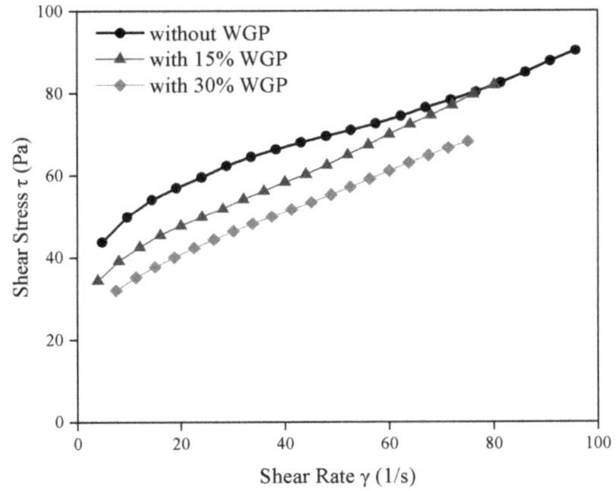

Figure 4. Shear stress-rate curves of fresh cement paste.

3.2. Hydration

Heat evolution in the first 168 h of plain paste, 15 wt.%, and 30 wt.% WGP-modified pastes is plotted in Figure 5. Two hydration parameters were analyzed: one is the heat flow per unit binder mass, and the other is the per unit cement mass (Figure 5b). Figure 5a reveals that the incorporation of WGP greatly decreases the heat flow rate and cumulative heat. Low activity of WGP at early ages is responsible for this diminishing trend, manifested by a lower heat evolution rate in pastes containing a higher amount of WGP. Such a retarding effect can also be reflected by the parameters of hydration peaks listed in Table 3. On the one hand, the hydration peaks appear later in WGP-modified pastes; 8.99 and 9.01 h in contrast with 8.69 h for the control. On the other hand, the intensity at the peaks decreases from 4.54 to 3.85 and 3.23 mW/g when 15% and 30% cement in weight are replaced by WGP. This retarding phenomenon in paste containing WGP from local waste glass bottles is similar to previous observations of other ground-glass powders [39,40]. The effective water-to-cement ratio increases when WGP is adopted, contributing to the delay of the hydration peak [41].

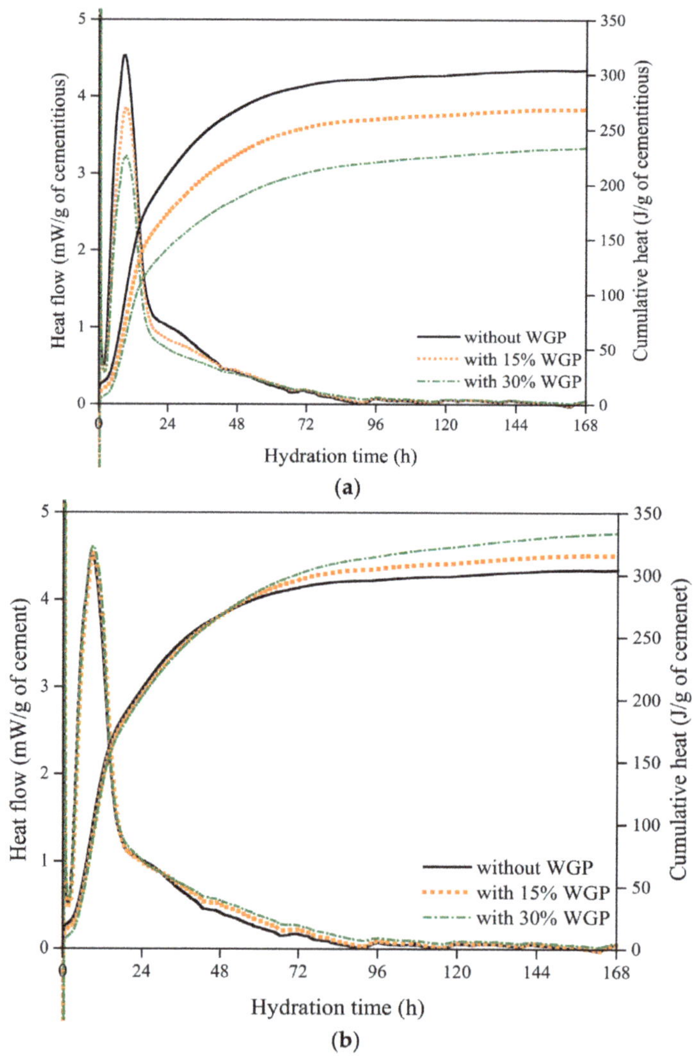

Figure 5. Heat evolution curves: (**a**) normalized heat evolution per unit binder in mass; (**b**) normalized heat evolution per unit cement mass.

Table 3. Mix proportions of cement pastes for heat hydration and parameters of hydration peaks.

Mixture	w/b	Water(g)	Cement (g)	WGP (g)	Hydration Peaks		Intensity of Hydration Peaks	
					Age (h)	Relative Value (%)	Intensity (mW/g)	Relative Value (%)
Plain paste	0.4	1.8827	4.7069	0.0000	8.69	1.000	4.54	1.000
15% GP	0.4	1.8827	4.0009	0.7060	8.99	1.035	3.85	0.848
30% GP	0.4	1.8827	3.2948	1.4121	9.01	1.037	3.23	0.711

When heat evolution rate per unit cement is discussed (Figure 5b), higher cumulative heat is observed in the presence of WGP after 48 h, which is also confirmed by Mirzahosseini and Riding [41]. More cement grains are exposed to free water thanks to the low water demand of glass beads, which can be concluded to be the dilution effect of WGP [37,42]. The decrease in alkali content in WGP-modified paste with a lower cement amount is also responsible for the heat increment since a high alkali liquid solution helps to shorten the induction period and promote tricalcium silicate (C_3S)/dicalcium silicate (C_2S) dissolution and CH formation [43]. Along with the hydration process, WGP particles gradually dissolve and produce an alkaline composition, accelerating the hydration process after 48 h. Du et al. [44] suggested that the accelerating mechanism of WGP in hydration might be the adsorption of calcium ions and the provision of growth sites for C-S-H nucleation. Na_2O in WGP serves as a catalyst for early C-S-H formation and development [45]. In addition, the filling effect of WGP with particle sizes less than 86μm would provide available space and surface for hydrate formation [46].

3.3. Strength Development

The cubic compressive strengths of all mixtures after standard curing for 7, 28, and 90 days are summarized in Figure 6. Strength decreases with an increment in WGP content, especially in the early ages. A reduction in cement amounts is responsible for this since hydration products are cut down. In the first seven days, the pozzolanic activity of WGP could not compensate for the decrease in strength due to the shortage of calcium hydroxide (CH). The average values of compressive strength after 90 days for WGP contents of 15–35 wt.% are close. Pastes containing 30 wt.% WGP exhibits the highest strength among all WGP-modified pastes, owing to its pozzolanic activity in the late hydration period [47]. This may be attributed to the pozzolanic reaction of WGP, which consumes CH in the cement paste, and the product generated fills the internal pores, thus reducing the strength loss rate.

Figure 7 further analyzes the strength growth rate of pastes with and without WGP. Herein, the strength growth rate is defined as the increase in the ratio of 28- or 90-day strength compared with 7-day strength. The growth rates in plain pastes at 28 and 90 days are close (37.3% and 41.9%), indicating a relatively weak development ability in strength. However, this situation is different for pastes containing WGP from waste glass bottles. Although the strength growth rate is lower at 28 days for paste with <25 wt.% WGP, larger values are observed for all WGP-modified pastes at 90 days. The ratios calculated even reach 100.80% and 101.96% for groups with 30 wt.% and 35 wt.% WGP, respectively. The secondary hydration reaction of WGP in late ages helps to improve mechanical properties, as also reported by other researchers [2,8].

As for the activity index of WGP, mainly three standards are quoted and compared. According to ASTM C311 [29], a 20% WGP replacement ratio is used, and its strength ratios at 7/28 days to plain pastes are calculated as 75.88% and 75.84%, slightly larger than the threshold of 75%. Based on BS EN 450-1 [31], the strength ratios of qualified pozzolanic WGP are 70% at 28 days and 80% at 90 days, with a WGP content of 25%. Herein, the average strengths of 25% WGP pastes are 45.79 and 57.70 MPa at 28 and 90 days, while plain cement pastes are 61.82 and 63.89 MPa. Ratios of 74.08% and 90.31% for 28 and 90 days meet the requirements of BS EN 450-1. When WGP increases to 30 wt.%, the activity index at the age of 28 days is evaluated as 69.23%, larger than the 65% required by GB 2847 [32]. Overall, WGP produced from waste glass bottles in the present study meets the basic requirement of pozzolanic materials for use in the cement industry in terms of activity index.

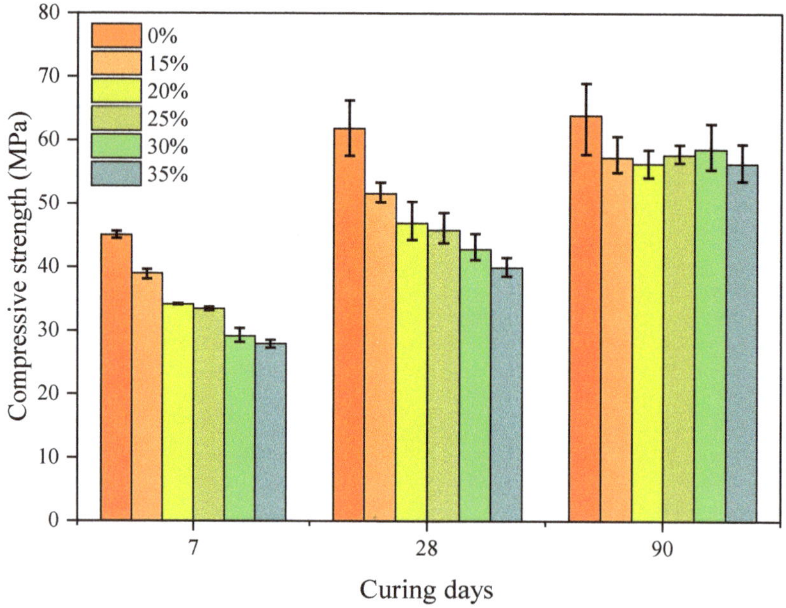

Figure 6. Compressive strength of pastes with and without WGP.

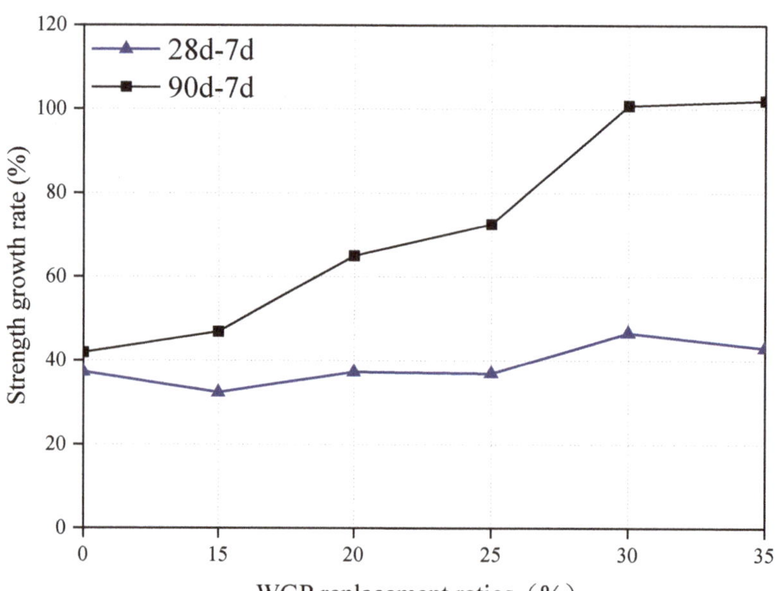

Figure 7. Strength growth rate of pastes with and without WGP.

3.4. XRD Analysis

XRD patterns of samples from plain pastes and 35 wt.% WGP-modified pastes at ages of 1, 7, 28, and 90 days are plotted in Figure 8. The main components in both pastes are CH, calcium carbonate ($CaCO_3$), dicalcium silicate (C_2S), tricalcium silicate (C_3S), quartz (SiO_2), ettringite ($3CaO{\cdot}Al_2O_3{\cdot}3CaSO_4{\cdot}32H_2O$), and other minerals. The quartz SiO_2 with diffraction peaks at an age of one day may be the impurity fraction contained in the cement itself. The hydration processes of C_3S and C_2S can be expressed in Equations (2) and (3), in which the hydration of C_2S is a bit slower and gives off less heat under the same conditions [48].

$$2C_3S + 6H \rightarrow C_3S_2H_3(C - S - Hgel) + 3CH + 120cal/g \quad (2)$$

$$2C_2S + 4H \rightarrow C_3S_2H_3(C - S - Hgel) + CH + 62cal/g \quad (3)$$

Herein, hyphens in C-S-H are used to represent the category of calcium silicate hydrated since the chemical composition of C-S-H is not constant and changes with hydration degree and original minerals. When hydration is completely accomplished, C-S-H can then be described as $C_3S_2H_3$, suitable for stoichiometric calculations [49]. Activated silica could react with the above hydration product CH, called secondary hydration or pozzolanic reaction, as shown in Equation (4):

$$CH + S \rightarrow C - S - Hgel \quad (4)$$

Figure 8a displays the hydration products at various curing ages for plain paste. The peaks of C_2S and C_3S gradually decrease as the hydration process proceeds, while the intensity of the hydration product CH slightly increases. When recycled WGP is adopted, the hydration degree of cement becomes higher, as depicted in Figure 8b, because the differences in endmembers of C_2S and C_3S are even greater in comparison with the control group, indicating an accelerating effect of WGP. Meanwhile, CH diffraction peaks exhibit a weakening trend in pastes with WGP. Herein, at one day, the difference between CH peaks of paste with and without WGP is not obvious. However, after 90 days of curing, CH peaks in WGP-modified paste are much smaller than those in the control group, implying consumption of CH in the presence of WGP. As cement hydration proceeds, CH content increases and the concentration of hydroxyl ions increases, prompting the pozzolanic reaction of WGP with amorphous silica [47]. Once consumption of CH gets larger than its generation from cement hydration, the total amount decreases, resulting in a weak intensity of CH endmembers. Such a chemical reaction would produce more hydration products and enhance the microstructure of pastes, which well explains the strength improvement of WGP-modified pastes at a later age.

3.5. CH Content

The pozzolanic activity of WGP could be reflected more specifically by the CH content change in pastes. Thermo-gravimetric parameters enable identification of physical phases and determination of CH content in hydrated pastes [48,50]. Based on the results of thermo-gravimetric tests (Figure 9a), cement itself contains CH and $CaCO_3$. Therefore, in order to minimize the effect of $CaCO_3$, a modified calculation method is proposed. The measured $CaCO_3$ content usually includes both raw material and CH carbonation, so it is essential to determine the CH and $CaCO_3$ content of the raw cement. $CaCO_3$ content measured typically includes both raw material and CH carbonation, which makes the determination of CH and $CaCO_3$ in the original cement critically necessary. The TG and DTG curves of cement, plain paste, and 35 wt.% WGP-modified paste are displayed in Figure 9. Generally, the heat absorption peak around 400 is caused by the dehydration decomposition of CH and 650 °C by $CaCO_3$. Based on the data from TG-DTG tests, the amount of CH can be quantitatively calculated as follows: The results are summarized in Table 4.

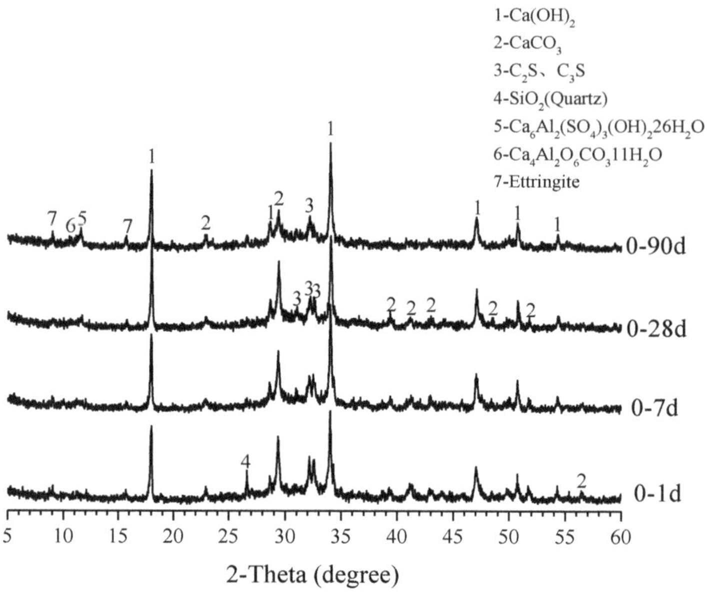

(**a**)

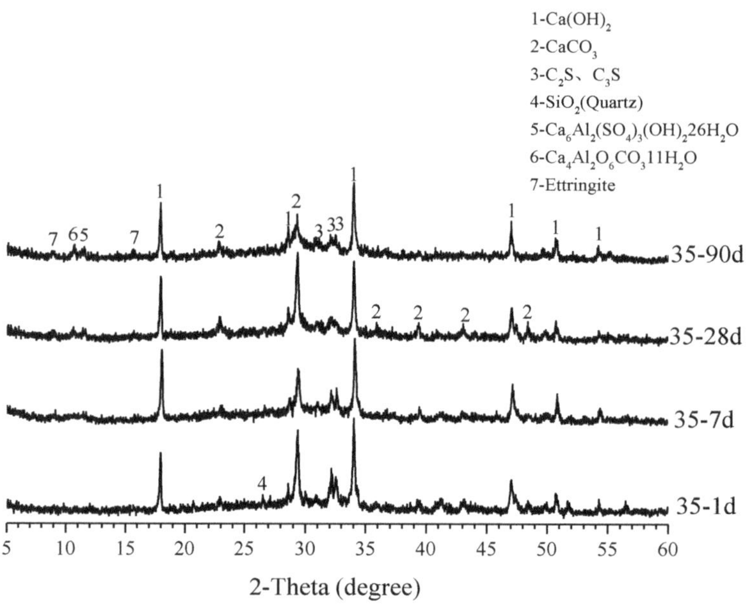

(**b**)

Figure 8. XRD patterns of (**a**) plain paste and (**b**) 35 wt.% WGP-modified paste.

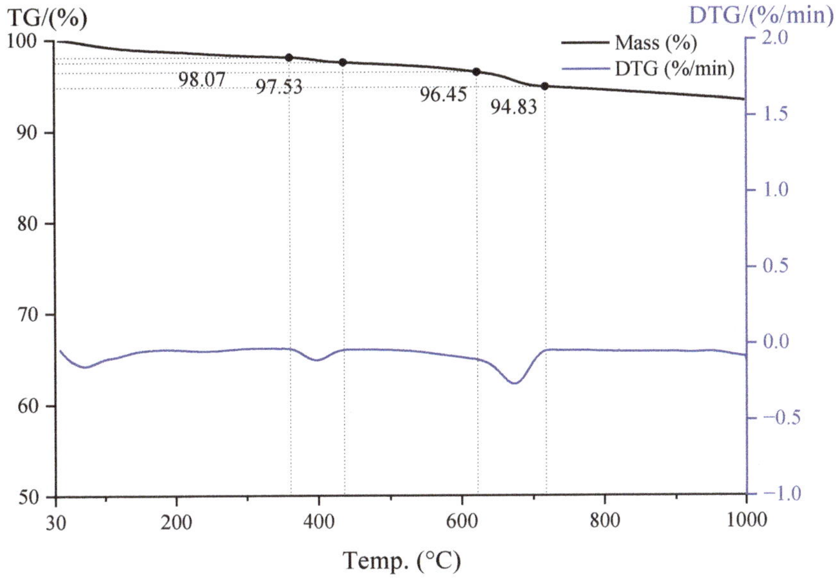

(**a**)

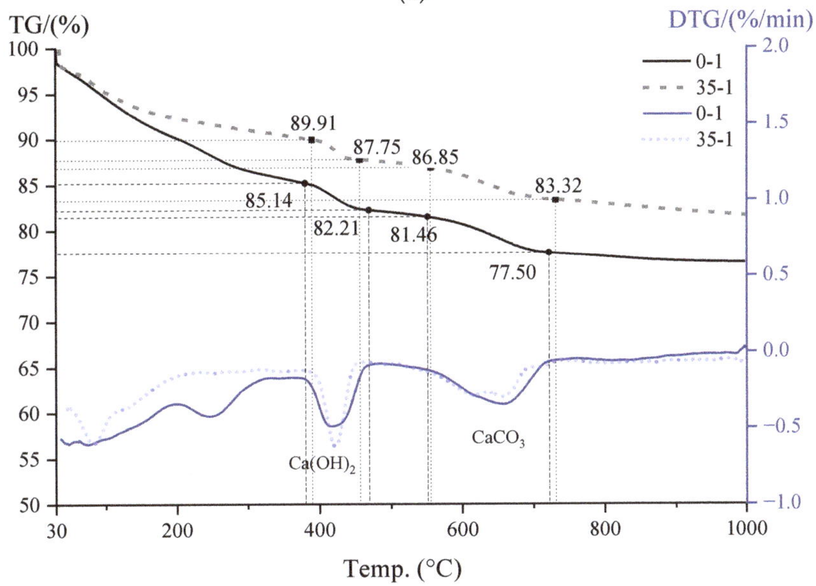

(**b**)

Figure 9. *Cont.*

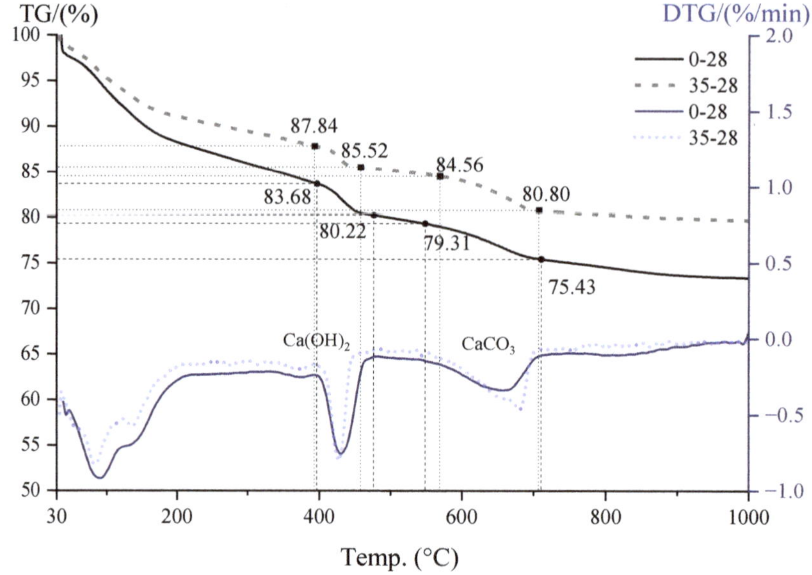

(c)

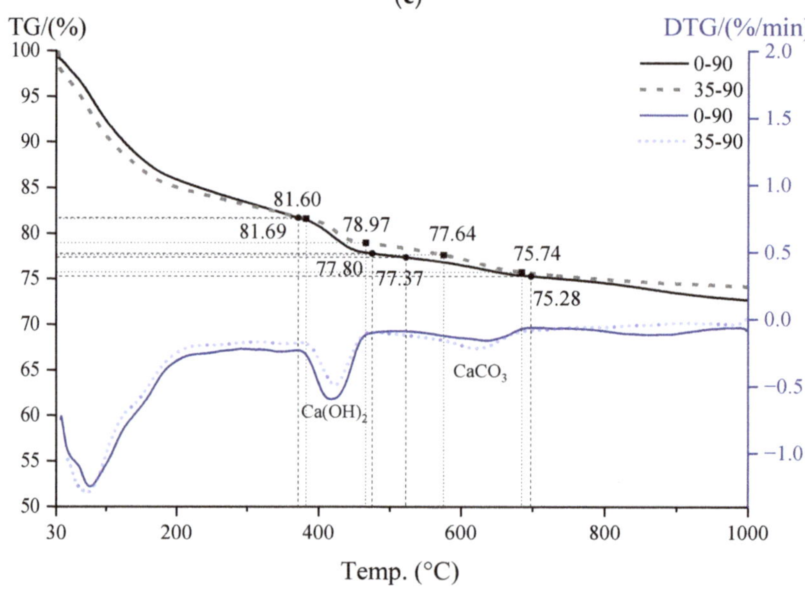

(d)

Figure 9. TG−DTG curves of the samples: (**a**) cement; (**b**) pastes at 1 day; (**c**) pastes at 28 days; (**d**) pastes at 90 days.

Table 4. Normalized CH content summarized.

Type	Age (Days)	Original Content		Carbonation Reaction Content		Total CH (%)	Normalized CH	
		CH (%)	$CaCO_3$ (%)	$CaCO_3$ (%)	CH (%)		Per 1 g Binder	Per 1 g Cement
Cement	-	2.22	3.68	-	-	-	-	-
Plain paste	1	12.05	9.00	5.32	3.94	13.76	0.1674	0.1674
35% WGP	1	8.88	8.02	5.63	4.17	11.60	0.1322	0.2034
Plain paste	28	14.22	8.82	5.14	3.80	15.81	0.1970	0.1970
35% WGP	28	9.54	8.55	6.15	4.55	12.65	0.1479	0.2275
Plain paste	90	16.07	4.75	1.07	0.79	14.64	0.1883	0.1883
35% WGP	90	10.81	4.32	1.93	1.42	10.79	0.1367	0.2103

Firstly, for cement, the weight loss around 400 °C caused by CH dehydration is $98.07\% - 97.53\% = 0.54\%$. According to the chemical equation shown in Equation (5), the amount of CH is $0.54\%/(18/74) = 2.22\%$, indicating a CH content of 2.22% in raw cement.

$$\begin{gathered} Ca(OH)_2 \rightarrow CaO + H_2O(\uparrow) \\ \text{Molecular weight (g/mol)}: \ 74 \ \ 18 \end{gathered} \tag{5}$$

The weight loss around 650 °C caused by $CaCO_3$ decomposition is $96.45\% - 94.83\% = 1.62\%$, based on Equation (6), $1.62\%/(44/100) = 3.68\%$ $CaCO_3$ exists in raw cement.

$$\begin{gathered} CaCO_3 \rightarrow CaO + CO_2(\uparrow) \\ \text{Molecular weight (g/mol)}: \ 100 \ \ 44 \end{gathered} \tag{6}$$

Secondly, Figure 9b shows the TG-DTG curves of plain paste and 35 wt.% WGP-modified paste after curing for 24 h. Similarly, the CH content of paste with WGP can be calculated as $(89.91\% - 87.75\%)/(18/74) = 8.88\%$, while the amount of $CaCO_3$ is $(86.85\% - 83.32\%)/(44/100) = 8.02\%$. Since raw cement contains 3.68% $CaCO_3$, the original amount of $CaCO_3$ is $3.68\% \times (100\% - 35\%) = 2.39\%$, and the amount generated by CH carbonation is $8.02\% - 2.39\% = 5.63\%$. The CH content in this section is then $5.63\%/(100/74) = 4.16\%$. On the basis of the 2.22% CH amount in the original cement, the total CH amount can be computed as $8.88\% + 4.16\% - 2.22\% \times (100\% - 35\%) = 11.60\%$. Figure 9c,d gives the total CH content of pastes with and without WGP at 28 and 90 days, summarized in Figure 10. CH content increases from 1 day to 28 days but decreases at 90 days for both series. Differences between pastes with and without WGP are 3.10%, 3.13%, and 3.87% for 1 day, 28 days, and 90 days, respectively, indicating evident pozzolanic activity of WGP in consuming CH in the samples, especially at later ages.

Thirdly, chemical-bonded water can be calculated right after decomposition of CH [51] as $100\% - 87.75\% = 12.25\%$ for 35 wt.% WGP-modified paste curing for one day, which can reflect the hydration degree of cement indirectly. Thus, normalized CH content per unit binder mass is calculated as $11.60\%/(100 - 12.25)\% = 0.13$, which means 1 g of binder produces 0.13 g of CH after one day of hydration. Furthermore, normalized CH content per unit cement mass can be worked out as $0.13/(100\% - 35\%) = 0.20$. Figure 11 displays the amounts of chemical-bonded water and normalized CH content. The amounts of chemical-bonded water in WGP-modified paste are lower than plain paste for all ages, whereas the growth rate is distinctly higher at 90 days. Compared with the samples without WGP, the CH content in the cementitious material containing 35% WGP decreased by 21.01%, 24.94%, and 27.41% at the curing ages of 1, 28, and 90 days, respectively. In addition, the normalized CH content per unit cement mass in Figure 11c confirms that WGP used can improve the hydration degree by 21.53%, 15.48%, and 11.68% at ages 1, 28, and 90 days, respectively, contributing to the late-age strength development of pastes.

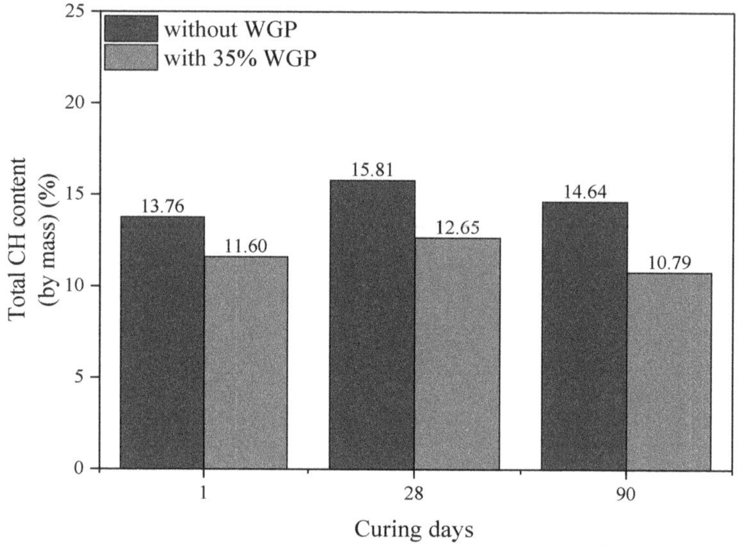

Figure 10. Total CH content of plain cement paste and 35% WGP-modified paste at different ages.

Figure 11. *Cont.*

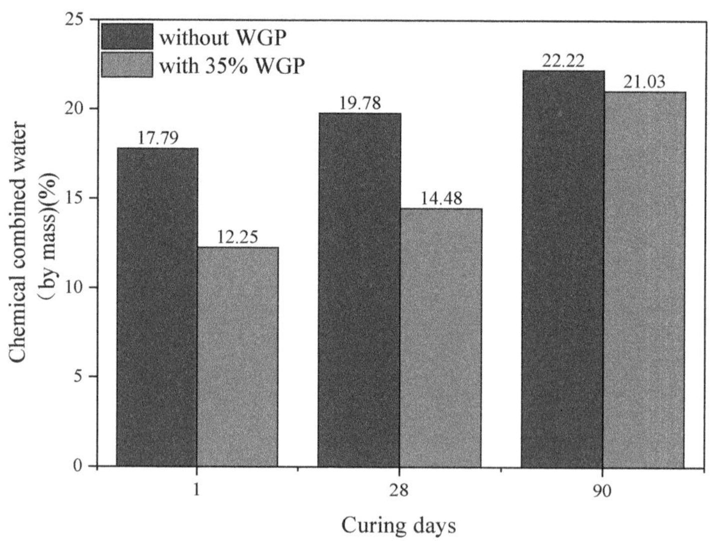

(**a**)

Figure 11. *Cont.*

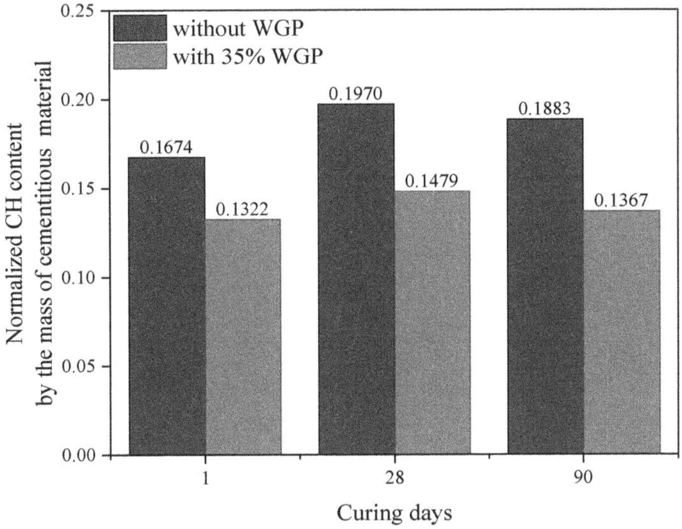

(b)

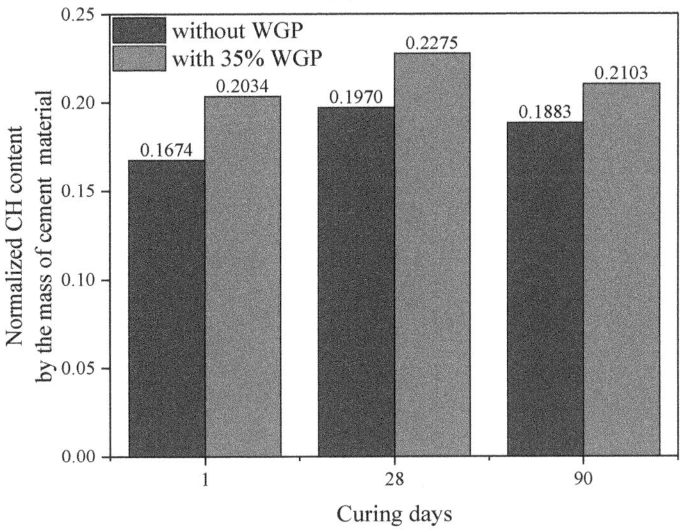

(c)

Figure 11. Amounts of chemical-bonded water and normalized CH content: (**a**) chemical-bonded water; (**b**) normalized CH content per unit binder mass; (**c**) normalized CH content per unit cement mass.

3.6. Microstructure

Micromorphologies of pastes containing WGP from waste glass bottles are shown in Figure 12 with various curing ages. WGP particles with diameters ranging from 10 to 80 μm are clearly photographed after mixing for 24 h (Figure 12a). Several hydrates appear on the surface of WGP particles, which are probably the nuclei of C-S-H, as suggested by Ouyang et al. [52]. Figure 12b shows the paste morphology after curing for seven days. Herein, the amount of total hydration products increases and the microstructure gets denser, with WGP particles surrounded by discontinuous hydration products. After curing for 28 days, glass powder grains are almost wrapped by granular nuclei growing in hydrates. The structure of the paste becomes more compact, as reflected by an increase in the size and quantity of hydrate sites. Interval pores between WGP and cement particles get smaller compared to those in the earlier ages, implying higher resistance to external forces and liquid transportation. At 90 days, WGP particles are closely covered by C-S-H gels, resulting in less surface area in the same field of view. Active silica is responsible for the microstructure change since it dissolves in pore solution and reacts with CH accompanied by cement hydration [1]. WGP particles provide nucleation sites for C-S-H gel and other hydration products. Based on the analysis through the entire hydration process, such a pozzolanic reaction is weak at early ages but becomes stronger after curing for 28 days and more, resulting in dense microstructure and further enhances the long-term properties of pastes.

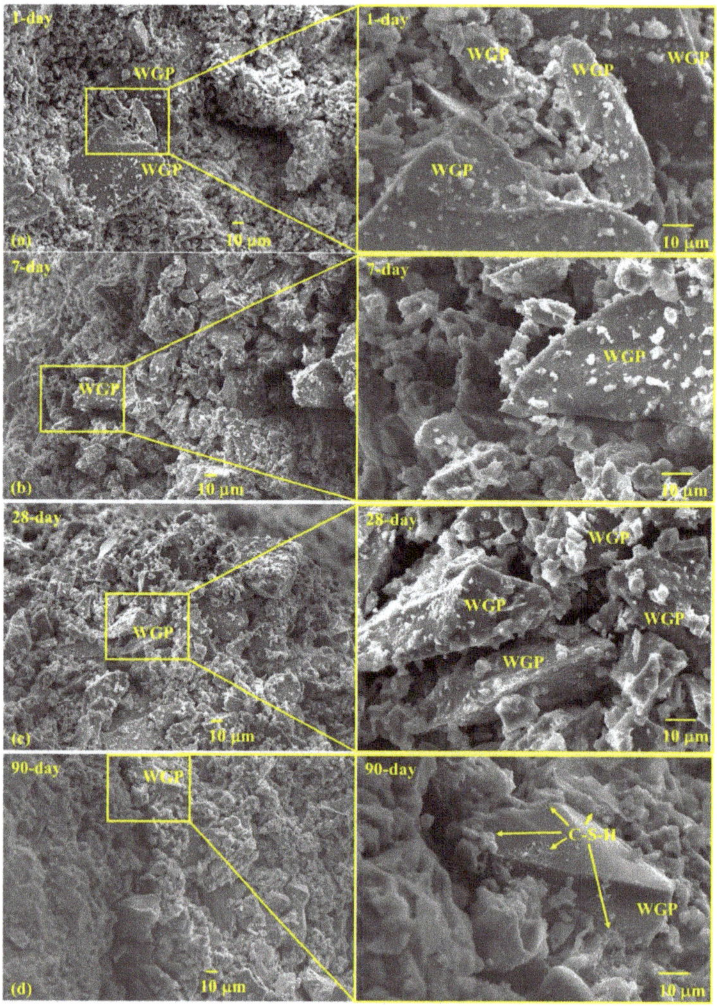

Figure 12. Micromorphologies of WGP-modified cement pastes: (**a**) 1 day; (**b**) 7 days; (**c**) 28 days; (**d**) 90 days.

4. CO_2 Emissions

The cement manufacturing process contributes greatly to CO_2 emissions, accounting for nearly a quarter of total CO_2 emissions [53]. WGP to replace Portland cement would lower the negative influence of cement-based materials on the environment. Herein, value engineering analysis (V) from the perspective of economics is adopted to analyze WGP impact by comparing the performance function (F) and carbon cost (C) of cement paste, expressed in Equation (7) [54].

$$V = \frac{F}{C} \quad (7)$$

This study focuses on the function of 90-day compressive strength and the carbon cost during the whole process of pastes with and without WGP. Generally, CO_2 emissions in the production of 1 ton of cement include those from raw material extraction, manufacturing, and transport (35.15 kg), decomposition (462.87 kg), calcination (505.92 kg), and product transportation (58.56 kg) [55]. To sum up, 1 ton of cement generates up to 1062.5 kg of CO_2. For WGP recycled from local waste glass bottles by crushing, grinding, and sieving, the main equipment used for the recycling process is ball mills and sieving machines. The consumption of electricity is recorded as 3 kW·h per ton of WGP. When coal-fired power is assumed, CO_2 production is 0.93 kg/(kW·h), which gives a CO_2 emission of 2.79 kg, according to Cabrera and Wolley [56].

The cost and function of plain paste and pastes with 15%, 20%, 25%, 30%, and 35% WGP in weight are listed in Table 5. For the control group, 1367 kg of cement is required for 1 m^3 of paste, and the emission of CO_2 reaches 1452.44 kg. When the cement is partially substituted by WGP from glass bottles, the cost greatly decreases based on the above analysis of CO_2 emissions. CO_2 emissions are taken as the C, i.e., the cost of carbon. In this paper, the compressive strength function is taken as F. For instance, 35 wt.% WGP-modified paste costs 945.42 kg, merely 65.1% of the plain paste, reducing it by 34.9%. And the value (F/C) increased from 0.044 to 0.060, increasing it by 26.7%. Although the incorporation of WGP may show a slight adverse effect on the strength function, a clear increasing trend in engineering value is detected, demonstrating the ecological effect of WGP in mitigating the emission impact of cement pastes. It shall be noted that the function here considers only the 90-day compressive strength of pastes; other parameters such as flexural strength, permeability, and carbonation resistance could be adopted for further analysis with higher accuracy.

Table 5. Estimation of environmental CO_2 emissions of pastes (1 m^3).

Groups	Content			Cost	Function	Value
	Cement (kg)	WGP (kg)	Water (kg)	CO_2 Emissions (kg)	Compressive Strength (MPa)	F/C
Plain paste	1367.00	0.00	546.80	1452.44	63.89	0.044
15% WGP	1161.95	205.05	546.80	1235.14	57.26	0.046
20% WGP	1093.60	273.40	546.80	1162.71	56.35	0.048
25% WGP	1025.25	341.75	546.80	1090.28	57.70	0.053
30% WGP	956.90	410.10	546.80	1017.85	58.61	0.058
35% WGP	888.55	478.45	546.80	945.42	56.36	0.060

5. Conclusions

In the present study, WGP recycled from local waste glass bottles is introduced in cement paste production. Its impact on the performance and carbon dioxide emissions of cement pastes were investigated. Based on experimental tests of rheology, heat of hydration, strength, and discussions of XRD, TG-DTG, and microstructure results, the following conclusions can be drawn:

- Recycling waste glass bottles into cement paste is technically feasible in terms of the mechanical properties required by relevant standards.

- WGP reduces the ecological impact of Portland cement by 34.9%, according to CO_2 emission analysis via value engineering analysis.
- The addition of 35% WGP increased cement hydration by 21.53%, 15.48%, and 11.68% at curing ages of 1, 28, and 90 days, respectively, compared to without WGP.
- The pozzolanic reaction of WGP consumes CH in cementitious systems, as confirmed by XRD patterns and CH content analysis from TG-DTG curves. Compared with the samples without WGP, the normalized CH content per unit binder mass containing 35% WGP decreased by 21.01%, 24.94%, and 27.41% at the ages of 1, 28, and 90 days, respectively.
- WGP particles provide nucleation places for hydration products, facilitating subsequent growth of C-S-H and strength growth in late ages.

Author Contributions: Conceptualization, F.R.; Methodology, F.R.; Formal analysis, Q.W.; Investigation, M.L.; Resources, Q.W.; Data curation, X.Z.; Writing—original draft, X.Z.; Writing—review & editing, Q.W. and J.S.; Supervision, Q.W.; Project administration, F.R.; Validation, J.S.; Funding acquisition, M.L. All authors have read and agreed to the published version of the manuscript.

Funding: This research was funded by National Natural Science Foundation of China (Nos. 52178125, 52008120), Yangcheng Scholar Project of Guangzhou Education Bureau (No. 202032849), and Natural Science Foundation of Guangdong Province (No. 2023A1515011624).

Institutional Review Board Statement: Not applicable.

Informed Consent Statement: Not applicable.

Data Availability Statement: All data, models, and code generated or used during the study appear in the published article.

Conflicts of Interest: The authors declare no conflict of interest.

References

1. Shi, C.; Wu, Y.; Riefler, C.; Wang, H. Characteristics and pozzolanic reactivity of glass powders. *Cem. Concr. Res.* **2005**, *35*, 987–993. [CrossRef]
2. Shao, Y.; Lefort, T.; Moras, S.; Rodriguez, D. Studies on concrete containing ground waste glass. *Cem. Concr. Res.* **2000**, *30*, 91–100. [CrossRef]
3. Gartner, E.; Hirao, H. A review of alternative approaches to the reduction of CO_2 emissions associated with the manufacture of the binder phase in concrete. *Cem. Concr. Res.* **2015**, *78*, 126–142. [CrossRef]
4. *197-1 BE*; Cement-Part 1: Composition, Specifications and Conformity Criteria for Common CementsCement-Part 1: Composition, Specifications and Conformity Criteria for Common Cements. British Standard Institution: London, UK, 2011.
5. Vereshchagina, T.; Kutikhina, E.; Solovyov, L.; Vereshchagin, S.; Mazurova, E.; Anshits, A. Hydrothermal Co-Processing of Coal Fly Ash Cenospheres and Soluble Sr(II) as Environmentally Sustainable Approach to Sr-90 Immobilization in a Mineral-like Form. *Materials* **2021**, *14*, 5586. [CrossRef]
6. Zhai, Q.; Kurumisawa, K. Effect of accelerators on Ca(OH)2 activated ground granulated blast-furnace slag at low curing temperature. *Cem. Concr. Compos.* **2021**, *124*, 104272. [CrossRef]
7. Gruber, K.A.; Ramlochan, T.; Boddy, A.; Hooton, R.D. Thomas MDA, Increasing concrete durability with high-reactivity metakaolin. *Cem. Concr. Compos.* **2001**, *23*, 479–484. [CrossRef]
8. Chen, Z.; Wang, Y.; Liao, S.; Huang, Y. Grinding kinetics of waste glass powder and its composite effect as pozzolanic admixture in cement concrete. *Constr. Build. Mater.* **2020**, *239*, 117876. [CrossRef]
9. Belouadah, M.; Rahmouni, Z.; Tebbal, N. Experimental characterization of ordinary concretes obtained by adding construction waste (glass, marble)—ScienceDirect. *Procedia Comput. Sci.* **2019**, *158*, 153–162. [CrossRef]
10. Yin, W.; Li, X.; Sun, T.; Chen, Y.; Xu, F.; Yan, G.; Xu, M.; Tian, K. Utilization of waste glass powder as partial replacement of cement for the cementitious grouts with superplasticizer and viscosity modifying agent binary mixtures: Rheological and mechanical performances. *Constr. Build. Mater.* **2021**, *286*, 122953. [CrossRef]
11. Pezeshkian, M.; Delnavaz, A.; Delnavaz, M. Development of UHPC mixtures using natural zeolite and glass sand as replacements of silica fume and quartz sand. *Eur. J. Environ. Civ. Eng.* **2019**, *25*, 2023–2038. [CrossRef]
12. Vaitkevicius, V.; Serelis, E.; Hilbig, H. The effect of glass powder on the microstructure of ultra high performance concrete. *Constr. Build. Mater.* **2014**, *68*, 102–109. [CrossRef]
13. Harbec, D.; Zidol, A.; Tagnit-Hamou, A.; Gitzhofer, F. Mechanical and durability properties of high performance glass fume concrete and mortars. *Constr. Build. Mater.* **2017**, *134*, 142–156. [CrossRef]

14. Shayan, A.; Xu, A. Performance of glass powder as a pozzolanic material in concrete: A field trial on concrete slabs. *Cem. Concr. Res.* **2006**, *36*, 457–468. [CrossRef]
15. Idir, R.; Cyr, M.; Tagnit-Hamou, A. Pozzolanic properties of fine and coarse color-mixed glass cullet. *Cem. Concr. Compos.* **2011**, *33*, 19–29. [CrossRef]
16. Pereira-de-Oliveira, L.A.; Castro-Gomes, J.P.; Santos, P.M.S. The potential pozzolanic activity of glass and red-clay ceramic waste as cement mortars components. *Constr. Build. Mater.* **2012**, *31*, 197–203. [CrossRef]
17. Lu, J.-X.; Zhan, B.-J.; Duan, Z.-H.; Poon, C.S. Using glass powder to improve the durability of architectural mortar prepared with glass aggregates. *Mater. Des.* **2017**, *135*, 102–111. [CrossRef]
18. Wang, S.; Baxter, L.; Fonseca, F. Biomass fly ash in concrete: SEM, EDX and ESEM analysis. *Fuel* **2008**, *87*, 372–379. [CrossRef]
19. Schwarz, N.; Neithalath, N. Influence of a fine glass powder on cement hydration: Comparison to fly ash and modeling the degree of hydration. *Cem. Concr. Res.* **2008**, *38*, 429–436. [CrossRef]
20. Yusuf, M.O.; Al-Sodani, K.A.A.; AlAteah, A.H.; Al-Tholaia, M.M.H.; Adewumi, A.A.; Bakare, A.O.; Usman, A.K.; Momohjimoh, I. Performances of the Synergy of Silica Fume andWaste Glass Powder in Ternary Blended Concrete. *Appl. Sci.* **2022**, *12*, 6637. [CrossRef]
21. Yusuf, M.O.; Al-Sodani, K.A.A.; Adewumi, A.A.; Alateah, A.H.; Al-Tholaia, M.M.H.; Shamsah, S.M.I.; Qazi, U.Y.; Alanazi, G.D.S. Microstructural Characteristics, Modeling ofMechanical Strength and Thermal Performance of Industrial Waste Glass Blended Concrete. *Appl. Sci.* **2022**, *12*, 8600. [CrossRef]
22. Mohammadyan-Yasouj, S.E.; Ghaderi, A. Experimental investigation of waste glass powder, basalt fibre, and carbon nanotube on the mechanical properties of concrete. *Constr. Build. Mater.* **2020**, *252*, 119115. [CrossRef]
23. Ramakrishnan, K.; Pugazhmani, G.; Sripragadeesh, R.; Muthu, D.; Venkatasubramanian, C. Experimental study on the mechanical and durability properties of concrete with waste glass powder and ground granulated blast furnace slag as supplementary cementitious materials. *Constr. Build. Mater.* **2017**, *156*, 739–749. [CrossRef]
24. *GB175;* Common Portland Cement, in Common Portland Cement. Ministry of Construction of China: Beijing, China, 2020.
25. Shayan, A.; Xu, A.M. Value-added utilisation of waste glass in concrete. *Cem. Concr. Res.* **2004**, *34*, 81–89. [CrossRef]
26. *ASTM C1866;* Standard Specification for Ground-Glass Pozzolan for Use in Concrete. ASTM International: West Conshohocken, PA, USA, 2020.
27. *ASTM C618;* Standard Specification for Coal Fly Ash and Raw or Calcined Natural Pozzolan for USE in concrete. ASTM International: West Conshohocken, PA, USA, 2012.
28. *196-1 BE;* Methods of Testing Cement-Part 1: Determination of Strength. British Standard Institution: London, UK, 2016.
29. *ASTM C311;* Standard Test Methods for Sampling and Testing Fly Ash or Natural Pozzolans for Use in Portland-Cement Concrete. ASTM International: West Conshohocken, PA, USA, 2017.
30. *GB/T 12957;* Test Method for Activity of Industrial Waste Slag Used As Addition to Cement. Ministry of Construction of China: Beijing, China, 2005.
31. *BS EN 450-1;* Fly Ash for Concrete-Part 1: Definition, Specifications and Conformity Criteria. British Standard Institution: London, UK, 2012.
32. *GB/T 2847;* Pozzolanic Materials Used for Cement Production. Ministry of Construction of China: Beijing, China, 2005.
33. *GB17671;* Method of testing cements-Determination of strength. Ministry of Construction of China: Beijing, China, 2020.
34. Lai, M.H.; Binhowimal, S.A.M.; Hanzic, L.; Wang, Q.; Ho, J.C.M. Cause and mitigation of dilatancy in cement powder paste. *Constr. Build. Mater.* **2020**, *236*, 117595. [CrossRef]
35. Wallevik, O.H.; Feys, D.; Wallevik, J.E.; Khayat, K.H. Avoiding inaccurate interpretations of rheological measurements for cement-based materials. *Cem. Concr. Res.* **2015**, *78*, 100–109. [CrossRef]
36. Jiang, Y.; Ling, T.C.; Mo, K.H.; Shi, C. A critical review of waste glass powder—Multiple roles of utilization in cement-based materials and construction products. *J. Environ. Manag.* **2019**, *242*, 440–449. [CrossRef]
37. Lu, J.-X.; Duan, Z.-H.; Poon, C.S. Combined use of waste glass powder and cullet in architectural mortar. *Cem. Concr. Compos.* **2017**, *82*, 34–44. [CrossRef]
38. Taha, B.; Nounu, G. Utilizing Waste Recycled Glass as Sand/Cement Replacement in Concrete. *J. Mater. Civ. Eng.* **2009**, *21*, 709–721. [CrossRef]
39. Liu, G.; Florea, M.V.A.; Brouwers, H.J.H. Performance evaluation of sustainable high strength mortars incorporating high volume waste glass as binder. *Constr. Build. Mater.* **2019**, *202*, 574–588. [CrossRef]
40. Lu, J.-X.; Duan, Z.-H.; Poon, C.S. Fresh properties of cement pastes or mortars incorporating waste glass powder and cullet. *Constr. Build. Mater.* **2017**, *131*, 793–799. [CrossRef]
41. Mirzahosseini, M.; Riding, K.A. Effect of curing temperature and glass type on the pozzolanic reactivity of glass powder. *Cem. Concr. Res.* **2014**, *58*, 103–111. [CrossRef]
42. Maraghechi, H.; Maraghechi, M.; Rajabipour, F.; Pantano, C.G. Pozzolanic reactivity of recycled glass powder at elevated temperatures: Reaction stoichiometry, reaction products and effect of alkali activation. *Cem. Concr. Compos.* **2014**, *53*, 105–114. [CrossRef]
43. Huang, L.; Yan, P. Effect of alkali content in cement on its hydration kinetics and mechanical properties. *Constr. Build. Mater.* **2019**, *228*, 116833. [CrossRef]
44. Du, H.; Tan, K.H. Waste Glass Powder as Cement Replacement in Concrete. *J. Adv. Concr. Technol.* **2014**, *12*, 468–477. [CrossRef]

45. Khmiri, A.; Chaabouni, M.; Samet, B. Chemical behaviour of ground waste glass when used as partial cement replacement in mortars. *Constr. Build. Mater.* **2013**, *44*, 74–80. [CrossRef]
46. Scrivener, K.L.; Juilland, P.; Monteiro, P.J.M. Advances in understanding hydration of Portland cement. *Cem. Concr. Res.* **2015**, *78*, 38–56. [CrossRef]
47. Liu, G.Y.; Fan, L.; Jin, D.Z.; Lu, R.Y.; Li, L.J. Progress in Stimulating Techniques and Mechanism of Activity of Glass Powder. *Sci. Technol. Eng.* **2016**, *16*, 152–194.
48. Shi, C.; Yuan, Q. *Methods for Testing and Analysis of Cement-Based Materials*; China Construction Industry Press: Beijing, China, 2018.
49. Mehta, P.K.; Monteiro, P. *Concrete: Microstructure, Properties, and Materials*; McGraw-Hill Education: New York, NY, USA, 2013.
50. Lian, H.Z.; Tong, L.; Chen, E.Y. *Phase Research Foundation of Building Materials*; Tsinghua University Press: Beijing, China, 1995.
51. Zhang, J.; Scherer, G.W. Comparison of methods for arresting hydration of cement. *Cem. Concr. Res.* **2011**, *41*, 1024–1036. [CrossRef]
52. Ouyang, X.; Koleva, D.A.; Ye, G.; van Breugel, K. Understanding the adhesion mechanisms between C S H and fillers. *Cem. Concr. Res.* **2017**, *100*, 275–283. [CrossRef]
53. Mahoutian, M.; Shao, Y. Production of cement-free construction blocks from industry wastes. *J. Clean. Prod.* **2016**, *137*, 1339–1346. [CrossRef]
54. Liu, G.-Y.; Gao, P.-F.; Li, D.-C.; Yu, X.-L. Economic analysis of glass powder concrete based on value engineering analysis. *Concrete* **2020**, *6*, 163–165+174.
55. Yu, H.; Yang, H.; Zhang, H.; Jie, Z. Research of Carbon Emission of Cement in Lifecycle. *Sichuan Build. Mater.* **2017**, *43*, 3.
56. Cabrera, J.G.; Woolley, G.R. Life cycle benefits of calcium silicate replacements. *Waste Manag.* **1996**, *16*, 215–220. [CrossRef]

materials

MDPI

Article

Effect of Carbonation Treatment on the Properties of Steel Slag Aggregate

Jian Ma [1,2], Guangjian Dai [1], Feifei Jiang [2,3,*], Ning Wang [2], Yufeng Zhao [2] and Xiaodong Wang [1]

1 College of Architecture and Civil Engineering, Jiangsu University of Science and Technology, Zhenjiang 212000, China; mj198004@mail.ustc.edu.cn (J.M.); daigj1230@163.com (G.D.); 18205255126@163.com (X.W.)
2 Suzhou Institute of Technology, Jiangsu University of Science and Technology, Suzhou 215600, China; 18862615220@163.com (N.W.); 17625216718@163.com (Y.Z.)
3 College of Civil Engineering, Nantong Institute of Technology (NIT), Nantong 226000, China
* Correspondence: 999620140019@just.edu.cn

Abstract: Steel slag is the waste slag generated after steel smelting, which has cementitious activity. However, untreated steel slag can damage the integrity of steel slag concrete due to its harmful expansion. This study prepared porous aggregates by mixing powdered steel slag, fly ash, and cement and carbonated them with CO_2 under high pressure conditions (0.2 MPa). The effect of carbonation on the performance of steel slag aggregate was studied using volume stability and crushing value. The effect of different carbonation conditions on the products was studied using X-ray diffraction (XRD) and thermogravimetric (TG) analyses, and the carbon sequestration efficiency of steel slag under different treatment methods was quantitatively evaluated. The research results indicate that untreated steel slag was almost completely destroyed and lost its strength after autoclave curing. With the increase in temperature and carbonation time, the performance of steel slag aggregate gradually improved and the pulverization rate, expansion rate, and crushing value gradually decreased. According to the experimental results of XRD and TG, it was found that the reaction between f-CaO (free CaO) and CO_2 in steel slag generated $CaCO_3$, filling the pores inside the aggregate, which was the internal reason for the improvement of aggregate performance. After comparison, the best carbonation method was maintained at 55 °C for 72 h. After carbonation, the steel slag aggregate had a pulverization rate of 2.4%, an expansion rate of 0.23%, a crushing value of 23%, and a carbon sequestration efficiency of 11.27% per unit weight of aggregate.

Keywords: steel slag; carbonation; expansion; volume stability; waste management; sustainability

1. Introduction

With the rapid development of global industrialization, the harm of global climate change is increasing day by day. With the development of ecological civilization construction in China, carbon emissions and carbon neutrality have received great attention. The Chinese government solemnly promised in 2020 to achieve the "Dual Carbon" Goal of carbon peaking by 2030 and carbon neutrality by 2060 [1,2]. Currently, relevant organizations and institutions around the world, including the World Meteorological Organization (WMO), the European Integrated Carbon Observing System (ICOS), and the China Meteorological Administration, are monitoring greenhouse gases, predicting climate change trends, and proposing carbon reduction measures. Reducing carbon emissions and protecting the environment has gradually become a human consensus [3–5].

As a major country in the steel industry, the output of China's iron and steel has steadily occupied the first place in the world for many years. With the rapid development of the steel industry, a large amount of steel slag has been generated during the production of coarse steel [6]. At the same time, the steel industry produces a significant amount of CO_2 in its production. It is estimated that its CO_2 emissions in 2019 were about 2.6 billion tons, accounting for 7% of global energy system emissions [7]. Because of the complex composition of steel slag, the utilization ratio of steel slag is always low. At present, the utilization ratio of steel slag in China is only about 30% [8] due to the complexity of steel slag compositions with a high concentration of f-CaO and f-MgO [9]. Steel slag is widely used in road construction [10]. Cansu Kurtulus et al. [11] successfully synthesized foam glass from ferrochrome slag and waste soda-lime glass for the first time. Because of the high output and low consumption of steel slag, a large amount of steel slag can not only occupy a lot of land resources but also cause pollution to the surrounding soil and air [12,13]. Some scholars have studied the reuse of steel slag, and now steel slag is mainly used as aggregate for roadbed landfill [14,15], agricultural fertilizer [16,17], and wastewater treatment [18,19], but they have not been used on a large scale.

Due to the fact that concrete is the most commonly used building material, some researchers have attempted to use steel slag to prepare steel slag concrete. Carlo Pellegrino [20] used steel slag to replace natural aggregate to prepare the concrete, as well as the mechanical properties and durability of steel slag concrete. Zhuang et al. [21] used steel-slag-modified ultrafine blast furnace slag as an admixture to prepare high-strength concrete and studied its ultra-high temperature and self-shrinkage issues; the results showed that steel slag can effectively retard the main hydration heat release peak, decrease the accumulated heat, and reduce the growth rate of self-shrinkage with regard to cement–ultrafine blast furnace slag composite binder. M. Fernández Bertos et al. [22] found that carbonation can improve the physical and chemical properties of steel slag aggregate and can also fix CO_2 in the steel slag, which provides a new way for the utilization of steel slag. Wang et al. [23] studied the compaction and carbonation of gypsum, steel slag, and water; based on microscopic analysis, the results demonstrated that gypsum can promote the compression strength of steel slag and the absorption of carbon dioxide, and the main hydration products are C-S-H phase and ettringite, while the main carbonation products are calcite and monocarbonate. Liu et al. [24] studied the effects of curing conditions on the mechanical properties of steel slag–cement cementitious materials at high temperatures and under carbonation. The results show that the case after carbonation can effectively prevent the corrosion of cement-based materials, improve the durability of cementitious materials, and enhance the compactness of cementitious materials. Shi et al. [25–27] studied the results of an experiment using carbonation curing to improve the performance of recycled aggregates to quantify the degree of CO_2 curing of recycled aggregates by assessing the percentage of carbonation of the aggregates. It was found that with the increase in carbonation curing time, the strength of the aggregate and the degree of CO_2 curing increased. After carbonation curing, the aggregate shows a more compact microstructure. The formation of calcium carbonate and silica gel greatly reduces the porosity; thus, the aggregate has higher strength. Mo Liwu et al. [28] used a 99.9% concentration of CO_2 to cure aggregates at a pressure of 0.1MPa and studied the carbonation technology for treating steel slag to prepare concrete, improving its strength and stability [29–31]. The abovementioned research has explored the potential application of steel slag in concrete, recognized the impact of steel slag stability, and demonstrated that carbonation curing of steel slag can improve the volume stability of steel slag. However, further research on how to improve carbonation efficiency and accurately calculate carbon sequestration is needed. And due to the density of steel slag being about 3.2 g/cm^3, which is much greater than the 2.5 g/cm^3 density of stones, the weight of steel slag concrete is greater than that of ordinary concrete, which increases the load on the structural foundation. The light weight of steel slag concrete is also an important factor hindering its engineering application.

This study uses a mixture of fly ash and steel slag powder to prepare a steel slag aggregate. On the one hand, porous fly ash provides a channel for CO_2 transport. On the other hand, lightweight fly ash reduces the weight of concrete. On this basis, the effects of different carbonation times and temperatures on carbonation products were studied using XRD and TG, and the mass of CO_2 absorbed by the steel slag aggregate per unit mass was calculated in detail. The research results have positive significance for scientific carbon sequestration and the resource utilization of steel slag.

2. Raw Materials and Test

2.1. Raw Materials

The cement was P.O 42.5 Portland cement produced by Zhangjiagang Conch Cement Co., Ltd in Zhangjiagang, China. The fly ash was Class I fly ash produced by Shenhua Huashou Power Co., Ltd in Shanghai, China. The steel slag was provided by the Shagang Group in Suzhou, China. The steel slag powder was prepared by pulverizing the steel slag with a ball mill through a 0.08 mm sieve with a specific surface area of 260 m^2/kg. Figure 1 shows the particle size distribution of steel slag after grinding. From Figure 1, it can be seen that the median particle size of powdered steel slag is 26.9 μm and the average particle size is 34.3 μm. The mineral composition of steel slag was tested using XRD. The results are shown in Figure 2. The chemical composition of raw materials is determined using the cement chemical analysis method, as shown in Table 1. Figure 2 shows that steel slag contains more CaO, followed by SiO_2 and MgO, indicating that steel slag contains a large amount of basic oxide which can absorb and react with CO_2.

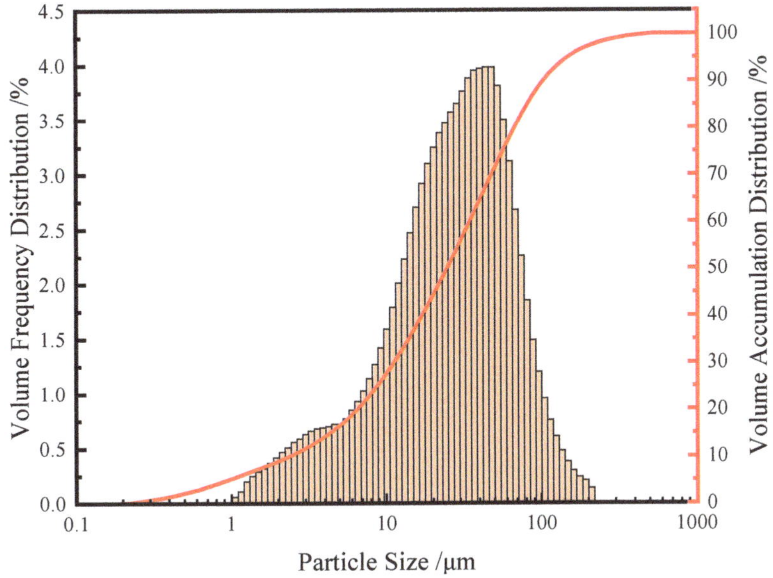

Figure 1. Particle size distribution of powdered steel slag.

2.2. Preparation

The steel slag powder was the main component of this experiment, which was obtained by grinding steel slag through a ball mill and passing through a 0.08 mm sieve. The mixture of the aggregate was composed of steel slag powder, cement, and fly ash, with a ratio of 5:3:1. The steel slag was used as the main raw material, cement was used as cementitious material, and fly ash was added to reduce the quality of the aggregate and make an aggregate with a porous structure. The fly ash has two main functions. The first effect is to reduce the weight of the aggregate, which can effectively reduce the load on the foundation of the building. The second function is that the porous fly ash provides a channel for the transport of CO_2, which can improve carbonation efficiency. The spherical steel slag aggregate was obtained by adding the abovementioned proportion into the pelletizer to mix evenly and spray-pelletizing. The particle size of the steel slag aggregate was 9–31.5 mm.

The aggregate was placed in the pre-curing condition for 7 days (relative humidity 55 ± 5%, 20 ± 2 °C) and the uncarbonated steel slag was obtained (Figure 3).

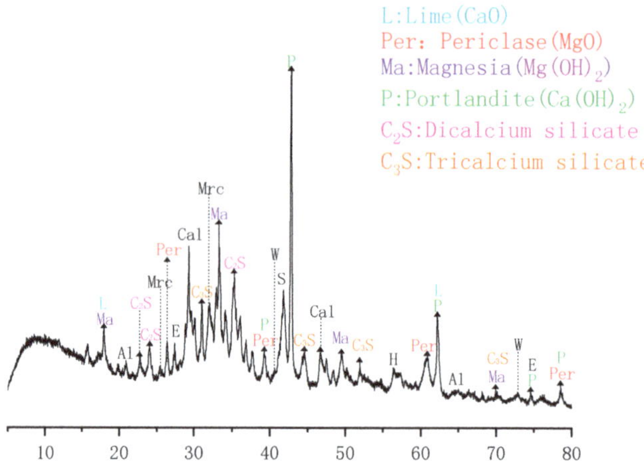

Per-MgO; L-CaO; Ma-Mg(OH)2; P-Ca(OH)2; Cal-CaCO3; Arg-CaCO3; S-SiO2; Mrc-FeS2; W-FeO; E-Aft

Figure 2. XRD pattern of steel slag.

Table 1. Chemical compositions of raw materials.

Raw Material	Mass Fraction									
	CaO	MgO	SiO_2	Al_2O_3	Fe_2O_3	K_2O	Na_2O	SO_3	LOSS	Total
Cement	63.95	2.40	19.40	4.76	2.94	0.66	0.12	2.59	2.83	99.65
Steel Slag	41.78	10.72	9.75	2.60	29.25	0.04	0.08	0.07	1.55	95.84
Fly Ash	3.96	1.06	48.82	27.84	6.10	1.14	0.40	0.94	6.13	96.39

Figure 3. Steel slag aggregate.

The pre-cured steel slag aggregate was placed in a custom-made carbonation box with a carbon dioxide concentration of 99.9% and an ambient pressure of 0.2 MPa for carbonation curing. The schematic diagram of the carbonation test is shown in Figure 4.

In order to determine the optimal carbonation time and carbonation effect at room and high temperatures, the carbonation temperature was set to 23 °C at normal temperature and 55 °C at high temperature for 3 h, 9 h, and 72 h, which were simplified as D1–D6, respectively. D0 was the blank control group without carbonation and granulation, as shown in Table 2.

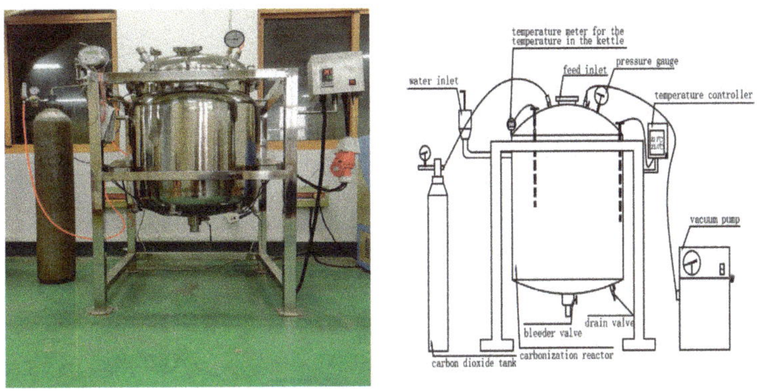

(**a**) (**b**)

Figure 4. Schematic diagram of carbonation test: (**a**) Photo of test; (**b**) Setup for split tensile strength test.

Table 2. Carbonation test grouping.

Test Group	D0	D1	D2	D3	D4	D5	D6
Carbonation temperature (°C)	-	23	23	23	55	55	55
Carbonation time (h)	-	3	9	72	3	9	72

2.3. *Characterizations*

2.3.1. Volume Stability

According to the specification of the steel slag stability test method (GB/T24175-2009) [32], the steel slag aggregate used in the test was determined using the method of water immersion expansion rate of steel slag as shown in Figure 5. The effect of different treatment methods on the expansibility of steel slag was evaluated using the expansion rate. In addition, the aggregates were autoclaved for 3 h at a temperature of 216 °C and pressure of 2 MPa. Its volume stability was determined by observing the damage of the steel slag aggregate particles.

(**a**) (**b**)

Figure 5. Measuring device for the swelling rate of immersion: (**a**) Photo of test; (**b**) Setup for split tensile strength test.

2.3.2. Crushing Value

The steel slag aggregate with a size range of 9.5–13.2 mm was utilized in the experiment, following the methodology for determining the crushing index value as stipulated in the test methods of highway engineering (JTG E42-2005) [33]. The mould containing the sample was placed on the press, and the pressure testing machine was started to reach a total load of 400 KN within 10 min, and the load was stabilized for 5 s. The finely crushed particles were removed by sieving with a diameter of 2.36 mm, and the mass left on the sieve was weighed to the nearest 1 g. The crushing index value Q was calculated according to Equation (1), accurate to 0.1%:

$$Q = \frac{M_1}{M_1 + M_2} \times 100\% \tag{1}$$

where M_1 is the fine particle mass with a diameter less than 2.36 mm; M_2 is the fine particle mass with a diameter greater than 2.36 mm.

2.3.3. Product Analysis

The mineral composition of steel slag samples before and after carbonation was analyzed using X-ray diffraction (Rigaku SmartLab (3), Tokyo, Japan); the pattern was acquired between 5° and 80° with steps of 0.02°; the scanning speed was 10 °/min, and the mineral composition and degree of carbonation of the phase were determined using a thermogravimetric differential scanning calorimeter (Netzsch STA 449,Germany). For the mass absorption formula of carbon dioxide after carbonation stated in Formula (2), X is the amount of carbon dioxide absorbed by steel slag aggregates, while the $CaCO_3$ formed is calculated from Formula (3) and Y is the amount of $CaCO_3$ formed in steel slag aggregates [34]:

$$X = \frac{G_1}{G_2} \times 100\% \tag{2}$$

$$Y = (G_3 - G_4) \times \frac{100}{44} \times 100\% \tag{3}$$

where G_1 is steel slag aggregate mass loss in 600–750 °C; G_2 is steel slag aggregate mass at 900 °C; G_3 is steel slag aggregate mass at 600 °C; G_4 is steel slag aggregate mass at 750 °C.

3. Results and Discussion

3.1. Volume Stability

The immersion expansion rate was measured for the steel slag aggregate, and the expansion rate of each group was plotted according to the results as shown in Figure 6; the expansion rate of steel slag aggregates with different carbonation times increases rapidly in the first three days but slowly thereafter whether at room temperature (23 °C) or a high temperature (55 °C). However, the expansion rate of uncarbonated D0 exceeds 2% at 7 d, which exceeds the prescribed limit. Except for group D1, which has an expansion rate of over 2% with a carbonation time of 6 h at 23 °C after 10 d, those of D2–D6 were 0.4%, 0.3%, 0.88%, 0.35%, and 0.23%, respectively, which was much less than that of the standard level.

The explanation for this phenomenon is that f-CaO content in the steel slag aggregate decreased obviously after the reaction at the high temperature and the long carbonation time, and the hydration product C-S-H gel was formed by C_2S and C_3S in the steel slag; this was equivalent to forming a protective shell on the surface of the steel slag, which hindered the further reaction between the steel slag and water and thus reducing the hydration reaction rate of the steel slag [35]. Furthermore, the expansion rate of the steel slag aggregate treated at 55 °C for 72 h is only 0.23%, which is 88.5% lower than the standard value and has the optimum expansion stability.

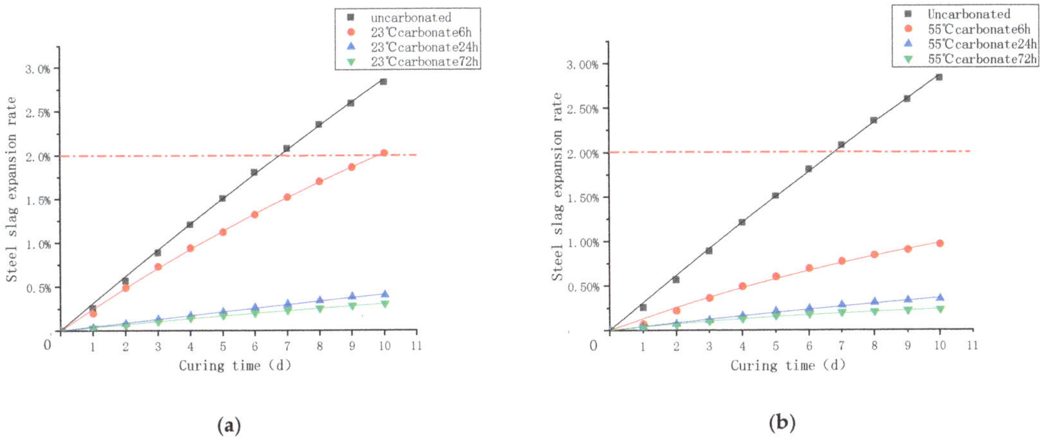

Figure 6. Daily expansion rate of steel slag aggregates. (**a**) carbonation temperature 23 °C; (**b**) carbonation temperature 55 °C.

As shown in Figure 7, the steel slag aggregate that has not undergone carbonation treatment is completely pulverized after the autoclaving test. Mainly due to the large amount of f-CaO, f-MgO in the aggregate undergoes hydration and expansion under saturated steam conditions, ultimately leading to the crushing and pulverization of the aggregate. After the carbonation treatment, the shattered ratio of the steel slag aggregate improved. With the increase in carbonation time and carbonation temperature, the shattered ratio of the steel slag aggregate also decreased. This is because when the steel slag aggregate undergoes carbonation treatment, f-CaO reacts with CO_2 and is consumed more without causing significant expansion. Under the carbonation curing condition of 55 °C for 72 h, the pulverization rate of steel slag aggregate was only 2.4% with almost no crushing (Table 3). This indicates that carbonation treatment can effectively improve the problem of poor volume stability caused by the expansion of steel slag aggregates. The prepared steel slag aggregate after carbonation treatment has a low expansion rate, good volume stability, can be better used in the building materials industry, especially in concrete and roadbed engineering, with good stability, and can be used to replace stones to a certain extent.

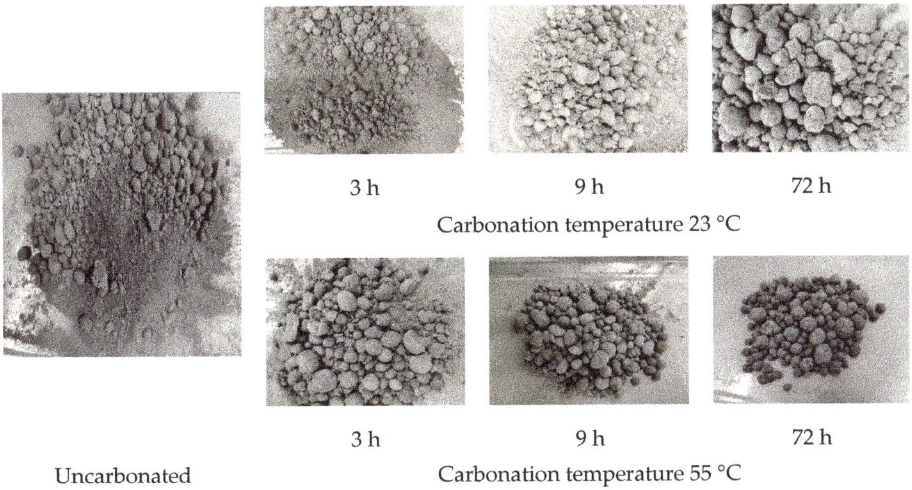

Figure 7. Photos of steel slag aggregates after the autoclaving test.

Table 3. The pulverization rate of steel slag aggregate.

Test Group	D0	D1	D2	D3	D4	D5	D6
pulverization rate (%)	55.8	17.7	9.8	4.4	11.2	3.5	2.4

3.2. *Crushing Value*

Figure 8 shows the crushing values of the steel slag aggregates with different carbonation times at 23 °C and 55 °C. At the same carbonation temperature, the crushing value of the aggregate decreases with the increase in carbonation time. The crushing values of the aggregates treated at a carbonation temperature of 55 °C are significantly lower than those at 23 °C. The crushing value of uncarbonated steel slag aggregate (D0) is 33.2%, which exceeds the limit for Grade III coarse aggregate in highway engineering (30%). When the carbonation temperature is 23 °C, only the crushing value of the steel slag aggregate carbonated for 72 h is less than 25%, satisfying the requirements for Grade III but not Grade II coarse aggregate yet. When the carbonation temperature increases to 55 °C, the crushing value of the steel slag aggregate cured for 72 h decreases to 23%, which is less than 25%, and reaches the standard requirement for Grade II coarse aggregate. Because the prepared steel slag aggregate had a porous structure and low strength, after carbonation curing, the generated calcium carbonate filled the pores, making the internal structure of the aggregate more dense, and the strength of the aggregate also increased.

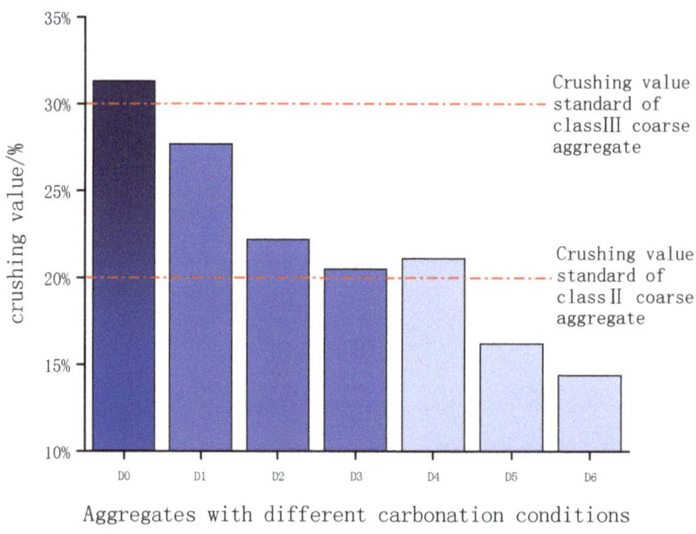

Figure 8. Crushing value of steel slag aggregates after carbonation treatment.

3.3. *Reaction Products*

As shown by the XRD pattern in Figure 9, the diffraction peak of C_2S decreases and the diffraction peak of C_3S and f-CaO disappears after carbonation as compared to the non-carbonated. In addition, the diffraction peak of $Ca(OH)_2$ decreased with the increase in carbonation time, while the diffraction peak of $CaCO_3$ was produced after carbonation and is obviously enhanced with the increase in carbonation temperature and carbonation time. The results confirmed that f-CaO, C_2S, C_3S, and $Ca(OH)_2$ in the steel slag aggregate reacted with CO_2 to form $CaCO_3$, as the following equations show [36]:

$$2(3CaO \bullet SiO_2) + CO_2 + 3H_2O \rightarrow 3CaO \bullet 2SiO_2 \bullet 3H_2O + CaCO_3 \quad (4)$$

$$2(3CaO{\bullet}SiO_2) + 3CO_2 + 3H_2O \rightarrow 3CaO{\bullet}2SiO_2{\bullet}3H_2O + 3CaCO_3 \quad (5)$$

$$CaO + H_2O \rightarrow Ca(OH)_2 \quad (6)$$

$$Ca(OH)_2 + CO_2 \rightarrow H_2O + CaCO_3 \quad (7)$$

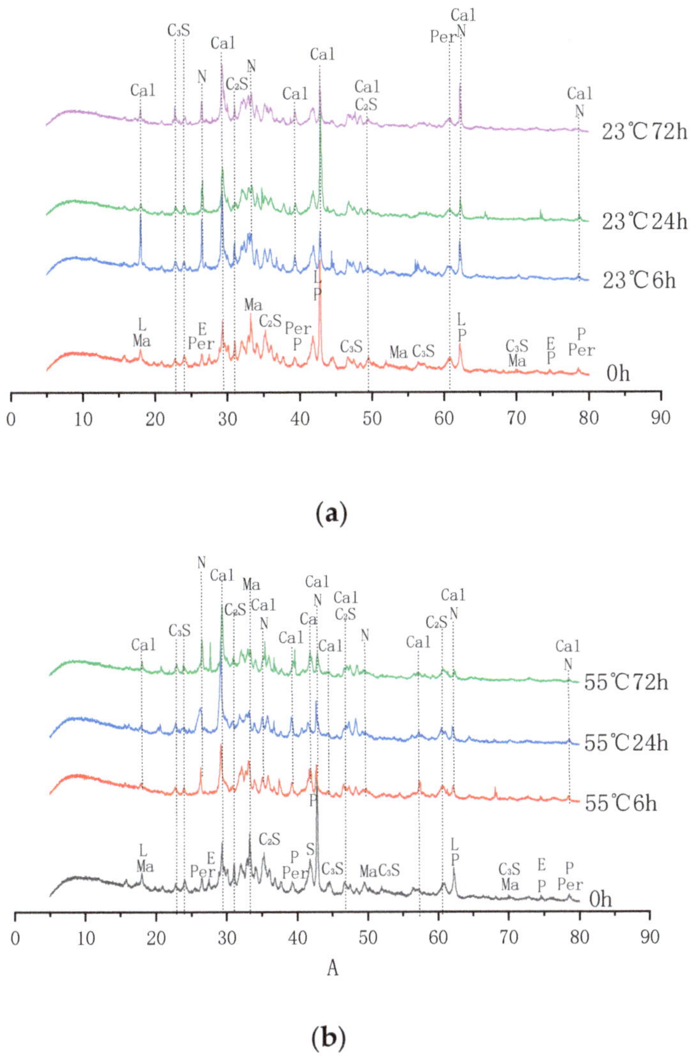

Figure 9. XRD analysis pattern of steel slag aggregates after carbonation curing. (**a**) carbonation temperature 23 °C; (**b**) carbonation temperature 55 °C.

The results confirmed that as the carbonation temperature and carbonation time increased, more of the reaction product $CaCO_3$ was generated and filled the pores in the aggregate, which enhanced the strength and decreased the crushing value of the aggregates.

In order to quantitatively analyze the CO_2 absorption capacity of the steel slag aggregates with TG/DSC, the results are shown in Figure 10.

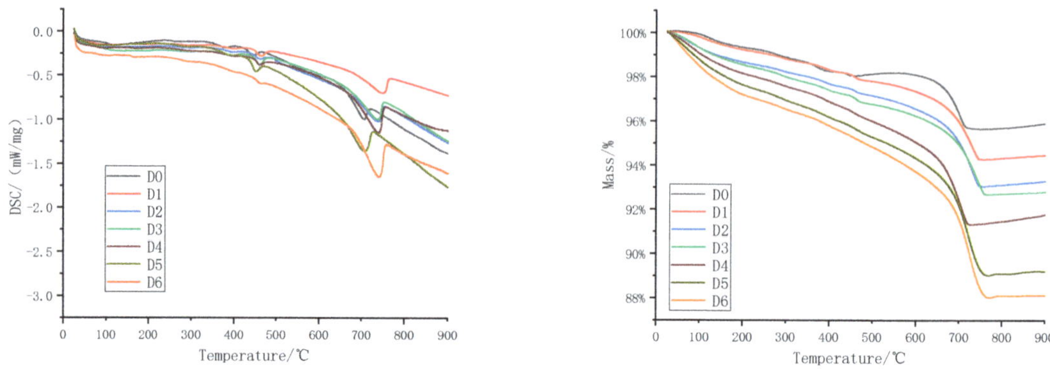

(**a**) (**b**)

Figure 10. TG/DSC curve of steel slag aggregates after carbonation. (**a**) DSC curve; (**b**) TG curve.

In Figure 10a, there are two distinct endothermic peaks, wherein the one within the temperature range of 400–450 °C can be ascribed to the decomposition of $Ca(OH)_2$ and the endothermic peak in the temperature range of 600–750 °C resulted from the decomposition of $CaCO_3$ [36,37]. The content of $Ca(OH)_2$ and the endothermic peak of $Ca(OH)_2$ decrease with the increase in carbonation temperature and carbonation time. In particular, the endothermic peak of the steel slag aggregate disappears almost completely after carbonation at 55 °C for 72 h, while the endothermic peak becomes more and more obvious in the temperature range of 600–750 °C, which indicates that more and more $CaCO_3$ was produced during carbonation curing.

Compared to other studies where steel slag was directly carbonated for 24 h, the carbon dioxide absorption was only 7.09 wt.% [36]. In this study, the carbon dioxide absorption of steel slag after carbonation treatment after grinding is much higher than the conclusion in the literature. The CO_2 absorption and $CaCO_3$ production of steel slag aggregate under carbonation curing were determined according to the TG/DSC curves listed in Table 4. It can be seen that the steel slag aggregate absorbs 6.76–9.04 wt.% CO_2 after carbonation. As the carbonation time increases, the amount of CO_2 absorption also increases. The steel slag aggregate absorbs an additional 7.4–8.74 wt.% CO_2 when the carbonation time increases from 0 to 72 h. When the carbonation time increases from 6 h to 72 h, the aggregate absorbs an additional 2.28 wt.% and 1.34 wt.% CO_2 at room temperature and high temperature, respectively. It can be seen that the carbonation efficiency of steel slag aggregate in 0–24 h is 1.89 times that of 24–72 h. This is due to the reaction of CO_2 with f-CaO, f-MgO, and so on during carbonation. Decreasing calcium content can take part in the reaction. In addition, with the formation of the carbonation product $CaCO_3$, the pores in the aggregate are filled up, remarkably boosting the enhancement of structural densification, which is unfavorable to the penetration of CO_2 gas and hinders the carbonation reaction; the efficiency of early carbonation is higher than that of late carbonation. From the tests above, not only will the carbonation temperature and time affect the carbonation efficiency of the steel slag aggregate but the composition and structure of the steel slag aggregate will also play a key role in carbonation. In addition, the steel slag aggregate shows different carbonation efficiencies at different carbonation periods.

Table 4. CO_2 absorption and $CaCO_3$ content in steel slag aggregate after carbonation curing.

Test Group	600–750 °C Mass Loss/mg	CO_2 Absorption/wt.%	$CaCO_3$ Content/wt.%
D0	0.46	2.53	1.05
D1	3.78	6.76	8.58
D2	5.52	7.57	12.55
D3	6.95	9.04	15.80
D4	6.76	9.93	15.36
D5	7.49	10.13	17.01
D6	9.13	11.27	20.74

4. Conclusions

This study not only reduced the weight of steel slag aggregate but also improved its carbonation efficiency by mixing powdered steel slag with fly ash. The reaction products of steel slag were analyzed using XRD and TG, and based on this, the carbon sequestration efficiency under different curing methods was calculated. The research results provide a solution for the resource utilization of steel slag and a theoretical calculation basis for scientifically reducing "carbon emissions". Based on the experimental results, the following conclusions can be drawn:

(1) Untreated steel slag contained a large amount of active CaO, with a pulverization rate of 55.8% and a crushing value of 33.2%. When used directly, it will cause harmful expansion and damage the integrity of the aggregate. Therefore, untreated steel slag cannot be used to prepare steel slag concrete.

(2) Under high temperatures and pressure, the steel slag could quickly react with CO_2 to generate $CaCO_3$, which filled the voids in the aggregate and improved its performance. This indicates that fly ash could not only reduce the weight of the aggregate but also provide channels for CO_2 and improve carbonation efficiency. This proves that the process of producing artificial aggregates by mixing powdered steel slag, fly ash, and cement in this study was reasonable. After comparison, the best carbonation method was to cure at 55 °C for 72 h. The pulverization rate of the carbonated steel slag aggregate was 2.4%, the expansion rate was 0.23%, and the crushing value was 23%, meeting the requirements of Class II aggregate.

(3) The carbon sequestration efficiency was closely related to temperature and time, and increasing the curing temperature and time could effectively improve the carbon sequestration efficiency of steel slag. When the curing temperature was 23–55 °C and the curing time was 6–72 h, the carbon sequestration efficiency was increased from 6.76% to 11.27% (in mass fraction).

Author Contributions: Methodology, J.M. and F.J.; Formal analysis, J.M. and G.D.; Investigation, F.J. and N.W.; Resources, F.J.; Data curation, G.D. and X.W.; Writing—original draft, N.W. and Y.Z.; Writing—review & editing, X.W. All authors have read and agreed to the published version of the manuscript.

Funding: This research was funded by Suzhou science and technology development plan, grant number (SNG2022015, SNG2021039, SS202109); Zhangjiagang science and technology development plan, (ZKYY2212).

Institutional Review Board Statement: Not applicable.

Informed Consent Statement: Not applicable.

Data Availability Statement: All data generated or analyzed in this research were included in this published article. Additionally, readers can access all data used to support conclusions of the current study from the corresponding author upon request.

Conflicts of Interest: The authors declare no conflict of interest.

References

1. Yao, H.; Xu, P.; Wang, Y.; Chen, R. Exploring the low-carbon transition pathway of China's construction industry under carbon-neutral target: A socio-technical system transition theory perspective. *J. Environ. Manag.* **2023**, *327*, 116879.
2. Ouyang, Y.; Guo, J. Carbon capture and storage investment strategy towards the dual carbon goals. *J. Asian Econ.* **2022**, *82*, 101527.
3. Naitza, L.; Cristofanelli, P.; Marinoni, A.; Calzolari, F.; Roccato, F.; Busetto, M.; Sferlazzo, D.; Aruffo, E.; Di Carlo, P.; Bencardino, M.; et al. Increasing the maturity of measurements of essential climate variables (ECVs) at Italian atmospheric WMO/GAW observatories by implementing automated data elaboration chains. *Comput. Geosci.* **2020**, *137*, 104432.
4. Lelandais, L.; Xueref-Remy, I.; Riandet, A.; Blanc, P.E.; Armengaud, A.; Oppo, S.; Yohia, C.; Ramonet, M.; Delmotte, M. Analysis of 5.5 years of atmospheric CO_2, CH_4, CO continuous observations (2014–2020) and their correlations, at the Observatoire de Haute Provence, a station of the ICOS-France national greenhouse gases observation network. *Atmos. Environ.* **2022**, *277*, 119020.
5. Li, S.; Zhong, Y.; Li, X.; Ayikan, M.; Deng, L.; Liang, M.; Liu, L. Background characteristics and influence analysis of greenhouse gases at Akedala regional background station in Northern Xinjiang. *Atmos. Pollut. Res.* **2022**, *13*, 101612.
6. Jiang, Y.; Ling, T.-C.; Shi, C.; Pan, S.-Y. Characteristics of steel slags and their use in cement and concrete—A review. *Resour. Conserv. Recycl.* **2018**, *136*, 187–197.
7. Villalva, A.J.E. Sustainability assessment in the steel industry using partial least squares—Structural equation model. *World Dev. Sustain.* **2023**, *2*, 100054.
8. Chen, J.; Xing, Y.; Wang, Y.; Zhang, W.; Guo, Z.; Su, W. Application of iron and steel slags in mitigating greenhouse gas emissions: A review. *Sci. Total Environ.* **2022**, *844*, 157041.
9. Wang, G.; Wang, Y.; Gao, Z. Use of steel slag as a granular material: Volume expansion prediction and usability criteria. *J. Hazard. Mater.* **2010**, *184*, 555–560.
10. Gao, W.; Zhou, W.; Lyu, X.; Liu, X.; Su, H.; Li, C.; Wang, H. Comprehensive utilization of steel slag: A review. *Powder Technol.* **2023**, *422*, 118449.
11. Kurtulus, C.; Kurtulus, R.; Kavas, T. Foam glass derived from ferrochrome slag and waste container glass: Synthesis and extensive characterizations. *Ceram. Int.* **2021**, *47*, 24997–25008. [CrossRef]
12. Liu, X.; Hu, M.; Ke, S.; Fu, C.; Guo, X.; Ye, X. A novel rammed earthen material stabilized with steel slags. *Constr. Build. Mater.* **2018**, *189*, 1134–1139. [CrossRef]
13. Chen, M.; Wen, P.; Wang, C.; Chai, Z.; Gao, Z. Evaluation of particle size distribution and mechanical properties of mineral waste slag as filling material. *Constr. Build. Mater.* **2020**, *253*, 119183. [CrossRef]
14. Pasetto, M.; Baliello, A.; Giacomello, G.; Pasquini, E. Sustainable solutions for road pavements: A multi-scale characterization of warm mix asphalts containing steel slags. *J. Clean. Prod.* **2017**, *166*, 835–843. [CrossRef]
15. Bieliatynskyi, A.; Yang, S.; Krayushkina, K.; Shao, M.; Ta, M. Study of the possibility of using phosphorous slags in road construction. *Eng. Sci. Technol. Int. J.* **2022**, *36*, 101262. [CrossRef]
16. Shin, J.; Kwak, J.; Kim, S.; Son, C.; Lee, Y.-G.; Kim, J.; Bae, S.; Park, Y.; Lee, S.-H.; Chon, K. Highly selective recovery of phosphate ions using a novel carbonaceous adsorbent synthesized via co-pyrolysis of spent coffee grounds and steel slags: A potential phosphatic fertilizer. *Chem. Eng. J.* **2023**, *451*, 138978. [CrossRef]
17. Velayudhan, A.M.; Singh, B.; Shrivastava, M.; Khandelwal, A.; Yadav, P.; Rohatgi, B.; Darjee, S.; Ramalingappa, P.L.; Singh, R. Development of low heavy metal—Linz-Donawitz slag for safe spinach cultivation. *Sustain. Chem. Environ.* **2023**, *1*, 100003. [CrossRef]
18. Sahu, J.N.; Kapelyushin, Y.; Mishra, D.P.; Ghosh, P.; Sahoo, B.K.; Trofimov, E.; Meikap, B.C. Utilization of ferrous slags as coagulants, filters, adsorbents, neutralizers/stabilizers, catalysts, additives, and bed materials for water and wastewater treatment: A review. *Chemosphere* **2023**, *325*, 138201.
19. Yu, Q.; Liu, G.; Shi, J.; Wen, T.; Li, L. Magnetic-activated carbon composites derived from sludge and steel converter slag for wastewater treatment. *J. Environ. Chem. Eng.* **2022**, *10*, 108553. [CrossRef]
20. Pellegrino, C.; Gaddo, V. Mechanical and durability characteristics of concrete containing EAF slag as aggregate. *Cem. Concr. Compos.* **2009**, *31*, 663–671. [CrossRef]
21. Zhuang, S.; Wang, Q.; Luo, T. Modification of ultrafine blast furnace slag with steel slag as a novel high-quality mineral admixture to prepare high-strength concrete. *J. Build. Eng.* **2023**, *71*, 106501. [CrossRef]
22. Fernández Bertos, M.; Simons, S.J.R.; Hills, C.D.; Carey, P.J. A review of accelerated carbonation technology in the treatment of cement-based materials and sequestration of CO_2. *J. Hazard. Mater.* **2004**, *112*, 193–205. [CrossRef] [PubMed]
23. Wang, X.; Ni, W.; Li, J.; Zhang, S.; Hitch, M.; Pascual, R. Carbonation of steel slag and gypsum for building materials and associated reaction mechanisms. *Cem. Concr. Res.* **2019**, *125*, 105893. [CrossRef]
24. Liu, Q.; Liu, J.; Qi, L. Effects of temperature and carbonation curing on the mechanical properties of steel slag-cement binding materials. *Constr. Build. Mater.* **2016**, *124*, 999–1006. [CrossRef]
25. Zhan, B.; Poon, C.S.; Liu, Q.; Kou, S.; Shi, C. Experimental study on CO_2 curing for enhancement of recycled aggregate properties. *Constr. Build. Mater.* **2014**, *67*, 3–7. [CrossRef]
26. Pan, X.; Shi, C.; Hu, X.; Ou, Z. Effects of CO_2 surface treatment on strength and permeability of one-day-aged cement mortar. *Constr. Build. Mater.* **2017**, *154*, 1087–1095. [CrossRef]
27. Lu, B.; Shi, C.; Hou, G. Strength and microstructure of CO_2 cured low-calcium clinker. *Constr. Build. Mater.* **2018**, *188*, 417–423. [CrossRef]

28. Liu, P.; Mo, L.; Zhang, Z. Effects of carbonation degree on the hydration reactivity of steel slag in cement-based materials. *Constr. Build. Mater.* **2023**, *370*, 130653. [CrossRef]
29. Liu, P.; Zhang, M.; Mo, L.; Zhong, J.; Xu, M.; Deng, M. Probe into carbonation mechanism of steel slag via FIB-TEM: The roles of various mineral phases. *Cem. Concr. Res.* **2022**, *162*, 106991. [CrossRef]
30. Mo, L.; Zhang, F.; Deng, M.; Jin, F.; Al-Tabbaa, A.; Wang, A. Accelerated carbonation and performance of concrete made with steel slag as binding materials and aggregates. *Cem. Concr. Compos.* **2017**, *83*, 138–145. [CrossRef]
31. Liu, P.; Zhong, J.; Zhang, M.; Mo, L.; Deng, M. Effect of CO_2 treatment on the microstructure and properties of steel slag supplementary cementitous materials. *Constr. Build. Mater.* **2021**, *309*, 125171. [CrossRef]
32. *GB/T 24175-2009.2009-06-25*; Test Method for Slag Stability. Standards Press of China: Beijing, China, 2009. (In Chinese)
33. *JTG E42-2005*; Test Methods of Aggregate for Highway Engineering. China Communications Press: Beijing, China, 2005. (In Chinese)
34. Yi, Y.-R.; Lin, Y.; Du, Y.-C.; Bai, S.-Q.; Ma, Z.-L.; Chen, Y.-G. Accelerated carbonation of ladle furnace slag and characterization of its mineral phase. *Constr. Build. Mater.* **2021**, *276*, 122235. [CrossRef]
35. Siyu, D.; Liao, H.; Cheng, F.; Tao, M. Effect of curing condition and carbonation enhancement on mechanical properties of fly ash—desulfurization gypsum—steel slag blocks. *J. CO_2 Util.* **2020**, *38*, 282–290.
36. Kravchenko, E.; Qin, C.; Lin, Z.; Ng, C.W.W. Effect of polyvinyl alcohol on the CO_2 uptake of carbonated steel slag. *Constr. Build. Mater.* **2023**, *375*, 130761. [CrossRef]
37. Lu, B.; He, P.; Liu, J.; Peng, Z.; Song, B.; Hu, X. Microstructure of Portland cement paste subjected to different CO_2 concentrations and further water curing. *J. CO_2 Util.* **2021**, *53*, 101714. [CrossRef]

MDPI AG
Grosspeteranlage 5
4052 Basel
Switzerland
Tel.: +41 61 683 77 34

Materials Editorial Office
E-mail: materials@mdpi.com
www.mdpi.com/journal/materials

www.ingramcontent.com/pod-product-compliance
Lightning Source LLC
Chambersburg PA
CBHW061838210926
51850CB00006B/923